Engineering the Bioelectronic Interface
Applications to Analyte Biosensing and Protein Detection

Engineering the Bioelectronic Interface
Applications to Analyte Biosensing and Protein Detection

Edited by

Jason Davis
Chemistry Research Laboratory, University of Oxford, Oxford, UK

RSCPublishing

ISBN: 978-0-85404-165-7

A catalogue record for this book is available from the British Library

© 2009 Royal Society of Chemistry

All rights reserved

Apart from fair dealing for the purposes of research for non-commercial purposes or for private study, criticism or review, as permitted under the Copyright, Designs and Patents Act 1988 and the Copyright and Related Rights Regulations 2003, this publication may not be reproduced, stored or transmitted, in any form or by any means, without the prior permission in writing of The Royal Society of Chemistry or the copyright owner, or in the case of reproduction in accordance with the terms of licences issued by the Copyright Licensing Agency in the UK, or in accordance with the terms of the licences issued by the appropriate Reproduction Rights Organization outside the UK. Enquiries concerning reproduction outside the terms stated here should be sent to The Royal Society of Chemistry at the address printed on this page.

Published by The Royal Society of Chemistry,
Thomas Graham House, Science Park, Milton Road,
Cambridge CB4 0WF, UK

Registered Charity Number 207890

For further information see our website at www.rsc.org

Preface

The interfacing of man-made electronics with redox proteins and enzymes not only tells us a great deal about the, often humbling, levels of sophistication active in biology, but also paths the way to utilising exactly this in derived sensory devices. Some of these have already had a profound impact on both clinical diagnostics and the quality of life enjoyed by those unfortunate enough to live with disease. Though much remains to be learnt about controlling and optimising these interfacial interactions, their potential utilisation is, if anything, growing.

This book outlines a selection of some of the recent advances made in controllably engineering the bioelectronic interface, establishing robust communication with proteins and enzymes, characterising these surfaces and potentially important derived clinical applications. It is not intended to be in any way exhaustive but to give readers a flavour of the possibilities, challenges, tools and potential applications.

The book begins with a discussion by Paul Bernhardt of the fascinating properties and chemical transformation abilities of molybdenum-containing oxidoreductase enzymes (with potential applications in environmental and clinical sensing). The demands associated with establishing reliable electrochemical communication with these molecules are those familiar to many in the bioelectrochemistry field. As such, this chapter sets the stage for some of the leaders in this field to discuss a range of interfacial design and analysis tools now available to the electrochemist. This begins in Chapter 2, with a discussion of scanning probe technology, its application to biomolecular imaging and the proximal probe resolution of single metalloprotein redox characteristics.

In Chapter 3 the group of Willner discuss their recent pioneering work in using molecular wires, electron relays, nanoparticles and novel reconstitution protocols in improving electrode-protein/enzyme coupling and the preliminary applications of this to amperometric biosensing and fuel cell technology. In a continuing theme, the group of James Rusling detail the considerable progress they have made in utilising carbon nanotube modified surfaces in biosensing configurations (including those which are vertically aligned) in Chapter 4. The unique electrical and topological characteristics of these structures facilitate a considerable increase in Faradaic current and analyte sensitivity over bare electrodes.

In Chapter 5, Heering and Canters describe the covalent or crystallised association of protein and enzymes and the information this provides about their natural electron transfer dynamics. These analyses highlight the often tremendously sensitive nature of these processes to interfacial interactions (information of great value in establishing comparable levels of communication at man-made surfaces). The importance of electrode surface preparation, pre-modification and the use of π-delocalised surface linkers in facilitating robust electron transfer are also discussed.

The Cytochrome P450 enzymes play a critical role in the human body by catalysing reactions involved in xenobiotic metabolism, biosynthesis of steroid hormones, oxidation of unsaturated fatty acids and stereo-specific and regio-specific metabolism. There exists considerable interest in both engineering their specific characteristics and interfacing them with electrode surfaces. In Chapter 6, Gilardi and Dodhia describe the fascinating catalytic properties of these molecules, their engineering by mutagenesis and progress made controllably interfacing with electrode surfaces (including the use of "Molecular Lego" approaches).

One's ability to precisely engineer the bioelectrochemical interface can be applied to the sensitive electronic detection of protein (arguably the most demanding and important medical challenge of the next decade). In Chapter 7 the Davis group survey the importance, requirements and methods of ultra-sensitive protein detection, a particular focus being the establishment of highly sensitive, specific, field effect detection of target protein in complex mixtures. Finally, in Chapter 8, the applications of biosensors are discussed from a clinical perspective by Ko Ferrigno, where the complexities and demands of screening cell and bodily fluid composition are, in particular, highlighted.

Jason J. Davis
Oxford

Contents

Chapter 1 **Communication with the Mononuclear Molybdoenzymes: Emerging Opportunities and Applications in Redox Enzyme Biosensors**   1
Paul V. Bernhardt

    1.1 Introduction – the Three Mo Enzyme Families   1
    1.2 Mechanism   2
    1.3 Amperometric Biosensors   3
    1.4 Emerging Applications of Mo Enzymes in Sensing   5
        1.4.1 Xanthine Oxidase Family   5
    1.5 Sulfite Oxidase Family   9
        1.5.1 Sulfite Oxidoreductase   10
    1.6 DMSO Reductase Family   15
        1.6.1 DMSO Reductase   15
        1.6.2 Nitrate Reductase   17
        1.6.3 Arsenite Oxidase   19
        1.6.4 Chlorate and Perchlorate Reductase   20
    1.7 Conclusions   20
    References   21

Chapter 2 **Scanning Probe Analyses at the Bioelectronic Interface**   25
Jason J. Davis, Ben Peters, Yuki Hanyu and Wang Xi

    2.1 Introduction   25
        2.1.1 Scanning Probe Microscopy   26
        2.1.2 SPM Applications at the Biomolecular Interface   35
        2.1.3 Summary   38

Engineering the Bioelectronic Interface: Applications to Analyte Biosensing and Protein Detection
Edited by Jason Davis
© 2009 Royal Society of Chemistry
Published by the Royal Society of Chemistry, www.rsc.org

2.2	Bioelectronic Analyses	39
	2.2.1 Electrode Surface Considerations	39
	2.2.2 AFM Imaging Case Studies	39
	2.2.3 The Direct Imaging of Electrochemistry and Enzyme Activity	41
	2.2.4 Spectroscopic Assessment Electrode-biomolecule Electronic Coupling	46
2.3	Summary	49
	References	50

Chapter 3 Electrical Interfacing of Redox Enzymes with Electrodes by Surface Reconstitution of Bioelectrocatalytic Nanostructures — 56
Itamar Willner, Ran Tel-Vered and Bilha Willner

3.1	Introduction	56
3.2	Reconstituted Enzyme Electrodes in Monolayer Configurations	59
3.3	Electrical Wiring of Redox Proteins with Electrodes by their Reconstitution on Cofactor-Functionalised Metallic Nanoparticles (NPs) or Carbon Nanotubes (CNTs)	63
3.4	Reconstitution of apo-Enzymes in Thin Films of Redox Polymers	70
3.5	Design of Electrically Contacted Enzyme Electrodes by the Crossing of Surface-confined Cofactor-enzyme Affinity Complexes	72
3.6	Reconstituted Enzyme Electrodes for Biofuel Cell Applications	82
3.7	Conclusions and Perspectives	89
	Acknowledgement	90
	References	90

Chapter 4 Single-wall Carbon Nanotube Forests in Biosensors — 94
James F. Rusling, Xin Yu, Bernard S. Munge, Sang N. Kim and Fotios Papadimitrakopoulos

4.1	Unique Properties of Carbon Nanotubes	94
	4.1.1 Introduction	94
	4.1.2 Electrocatalytic Properties	96
4.2	Biosensors Using Non-oriented Carbon Nanotube Electrodes	96
4.3	Biosensors Utilising Vertically Aligned Carbon Nanotube Forests	99

		4.3.1	CNT Forest Fabrication	99

	4.3.1 CNT Forest Fabrication	99

4.3.1 CNT Forest Fabrication 99
4.3.2 Biosensor Applications of SWNT Forests 107
4.4 Outlook for the Future 112
References 113

Chapter 5 Activating Redox Enzymes through Immobilisation and Wiring 119
H.A. Heering and G.W. Canters

5.1 Introduction 119
5.2 Protein Complexes 120
    5.2.1 Co-crystallisation 120
    5.2.2 Covalent Complexes 122
5.3 Electron Transfer at Electrodes 126
    5.3.1 Voltammetry 128
    5.3.2 Chronoamperometry 128
5.4 Surface Preparation 132
    5.4.1 Carbon 132
    5.4.2 Gold 134
    5.4.3 Other Methods 134
5.5 Immobilisation 136
    5.5.1 Direct Immobilisation 137
    5.5.2 Wires 139
    5.5.3 Wiring Proteins 143
5.6 Conclusion 146
References 146

Chapter 6 Cytochromes P450: Tailoring a Class of Enzymes for Biosensing 153
Vikash R. Dodhia and Gianfranco Gilardi

6.1 Introduction 153
6.2 Structure-function of Bacterial and Human Cytochromes P450 155
6.3 The Need for Electrons: the Cytochrome P450 Catalytic Cycle 158
6.4 Human Cytochromes P450 and Drug Metabolism 161
6.5 Protein Engineering of P450s to Improve or Expand their Catalytic Properties 165
    6.5.1 Directed Evolution of Cytochrome P450 Enzymes 166
    6.5.2 Rational Design of Cytochrome P450 Enzymes 167
6.6 Interfacing Cytochromes P450 to Electrodes 171
    6.6.1 Immobilisation on Unmodified Electrodes 172
    6.6.2 Immobilisation with Surfactants, Polymers and Gold Nanoparticles 173

	6.6.3	Immobilisation by Covalent Linkage on Gold Electrodes: Use of Spacers	178
	6.6.4	Protein Engineering to Control Protein Immobilisation and Catalytic Turnover on Electrode Surfaces	180
6.7	Conclusions		185
References			186

Chapter 7 Label-free Field Effect Protein Sensing   193
Jan Tkac and Jason J. Davis

7.1	Interfacial Protein Detection		193
7.2	Protein Microarrays		194
	7.2.1	Array Substrates	194
	7.2.2	Surface Chemistry and Immobilisation	196
	7.2.3	Capture Biomolecules	198
	7.2.4	Detection Tools	200
	7.2.5	Ultrasensitive Protein Detection	204
7.3	Label-free Field Effect Protein Detection		206
	7.3.1	Field Effect Transistor (FET) based Protein Sensing	207
	7.3.2	Capacitance/Impedance Label-free Protein Sensing	209
	7.3.3	Nanoscale Devices for Label-free Field Effect Protein Sensing	213
7.4	Conclusions		215
References			216

Chapter 8 Biological and Clinical Applications of Biosensors   225
Paul Ko Ferrigno

8.1	Biosensing for Pure Biological Research		225
	8.1.1	The Challenges of "Omics" and "Systems" Approaches	225
	8.1.2	Biological Complexity	226
	8.1.3	The Types of Device Required	230
8.2	Biosensing for Clinical Applications		231
	8.2.1	The Clinical Problems – Diagnosis, Prognosis, Personalised Medicine	231
	8.2.2	Biosensors for Clinical Applications	239
8.3	Further Reading		240
References			240

Subject Index   243

CHAPTER 1

Communication with the Mononuclear Molybdoenzymes: Emerging Opportunities and Applications in Redox Enzyme Biosensors

PAUL V. BERNHARDT

Centre for Metals in Biology, School of Chemistry and Molecular Biosciences, University of Queensland, Brisbane 4072, Australia

1.1 Introduction – the Three Mo Enzyme Families

The mononuclear Mo-enzymes are remarkable in their coherence of active site structure and function yet equally interesting in the diversity of substrates that they are capable of oxidising or reducing. Apart from the well-studied enzyme nitrogenase, where the Mo ion is found within a S-bridged cluster of metals including Fe, all other enzymes containing Mo bear a single metal at the active site.[1] A recent exception to this may be the novel Mo enzyme CO dehydrogenase, where a Cu ion shares a sulfide bridging ligand with the Mo ion at the active site.

All known enzymes from this family bear either one or two bidentate pterin-dithiolene (molybdopterin, MPT) ligands bound to the Mo ion at the active site (Figure 1.1). Hille proposed[2] a classification of this group of enzymes into three

Engineering the Bioelectronic Interface: Applications to Analyte Biosensing and Protein Detection
Edited by Jason Davis

Published by the Royal Society of Chemistry, www.rsc.org

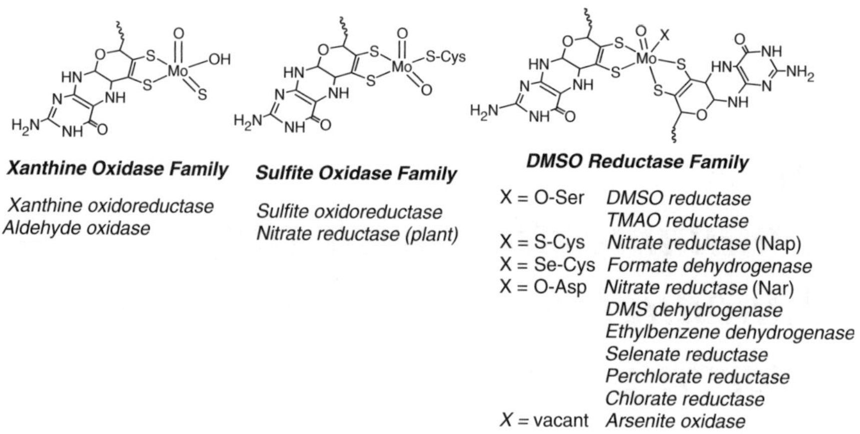

Figure 1.1 Active site structures of the three mononuclear Mo enzyme families.

Xanthine Oxidase Family
Xanthine oxidoreductase
Aldehyde oxidase

Sulfite Oxidase Family
Sulfite oxidoreductase
Nitrate reductase (plant)

DMSO Reductase Family

X = O-Ser DMSO reductase
           TMAO reductase
X = S-Cys Nitrate reductase (Nap)
X = Se-Cys Formate dehydrogenase
X = O-Asp Nitrate reductase (Nar)
           DMS dehydrogenase
           Ethylbenzene dehydrogenase
           Selenate reductase
           Perchlorate reductase
           Chlorate reductase
X = vacant Arsenite oxidase

families based on the coordination environment of the metal as shown in Figure 1.1.

At this time, enzymes from the DMSO reductase family (the most diverse of all) have only been found in bacteria and archea whilst enzymes from the other two families are found in all forms of life.

1.2 Mechanism

Although subtle differences exist in the mechanism of the mononuclear Mo enzymes, as a starting point, the reactions catalysed by this enzyme superfamily can be generalised by eqn (1) written in either the forward (reductase) or reverse (oxidase/dehydrogenase) direction. The substrates represented by the generic symbols Z and ZO are apparent from the names of the respective oxidases/dehydrogenases (Z) or reductases (ZO) shown in Figure 1.1.

$$ZO + 2e^- + 2H^+ \rightarrow Z + H_2O \tag{1}$$

The use of Mo enzymes in electrochemically driven (amperometric) biosensors relies on connecting a working electrode with the redox active species involved in the catalytic cycle *i.e.* the enzyme and/or its substrates and products. The electrons required to sustain catalysis are provided or accepted by the electrode rather than the enzyme's natural cosubstrate. The various ways in which this can be done are summarised in the following section, but the most relevant point is that the Mo ion at the active site always cycles between its Mo^{VI} and Mo^{IV} oxidation states during catalysis and an O-donor ligand (oxo or hydroxo) is exchanged with the substrate during turnover. The Mo^{VI} form is the catalytically active form of the oxidases/dehydrogenases whilst the reductases must be reduced to Mo^{IV} before turnover can commence. The Mo^{V} form is a thermodynamically stable intermediate in most, but not all, cases, but it is incapable of turning over substrates in either direction. However, it may become an important rate-limiting intermediate in some cases.

Ligands bonded to the Mo ion are activated by coordination to perform some remarkable bond-breaking and formation reactions that otherwise do not occur in the absence of the enzyme. In the well-studied xanthine oxidoreductases, a hydroxo ligand participates in a base-assisted nucleophilic attack at C-8 of xanthine coupled with a hydride abstraction by the sulfido ligand (Figure 1.2A).[3] This mechanism is significantly different from that seen in enzymes from the sulfite oxidase (Figure 1.2B) and DMSO reductase (Figure 1.2C) families where an oxo ligand is exchanged directly between the Mo ion and substrate during turnover.

1.3 Amperometric Biosensors

The development of enzyme-based biosensors in general has evolved over recent times as methods for addressing the active sites of enzymes have become better understood. Initially, enzyme electrochemistry relied upon the voltammetric detection of either the product or cosubstrate (so-called first-generation biosensors, Figure 1.3). The most common analyte that has been detected in this way is hydrogen peroxide, a typical product of oxidase enzyme turnover where the cosubstrate dioxygen is reduced in a two-electron proton-coupled reaction by the enzyme after substrate turnover. Alternatively the product itself may be electroactive but this is exceptional. A complementary approach is to monitor the depletion of cosubstrate, *e.g.* dioxygen, during turnover. However, this approach has limitations as variations in dioxygen concentrations may result from changes to the solution during analysis, *e.g.* temperature, stirring *etc.*, and thus give false readings.

Second generation biosensors removed the natural cosubstrate from the system altogether, replacing it with a small molecule electron transfer mediator *e.g.* a redox active coordination compound or organic molecule. This approach has its roots in traditional enzyme solution assays where a mediator is oxidised or reduced chemically rather than electrochemically to drive the catalytic reaction. In electrochemistry, the mediator serves the dual purpose of (i) undergoing homogeneous electron transfer with the enzyme to restore it to its active form following turnover and (ii) undergoing heterogeneous electron transfer with the working electrode to provide the current that quantifies the enzyme-substrate reaction. This approach underpins most commercial enzyme biosensors to date including the glucose oxidase biosensor,[4] which utilised a ferrocenium mediator (rather than dioxygen) as its artificial cosubstrate. The use of redox active polymers adsorbed on the electrode also comes under this classification. In all cases it should be emphasised that the currents observed are due to the mediator and they appear at the formal potential of the mediator and not of the enzyme. Ideally the redox potential of the mediator is in the vicinity of that of the active site. This avoids excessively large overpotentials which may lead to non-specific redox reactions with species in the sample other than the substrate.

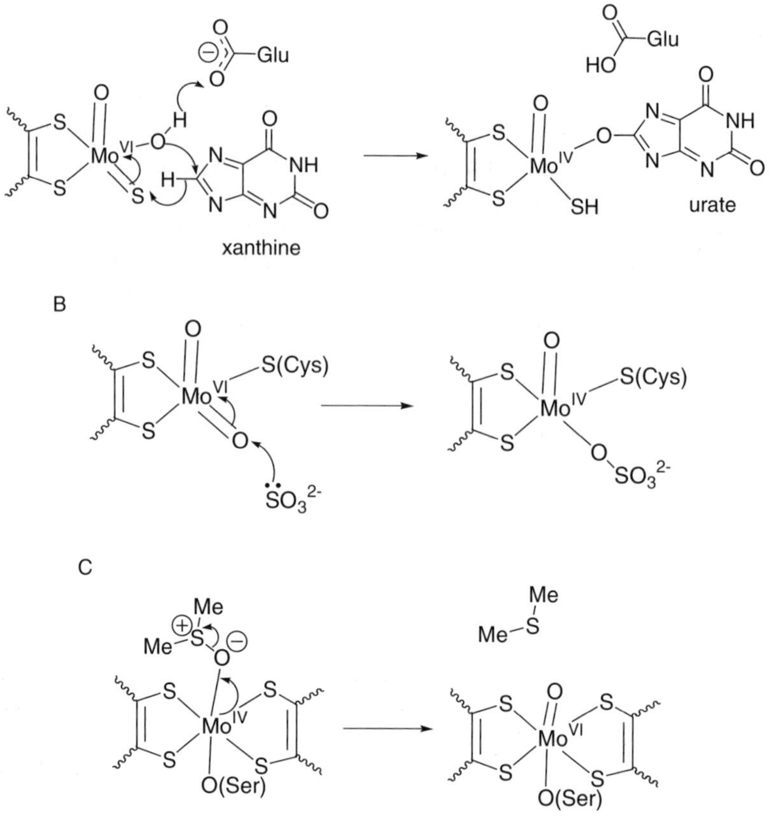

Figure 1.2 The currently accepted mechanisms for substrate turnover at the active sites of enzymes from the (A) xanthine oxidase; (B) sulfite oxidase and (C) DMSO reductase families. Similar mechanisms may be derived for other members of each enzyme family.

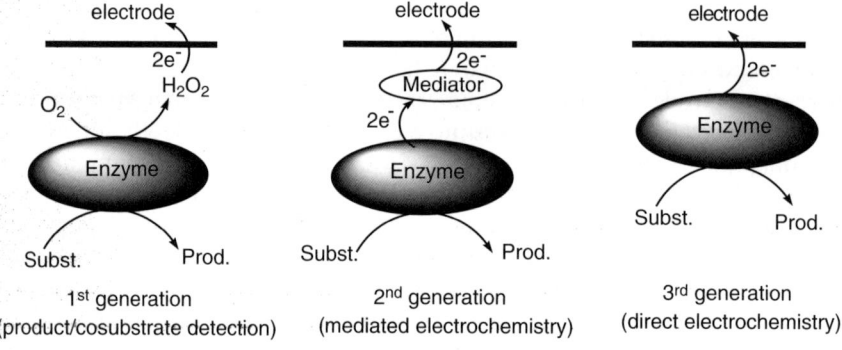

Figure 1.3 The three generations of electrochemical enzyme biosensors.

The final approach (third generation) is to remove all cosubstrates from the system (natural or artificial) and to achieve direct electron exchange between the enzyme and the electrode. Although this has yet to be applied in a commercial biosensor, it offers significant advantages over (ternary) mediated systems, which each require a certain artificial electron relay tailor-made for the enzyme in question. The challenges in achieving direct enzyme electrochemistry are many.[5–7] These include avoiding enzyme denaturation of the necessarily electrode-adsorbed enzyme whilst ensuring efficient electronic communication between the active site (or other redox cofactors within the enzyme) and the electrode. However, significant progress has been made in the last 15 years or so and a number of robust enzyme electrode systems have been reported which require no mediators. In the absence of mediators, which mask electron transfer events with the enzyme, some interesting mechanistic studies have been possible which provide new insight to electron transfer pathways in complex enzyme systems.

The following section will review the evolution of Mo enzyme based electrochemical biosensors over recent times with examples of all three types of biosensor being covered. The ordering of sections follows that of the three enzyme sub-families.

1.4 Emerging Applications of Mo Enzymes in Sensing

1.4.1 Xanthine Oxidase Family

1.4.1.1 Xanthine Oxidoreductase

Xanthine oxidase from bovine milk is the most studied of all mononuclear Mo enzymes. The volume of literature on this enzyme alone and its employment in enzyme electrode biosensors far outweighs all other Mo enzymes accordingly. Xanthine oxidase primarily catalyses the oxidation of xanthine to uric acid as part of the process of purine metabolism, but it also is capable of oxidising hypoxanthine to xanthine (Figure 1.4).

Regardless of their origin, all xanthine oxidase/dehydrogenase enzymes contain an active site comprising a single bidentate molybdopterin chelate, an equatorial terminal sulfido, an axial oxo and an equatorial hydroxo/aqua ligand depending on pH (Figure 1.1). The enzymes from this family bear three additional redox cofactors comprising two [2Fe-2S] clusters and an FAD cofactor. Crystal structures of various xanthine oxidoreductases[8,9] have shown that electrons flow along the pathway Mo→[2Fe-2S]→[2Fe-2S]→FAD. The FAD cofactor is oxidised by either $NADP^+$ or dioxygen depending on whether it is present in its dehydrogenase or oxidase form.

Like many other oxidase enzymes, hydrogen peroxide is a product of substrate turnover in xanthine oxidase when dioxygen is the cosubstrate. The electroactivity of H_2O_2 enables its voltammetric detection and provides a method for monitoring turnover without requiring direct or mediated electron transfer with the enzyme itself. A wide variety of electrode systems have been

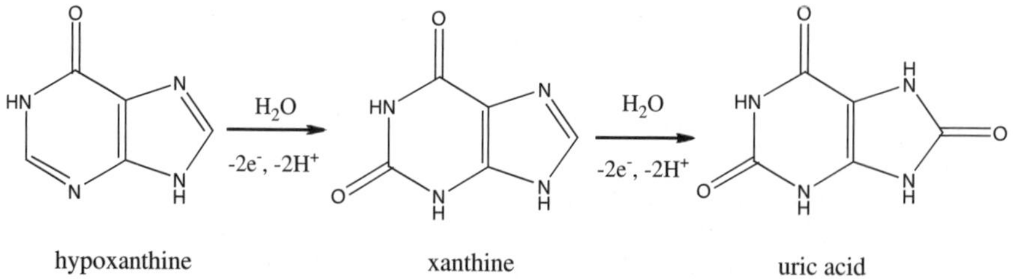

Figure 1.4 The substrates and products of xanthine oxidase activity.

described that utilise immobilised xanthine oxidase to produce H_2O_2 as an electrochemically detectable product or alternatively to monitor the depletion of cosubstrate dioxygen.

Hypoxanthine in particular is an analyte of interest as its presence is an indicator of spoilage in otherwise fresh fish[10,11] and beef products.[12] Sol-gel methods were used to produce an electrode coated with a silica-graphite matrix in which xanthine oxidase is entrapped.[13] Following substrate (hypoxanthine) turnover, the H_2O_2 produced is detected electrochemically by poising the electrode at 0.58 V vs. Ag/AgCl. Alternatively the complementary consumption of dioxygen may be determined at low potential voltammetrically (mediated by viologens) or with an oxygen electrode. In H_2O_2 production mode the sensor maintained a linear current/concentration response up to 500 µM hypoxanthine. Above this substrate concentration the response saturated following Michaelis–Menten kinetics ($K_{M,app}$ 450 µM). A lower detection limit of 1.3 µM hypoxanthine was reported.[13] The high surface area and biocompatibility of the silica matrix was found to be ideal for encapsulating the enzyme/graphite composite. Furthermore, the mild synthetic sol-gel techniques enabled the enzyme to be incorporated during the synthesis of the matrix.

Screen-printed xanthine oxidase electrodes have also been reported using a vast array of mediators including metal oxides (RuO_2, $Pd-IrO_2$) within the matrix to lower the overpotential for H_2O_2 oxidation.[14,15] Even more elaborate bi-enzyme systems have been developed comprising both xanthine oxidase and peroxidase (an enzyme that reduces H_2O_2 to water) where the H_2O_2 produced by xanthine oxidase turnover is quantified by the current produced through its ferrocyanide-mediated reduction by peroxidase[16] thus enabling the bi-enzyme system to function at a low potential (ca. −100 mV vs. Ag/AgCl) and minimising possible interferences from oxidation of other analytes at higher potentials.

A miniaturised xanthine oxidase electrode has been developed for monitoring the concentrations of hypoxanthine in myocardial cell culture media.[17] Purines are also associated with signalling in the nervous system and multi-enzyme electrodes (including xanthine oxidase) have been developed to monitor the local changes in purine concentrations *in vivo*.[18] The enzyme purine nucleoside phosphorylase (NP) catalyses the phosphorylation of inosine (by phosphate) to release hypoxanthine and ribose-1-phosphate (Scheme 1.1). The stoichiometry of the overall reaction coupled with xanthine oxidase activity

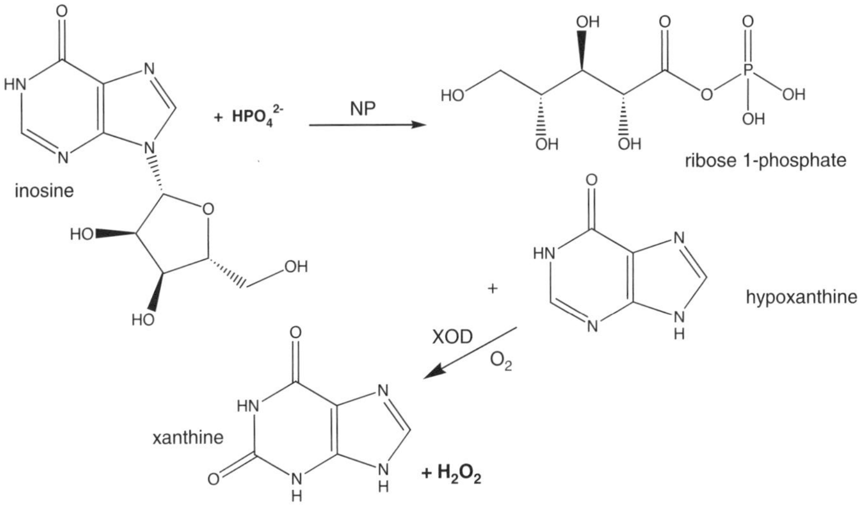

Scheme 1.1

means that one equivalent of H_2O_2 is produced for every (hydrogen)phosphate anion present.[19]

A number of groups have developed amperometric bi-enzyme systems of this type.[20–22] Haemmerli et al.[21] reported a bi-enzyme (NP/xanthine oxidase) system which exhibited a linear response up to 250 μM phosphate. Various ratios of the two enzymes were investigated and the most ideal was found to be 10 : 1 NP : xanthine oxidase. This novel approach provides a viable alternative to otherwise tedious wet chemical (colorimetric) methods for phosphate determination.

Purine analysis can also be achieved with this system. For example, the concentration of inosine (the cosubstrate, with phosphate, in eqn (2a)) can be determined.[23,24] The so-called hypoxanthine ratio ([hypoxanthine]/([hypoxanthine] + [inosine] + [inosine-monophosphate])[25] is an important parameter in the analysis of fresh fish as hypoxanthine is a product of nucleotide degradation. Like hydrogen peroxide, uric acid (the product of enzymatic xanthine oxidation, Figure 1.4) is electroactive and it can also be detected electrochemically thus providing a method for quantifying xanthine oxidase turnover.[26]

Electron transfer mediators can be used to great effect in providing a link between the electrode and the enzyme cofactors. There are many approaches that may be taken. Conducting polymers such as poly-*p*-benzoquinone and poly(mercapto-*p*-benzoquinone) have been used to provide a redox active matrix in which xanthine oxidase can be both immobilised and addressed electrochemically.[27] The well-studied Os-pyridine-based hydrogel polymers developed by the Heller group[28] effectively mediate electron transfer in horseradish peroxidase (HRP). This polymer has been cast on a glassy carbon electrode and coupled with xanthine oxidase which enables the production of H_2O_2 from (hypo)xanthine turnover to be monitored electrochemically *via* the

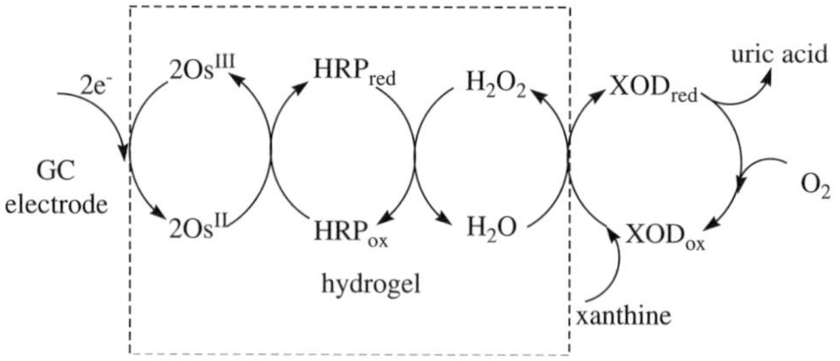

Scheme 1.2

Os-mediated reduction of HRP within the polymeric hydrogel.[29] This is illustrated in Scheme 1.2.

The benefit of using Os-mediated reduction of HRP is that the biosensor functions at a low potential (ca. 0 mV vs. Ag/AgCl) relative to that required for the direct oxidation of H_2O_2 (> 600 mV), where other species may be oxidised non-specifically as well. The sensor responds to hypoxanthine in a continuous flow system (linear response up to 80 μM with a detection limit of 0.2 μM) and also xanthine but is unaffected by potential interference from glutamate, lactate, glucose and glutathione. Ascorbate interference (significant at a working potential of 0 mV) was negligible when the working electrode was poised at −200 mV vs. Ag/AgCl. Otherwise redox inert polymers such as Nafion® can be made electroactive by the inclusion of small cationic mediators such as methyl viologen, which ion-exchanges within the anionic polymer. Xanthine oxidase adsorbed on such a viologen-modified polymer (cast on a glassy carbon electrode) produces a hypoxanthine biosensor[30] that functions in O_2-consumption mode. In this case the viologen mediates the reduction of O_2 and lowers the overpotential for reduction by about 100 mV relative to direct oxygen reduction at the electrode.

High potential redox mediators, such as ferrocenium derivatives, have been very effective in acting as artificial electron acceptors for a range of oxidase enzymes; glucose oxidase being the most famous example. Similarly, hydroxymethylferrocenium can accept electrons from xanthine oxidase (replacing dioxygen) to produce a mediated amperometric hypoxanthine biosensor. In this case the enzyme was entrapped within a membrane covering the carbon-paste working electrode.[31] A wide linear range (up to 700 μM hypoxanthine with a detection limit to 0.6 μM) was reported.

Ironically, despite its intensive investigation by enzymologists and spectroscopists for almost 50 years, it is only recently that a direct electrochemical study of a xanthine oxidoreductase was reported, namely the bacterial xanthine dehydrogenase from R. capsulatus.[32] Non-turnover signals from all cofactors were seen and EPR potentiometry was also undertaken to resolve the potentials of the cofactors. Pronounced catalytic voltammetry was seen in the presence of xanthine. The bell-shaped pH profile of the catalytic wave mirrored that seen in solution. A high pH deprotonation of xanthine (pK_a 7.7) and a low potential protonation of a glutamate residue (pK_a 6) essential for base catalysis each switch off catalysis. An unusual feature of this study was that the potential at which catalysis was observed was *ca.* 600 mV more positive than that of the highest potential cofactor (FAD). Essentially the same potential "delay" in catalysis has been seen for bovine milk xanthine oxidase when immobilised on a pyrolytic graphite electrode.[33]

1.4.1.2 CO Dehydrogenase

A recent addition to the Mo enzyme family is the novel CO dehydrogenase from the bacterium *Oligotropha carboxidovorans*, which catalyses the oxidation of CO to CO_2.[34] Its crystal structure first appeared to reveal an unusual Mo-S-Se group at the active site[35] but this interpretation was later revisited. Coupled with spectroscopic measurements, an equally novel Mo-S-Cu moiety,[36] was identified instead of the terminal Mo=S group typically found in enzymes from the xanthine oxidase family (Figure 1.1). Given its dinuclear active site CO dehydrogenase, strictly speaking, does not belong in the mononuclear enzyme family at all. Notwithstanding, CO dehydrogenase possesses a single molybdopterin ligand bound to the Mo ion and two [2Fe-2S] clusters in addition to an FAD cofactor like all other members of the xanthine oxidase family and on this basis it appears to belong within this grouping.

The importance of quantitatively detecting CO, a potentially lethal gas, in the home and in the field is a significant driving force for the development of a CO dehydrogenase biosensor. Very little electrochemical work has been performed on this enzyme. The most significant appeared more than 25 years ago, well before its novel structure was known, when ferricyanide and ferrocenium-mediated catalytic electrochemistry of CO dehydrogenase was reported.[37] More studies with this remarkable addition to the Mo enzyme family are eagerly awaited.

1.5 Sulfite Oxidase Family

The sulfite oxidases and dehydrogenases dominate this Mo-enzyme family;[2] a plant nitrate reductase being the only other member of this sub-grouping that has been characterised to date.[38] Sulfite oxidoreductases are found in most forms of life and there now exist structurally characterised examples from animals,[39,40] plants[41] and most recently bacteria.[42] In humans they catalyse the last stage in the degradation of S-containing amino acids where sulfite is oxidised to sulfate. The eukaryotic sulfite oxidases can donate electrons to dioxygen (eqn (2a)) or cytochrome *c* (eqn (2b)) whilst the dehydrogenases only use cytochrome *c* as their electron acceptor.

$$SO_3^{2-} + O_2 + H_2O \xrightarrow{\text{sulfite oxidase}} SO_4^{2-} + H_2O_2 \quad (2a)$$

$$SO_3^{2-} + 2cyt.\ c_{ox} + H_2O \xrightarrow{\text{sulfite oxidase/dehydrogenase}} SO_4^{2-} + 2cyt.\ c_{red} + 2H^+ \quad (2b)$$

In animal and bacterial sulfite oxidoreductases, the Mo active site is coupled with a heme cofactor that accepts electrons in sequence from the Mo^{IV} ion after turnover and passes them on to a c-type cytochrome in solution. In plants it appears that the heme cofactor is absent[41] and the Mo ion donates electrons directly to dioxygen.

1.5.1 Sulfite Oxidoreductase

The first crystal structure of a sulfite oxidoreductase to appear was that of the enzyme from chicken liver.[39] This structure presented a mechanistic problem in that the heme and Mo cofactors (supposedly electron partners) were separated by 33 Å; well in excess of the distance that electrons are known to tunnel between redox centres.[43] Subsequently it has emerged from spectroscopic studies of electron transfer[44,45] that the crystallographically characterised conformation of chicken liver sulfite oxidase is not that of the active enzyme. Instead, a polypeptide "hinge" enables the heme cofactor to "swing" around from this remote location to be in proximity with the Mo active site such that intramolecular electron transfer is then possible. The initial uncertainty surrounding the active conformation of chicken liver sulfite oxidase has been resolved by the crystal structure of a bacterial sulfite dehydrogenase[42] where the Mo and heme cofactors are adjacent as expected (Figure 1.5). This structure serves as a good model for the chicken liver and human enzyme in their active conformations, which so far have eluded identification by X-ray crystallography.[39,40]

The substrate sulfite is of great importance to the food and beverage industry as it is used extensively as a preservative due to its antioxidant and antimicrobial properties.[46] The concentration of added sulfites is regulated as allergies to this chemical in some people can present a serious health risk.[47] However, the analytical determination of sulfite using wet chemical methods is not a simple procedure and the development of alternative procedures that can directly determine sulfite in foods and beverages is an important goal. Although the direct electrochemical oxidation of sulfite to sulfate can in principle be performed, the high potential that is required becomes problematic due to the simultaneous oxidation of species such as ascorbate and polyphenols, which are also often present in food and beverage samples. The development of enzyme-based (sulfite oxidase) biosensors offers a great improvement in selectivity.

The commercial availability of chicken liver sulfite oxidase has led to its dominance of the amperometric sulfite biosensor literature. In principle, the oxidase activity of the enzyme (in producing electroactive H_2O_2) can be exploited to develop a first generation biosensor where hydrogen peroxide is detected amperometrically. However, a major problem to be overcome with this approach is that H_2O_2 and sulfite are each oxidised at similar potentials on bare electrodes (*ca.* 400 mV *vs.* Ag/AgCl at neutral pH). The use of electropolymerised polytyramine to entrap sulfite oxidase has proven a successful solution to this problem and a first-generation sulfite biosensor that detects

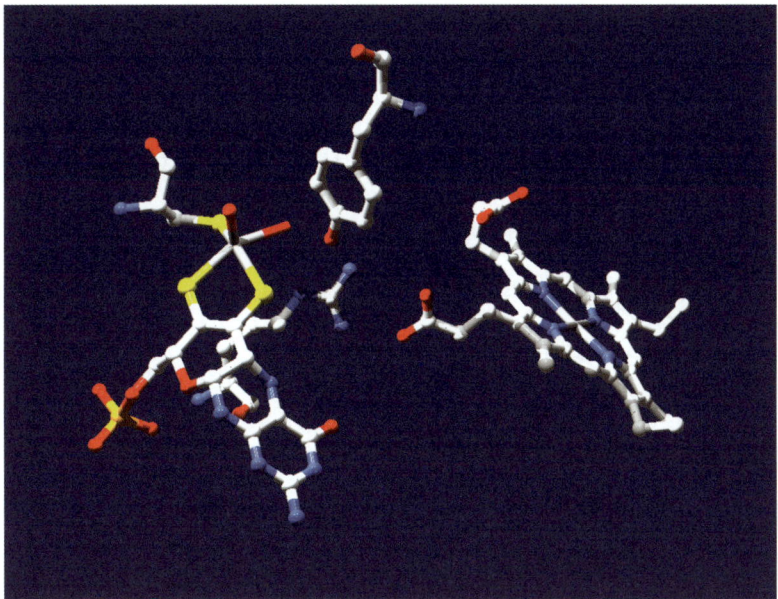

Figure 1.5 The proximity of the Mo (left) and heme (right) cofactors in *S. novella* sulfite dehydrogenase as determined by X-ray crystallography. Coordinates taken from the Brookhaven Protein Data Bank (of structure published in ref. 42) and rendered with PovRay vers. 3.5.

H_2O_2 produced as a result of enzyme turnover (eqn (2a)) has been reported.[48] Direct oxidation of sulfite (Figure 1.6, curve a) is seen on a bare GC electrode but this response is completely suppressed when the polytyramine film is cast on the same electrode (Figure 1.6, curve b). Introduction of sulfite oxidase and sulfite (Figure 1.6, curve d) generates a pronounced wave due to electrochemical H_2O_2 oxidation produced from enzyme turnover. This biosensor gave a linear response up to 300 µM sulfite.

An Hg-film coated glassy carbon electrode was used in oxygen depletion mode to monitor sulfite oxidase activity.[49] The Hg film significantly lowered the overpotential for O_2 reduction at the electrode. Entrapment of sulfite oxidase within electropolymerised polypyrrole films has also been reported.[50] Additives such as dextran[51] were shown to improve mechanical stability of the enzyme-polymer film.

The problems associated with accurately measuring oxygen consumption or detecting H_2O_2 amperometrically in the presence of sulfite are avoided if mediators are used to accept electrons from sulfite oxidase instead of dioxygen. The electron transfer protein cytochrome *c* is an ideal choice (Scheme 1.3) as it is known to be a physiological electron partner of sulfite oxidase and its direct electrochemistry at a number of different electrode surfaces has been intensively studied. Under anaerobic conditions, no H_2O_2 is produced from enzyme turnover and the anodic current is merely that of the cytochrome.

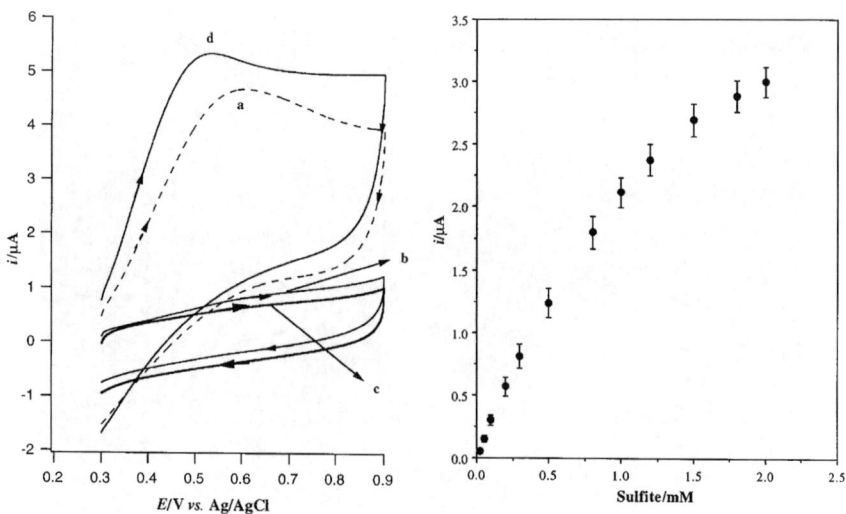

Figure 1.6 (left) Cyclic voltammograms for (a) 1.0 mM sulfite on a bare GC electrode, (b) 1.0 mM sulfite on a GC/polytyramine electrode, (c) blank phosphate buffer solution on a GC/poltyramine electrode and (d) 1.0 mM sulfite, 2 units mL^{-1} sulfite oxidase in the solution on a GC/polytyramine electrode and (right) current response as a function of sulfite concentration. Reproduced from ref. 48 with permission from the Royal Society of Chemistry.

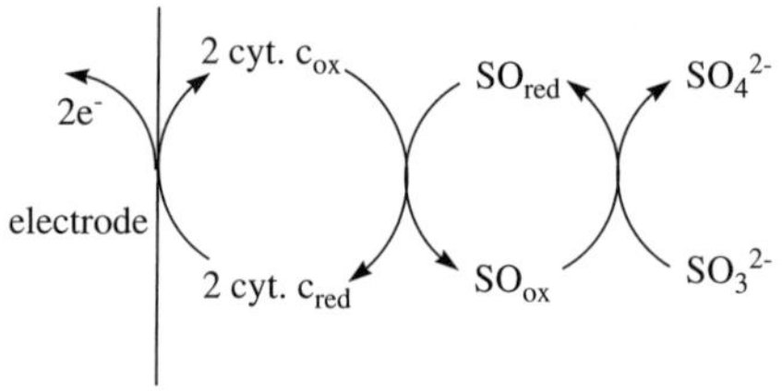

Scheme 1.3

A comprehensive study by Ferapontova and coworkers[52] showed that horse heart cytochrome c at a 1 : 1 mercaptoundecanoic acid and mercaptoundecanol modified Au electrode is capable of mediating sulfite oxidase electron transfer (Figure 1.7). The electrochemistry of an almost equimolar mixture of cytochrome c and sulfite oxidase (Figure 1.7, curves 1 and 3 in inset) is characteristic of diffusion-controlled voltammetry of the cytochrome, indicating only weak association with the thiol-modified Au electrode. Loss of the proteins from the electrode surface into the bulk was prevented by their entrapment beneath a dialysis membrane ensuring all proteins remained in proximity to the electrode. In the presence of sulfite the reversible peak-shaped response of the cytochrome transforms into a sigmoidal wave (curve 2) where the amplification of current is

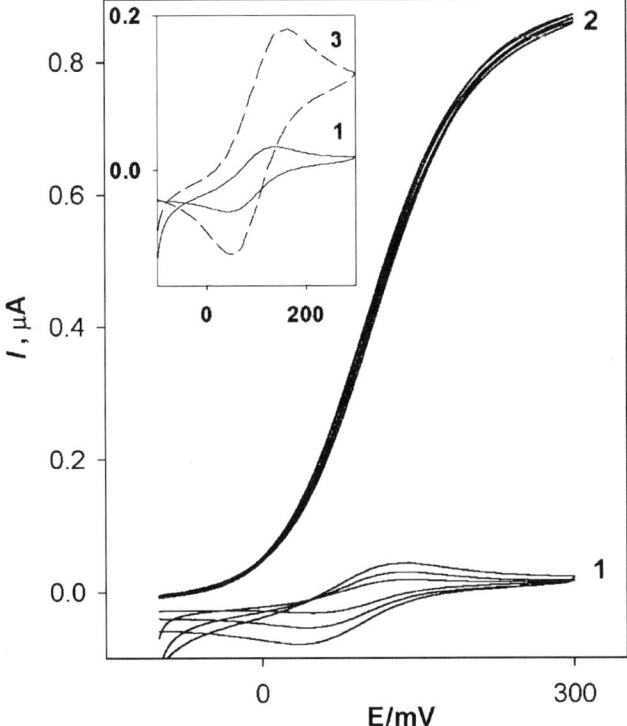

Figure 1.7 Cyclic voltammograms of a sulfite oxidase (36 μM)/cytochrome c (30 μM) mixture entrapped between a thiol-modified Au electrode and a membrane (1) in the absence and (2) in the presence of sulfite (3.3 mM) (0.1 M Tris-HCl, pH 7.4). Note the potentials on the horizontal axes are *versus* Ag/AgCl. Reprinted with permission from ref. 52. Copyright 2003 American Chemical Society.

due to recycling of reduced cytochrome c, which accepts electrons from sulfite oxidase following turnover. A very similar study using human sulfite oxidase has appeared more recently and utilising the same thiol-modified Au procedure as Ferapontova *et al.* and with cytochrome c acting as a mediator.[53]

Screen-printed carbon electrode composites containing both sulfite oxidase and cytochrome c have been reported for the purpose of analysing gaseous SO_2 (dissolved in water).[54] The sensor operated at 300 mV *vs.* Ag/AgCl and exhibited a linear response in the range 4–50 ppm sulfite.

Artificial mediators have also been used to accept electrons from sulfite oxidase during catalysis. Murray and coworkers reported a series of papers where a number of different high-potential mediators were used.[55–58] Given the nature of the experiment, the intrinsic catalytic constants were hidden by the use of mediators, but information regarding the relative rates of bimolecular (mediator-enzyme) electron transfer, modelled with Marcus theory,[56] and intramolecular (Mo-heme) electron transfer[58] were elucidated. Interestingly when the heme domain was cleaved from the enzyme, the mediator was unable to oxidise the reduced Mo active site thus indicating that the heme moiety was the site at which intermolecular electron transfer took place.[56]

Towards third-generation biosensors, the direct electrochemistry of sulfite oxidoreductases has been a more recent development. Direct electrochemistry of the bacterial sulfite dehydrogenase from *S. novella* adsorbed on an edge plane pyrolytic graphite electrode has been reported[59] where non-turnover responses from both the Mo and heme cofactors were identified. A sigmoidal catalytic oxidation wave was seen upon addition of sulfite to the cell (Figure 1.8) at a potential corresponding to the heme potential (and *ca.* 200 mV lower than that due to direct sulfite oxidation).

This represents an ideal example of direct enzyme electrochemistry where the potential at which a catalytic current emerges is that of the enzyme rather than of a mediator. The apparent Michaelis constant derived from the concentration dependence of the steady state limiting current ($K_{M,app} = 26\,\mu M$) was similar to that obtained from solution studies. It is notable that the absence of mediators and the fact that the enzyme is adsorbed on an electrode significantly increases the sensitivity whilst narrowing the linear response of the enzyme electrode to around 10 μM sulfite. Elliot *et al.* reported direct electrochemistry of chicken liver sulfite oxidase adsorbed on a pyrolytic graphite electrode. The only non-turnover signal in the absence of substrate was assigned to the heme which was replaced by sigmoidal catalytic oxidation wave in the presence of sulfite.[60] Ferapontova *et al.* reported a similar activity with chicken sulfite oxidase but used self-assembled monolayers of long-chain alkanethiols bound to an Au working electrode.[52] Interestingly, the electrocatalytic activity was very

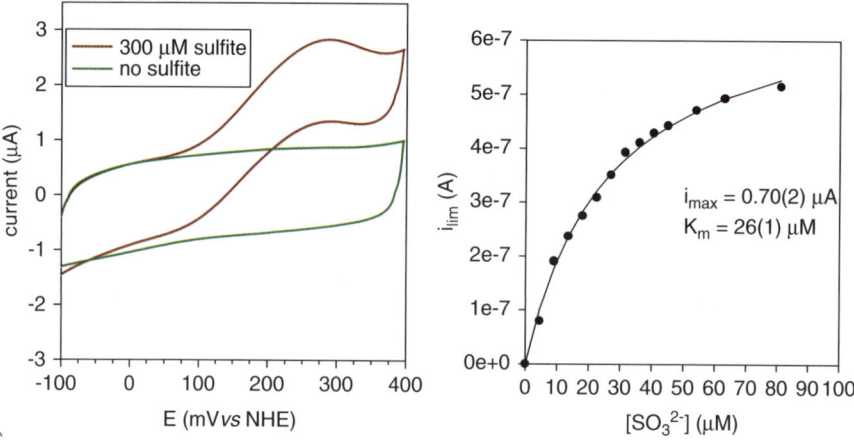

Figure 1.8 (left) Cyclic voltammograms of *S. novella* sulfite dehydrogenase (pH 8.0) in the absence and presence of sulfite; (right) sulfite concentration dependence of the steady state current. Note the potentials are *versus* the normal (standard) hydrogen electrode. Reprinted with permission from ref. 59. Copyright 2003 American Chemical Society.

sensitive to both the terminal group on the alkanethiol and the length of the chains comprising the self-assembled monolayer.

1.6 DMSO Reductase Family

Unlike the xanthine oxidase and sulfite oxidase families, which contain very few members, the DMSO reductase family boasts an impressive array of enzymes that act upon a wide variety of inorganic and organic substrates. This family perhaps offers the most opportunities for the development of novel biosensors for challenging (and otherwise chemically inert) substrates. This is a rapidly developing field and new members of the DMSO reductase family are continuing to be identified each year.

1.6.1 DMSO Reductase

The parent enzyme from this family has been isolated from three different bacteria (*E. coli*, *R. capsulatus* and *R. sphaeroides*). The *E. coli* DMSO reductase is a complex membrane-bound enzyme which bears five Fe-S clusters in addition to its Mo active site. It obtains its electrons from the quinone pool. The two *Rhodobacter* enzymes are periplasmic, very similar to each other and contain no other redox active cofactors. They are reduced by a multi-heme containing electron partner.

Regardless of these differences, all catalyse the reduction of DMSO to dimethyl sulfide (DMS) where the substrate is the terminal electron acceptor for anaerobic respiration. As well as being present in seawater (and an important component of the global sulfur cycle) DMSO is found in food and beverages. Its reduction to highly volatile DMS is responsible for the unpleasant odour of otherwise pure DMSO. The analytical determination of DMSO is complicated by its chemical inertness and its high miscibility with most solvents, making separation and isolation by extraction very difficult.

The highly selective reduction of DMSO by DMSO reductase offers a direct way of determining this compound analytically in the presence of other species. DMSO reductase may accept electrons from the reduced methyl viologen radical cation ($MV^{+\cdot}$) in solution assays and this has been adapted for the electrochemically driven catalysis of the enzyme from *R. sphaeroides* (Scheme 1.4) in the

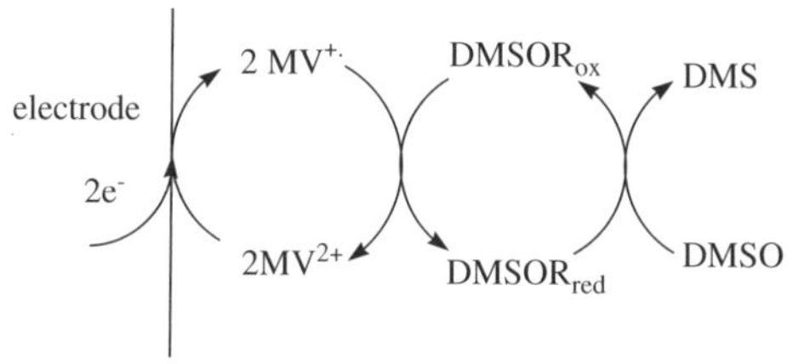

Scheme 1.4

presence of the viologen as the electrochemically detectable species. Apart from DMSO itself, enantioselective reduction of various chiral sulfoxides was demonstrated with very high enantiomeric excess.[61] This has been further developed into a biosensor for DMSO in solution.[62] A linear current response was found up to 6 mM DMSO using methyl or benzyl viologen as the mediator. Figure 1.9 (left) shows the transformation of the first electron reduction of methyl viologen ($MV^{2+/+ \cdot}$) from a peak-shaped reversible response to a sigmoidal steady state wave where the dication is regenerated. The second wave at lower potential is due to the $MV^{+ \cdot /0}$ couple.

The first direct (unmediated) electrochemical investigation of a mononuclear Mo enzyme comprised a voltammetric study of DMSO reductase from *E. coli* (DmsABC).[63] No response from any of the redox active cofactors was seen in the absence of substrate but, upon addition of DMSO, a cathodic catalytic wave emerged. Curiously, the catalytic voltammogram at pH 8.9 (Figure 1.10) is peak shaped instead of the ideally sigmoidal profile of a steady state electrochemical system more apparent at pH 7.0.

This behaviour was rationalised by a model whereby protonation steps at the active site accompanying reduction ($Mo^{VI} \rightarrow Mo^{V} \rightarrow Mo^{IV}$) become rate limiting in the catalytic cycle. Central to this hypothesis is the assumption that the Mo^{V} form undergoes protonation much more rapidly than Mo^{IV}. At low potentials (high driving force) and high pH the Mo^{IV} form is generated rapidly but catalysis is limited by a slow protonation reaction on the Mo ion. Conversely at intermediate potentials, the intermediate Mo^{V} form accumulates and undergoes rapid (non-rate-limiting) protonation before further reduction to the active Mo^{IV} state and the overall reaction proceeds more quickly. Non-physiological oxidation of PMe_3 to $OPMe_3$ was also seen.

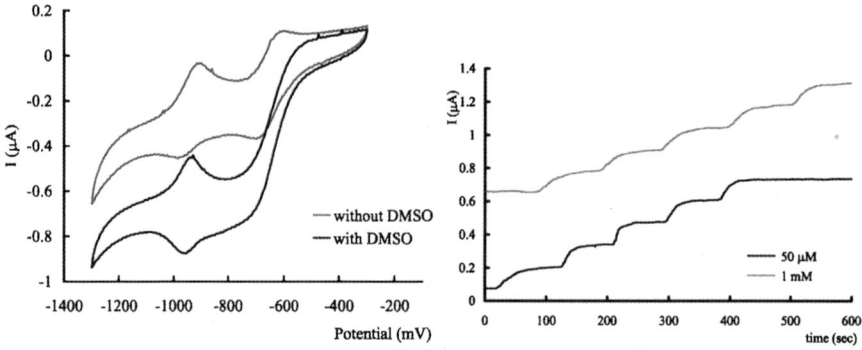

Figure 1.9 (left) Cyclic voltammograms of DMSO reductase/methyl viologen (50 mM) without (light grey) and with DMSO (50 mM, dark grey). The enzyme was coupled with bovine serum albumin using glutaraldehyde and adsorbed on a glassy carbon electrode: pH 6.5, sweep rate 150 mV s^{-1}, potentials *versus* Ag/AgCl; (right) Amperometric MV^{2+} catalytic reduction currents at −750 mV *vs*. Ag/AgCl upon successive additions of DMSO (10 μL, 50 mM). The light-grey line is with 1 mM MV^{2+} and the dark-grey line is with 50 μM MV^{2+}. Reprinted from ref. 62 with permission from Elsevier.

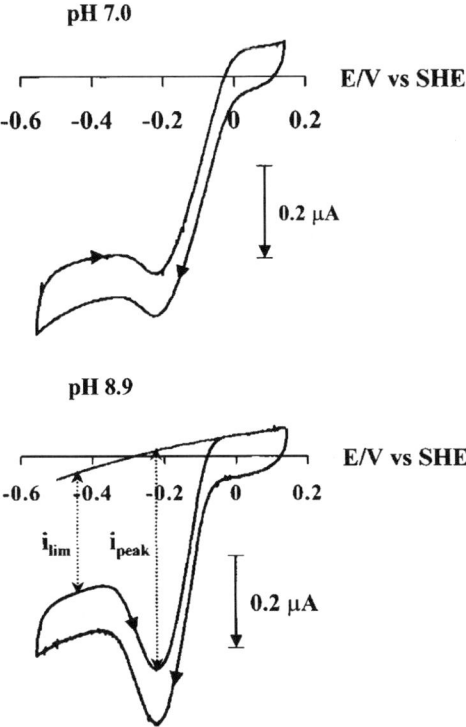

Figure 1.10 Baseline-subtracted cyclic voltammograms of DMSO reductase adsorbed on an edge plane pyrolytic graphite electrode at pH 7.0 and pH 8.9 in the presence of 20 mM DMSO; scan rate 5 mV s^{-1}. Reprinted with permission from ref. 63. Copyright 2001 American Chemical Society.

The periplasmic DMSO reductase from *Rhodobacter capsulatus* (DorA) is structurally unrelated to membrane-bound DmsABC except that they are believed to share the same active site. This enzyme (and the highly homologous enzyme from *R. sphaeroides*) possesses no other cofactors and is thus an attractive choice for spectroscopists where competing optical and EPR signals from Fe-S clusters and heme moieties do not mask those of the Mo cofactor. In fact this simplicity enabled clear resolution, for the first time, of non-turnover Mo$^{VI/V}$ and Mo$^{V/IV}$ signals from a Mo enzyme active site (in the absence of substrate).[64] Optical spectroelectrochemistry was also employed to observe the enzyme in its fully oxidised and reduced forms. Upon addition of DMSO, a catalytic wave was seen.

1.6.2 Nitrate Reductase

Like DMSO reductase, nitrate reductase can be found in distinctly different forms; a soluble periplasmic form (Nap) and a membrane-bound form (Nar) to name but two. Crystal structures of both NapAB[65,66] and NarGH(I)[67–69] have appeared recently. The NarGHI system is a membrane-bound quinol nitrate oxidoreductase complex comprising a number of redox active cofactors in each subunit. The NarI sub-unit is membrane bound and contains two hemes which receive electrons from the quinol pool, but this sub-unit may be separated to leave a catalytically competent and soluble NarGH dimer. The NarG component contains the Mo cofactor and a recently identified Fe-S cluster, while the NarH sub-unit bears a number of Fe-S clusters. Butt and coworkers have reported studies on the soluble NarGH enzyme from *Paracoccus pantotrophus*[70,71] where catalytic reduction ($NO_3^- \rightarrow NO_2^-$) waves were seen in the presence of substrate. At low substrate concentrations, a peak appears in the catalytic wave. The authors suggest that rate-limiting substrate binding to the Mo^V form is more rapid than to Mo^{IV} (a similar argument to the protonation model proposed by Heffron *et al.* for DMSO reductase (Figure 1.10),[63] except that no substrate concentration-dependence of the catalytic waveform was seen in that case. However, given the number of other redox centres in the enzyme it is also possible that a different redox event is responsible for the attenuation in catalytic current at lower potentials. The NarGH enzyme from *E. coli* has more recently been investigated[72] and again complex behaviour is seen comprising low- and high-potential catalytic waves. At micromolar nitrate concentrations a peak is seen in the higher potential component of the catalytic wave, which is then dominated by a more intense lower potential wave at millimolar concentrations of substrate. The high-potential component is pH dependent while the lower component is not. The conclusions of this study again implicate the Mo^V form as a more effective Lewis acid in being able to bind nitrate more rapidly and more tightly than the Mo^{IV} form. It should be reiterated that the catalytic cycle cannot finish until the tetravalent oxidation state is reached but substrate binding emerges as a rate-limiting event in this system.

The periplasmic (Nap) nitrate reductases have received comparatively little attention from electrochemists. They are quite different from the NarGHI nitrate reductases both in the active site structure (a cysteine is coordinated to the Mo instead of an aspartate) and their cofactors. The NapA sub-unit contains the Mo active site and a [4Fe-4S] cluster while the NapB sub-unit has two heme cofactors. Frangioni *et al.* have communicated their results from the direct protein film voltammetry of NapAB from *R. sphaeroides*.[73] Once again, despite the significant differences in active site structure and cofactors, a peak in the ideally sigmoidal catalytic rotating disk voltammogram was seen across a wide range of nitrate concentrations. The authors refer to more complicated behaviour at high nitrate concentrations but no further analysis or data were given.[73]

Reports of mediated nitrate reductase voltammetry appeared earlier than the above-mentioned direct electrochemistry. An early paper used a methyl-viologen-appended poly-pyrrole matrix within which nitrate reductase (*E. coli*) was entrapped.[74] The enzyme/monomer mixture was pre-adsorbed on the electrode to ensure a significant amount of enzyme remained within the poly-pyrrole matrix. A later study from the same group utilised a clay-modified electrode as a template for electropolymerisation which improved the current response.[75] The need for low-potential mediators was established in separate investigations with nitrate reductase enzymes from different organisms. Low-potential viologens and bromophenol dyes (red and blue) are particularly effective whilst higher-potential dyes (*e.g.* indigo sulfonates, with potentials closer to the Mo$^{VI/V}$ and Mo$^{V/IV}$ redox couples) are ineffective.[76,77]

1.6.3 Arsenite Oxidase

Arsenic poisoning from anthropogenic sources is a major health problem in some parts of the world. The most toxic form is trivalent As (which exists naturally as arsenite (As(OH)$_3$). A number of bacteria can defend themselves from arsenite poisoning by using the Mo enzyme arsenite oxidase which oxidises arsenite to the less toxic arsenate (AsO$_4^{3-}$). One of these (*Rhizobium sp. str.* NT-26) is unique in being able to draw energy from arsenite oxidation through its own arsenate oxidase. Direct electrochemistry of this enzyme adsorbed on a pyrolytic graphite electrode has been reported, where a catalytic current in the presence of arsenite is seen (Figure 1.11).[78] The pH optimum and kinetic constants mirrored those seen in solution and the enzyme exhibits stability on an electrode for periods of days. This study laid the foundation for the use of a carbon-nanotube-arsenite-oxidase modified electrode that was used to detect arsenite in a number of water samples.[79] A similar but distinct arsenite oxidase (from *A. faecalis*) is also electroactive and appears to be oxidised in a

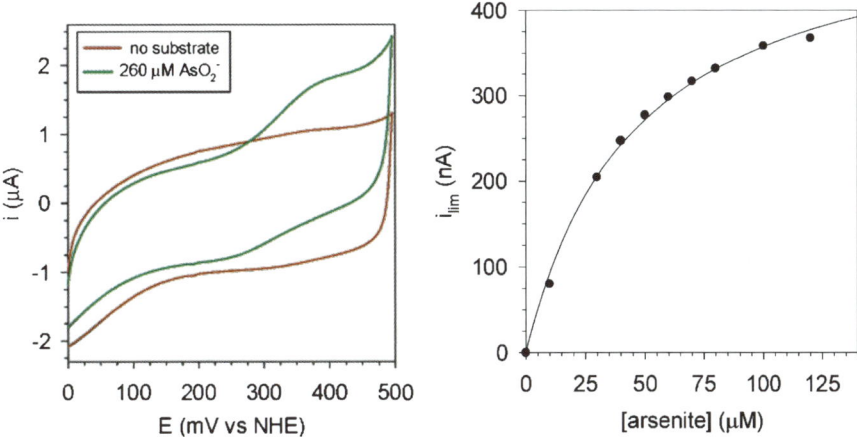

Figure 1.11 (left) Catalytic direct electrochemistry of arsenite oxidase in the absence (red curve) and in the presence (green curve) of arsenite; (right) arsenite concentration dependence of the steady state catalytic current ($K_{M,app}$ 46 μM at pH 5.6). Reprinted with permission from ref. 78. Copyright 2006 American Chemical Society.

cooperative 2-electron reaction (Mo^{IV} to Mo^{VI})[80] in the absence of substrate making it unusual compared with most Mo enzymes.

1.6.4 Chlorate and Perchlorate Reductase

There are bacteria that are capable of respiring on oxoanions such as perchlorate (ClO_4^-) or chlorate (ClO_3^-). The overall picture is quite complex.[81,82] Some bacteria may reduce either anion for respiration by utilising distinctly different (but related) Mo enzymes perchlorate reductase[83] and chlorate reductase.[84] The analytical detection of perchlorate is difficult as the anion, although in a very high oxidation state and a potentially strong oxidant, is rather inert in solution. Immobilisation of perchlorate reductase in a Nafion® film cast on a glassy carbon electrode has recently enabled the construction of a functional perchlorate biosensor.[85] Although chlorate reductase electrochemistry has not been investigated to date, chlorate can be catalytically reduced by the Mo enzyme nitrate reductase (NarGH)[71] in a non-physiological but important reaction.

1.7 Conclusions

Studies of mononuclear Mo enzymes received a tremendous boost with the publication of the first crystal structures of these enzymes a little more than 10 years ago,[86] which provided a clearer view of the organisation of electron transfer within these mostly complex metalloenzymes than was ever possible before. Since then the combined efforts of microbiologists, biochemists, geneticists, structural biologists, spectroscopists and, most recently, electrochemists have unearthed an ever-expanding family of enzymes whose substrates span a remarkably wide range of species, both organic and inorganic, yet who share essentially the reaction stoichiometry (eqn (1)). The observation of some unusual electrochemical behaviour, particularly the attenuation of activity at high overpotential, is remarkable but has now been observed in a number of Mo enzymes (mostly from the DMSO reductase family). Whether this non-classical electrochemical behaviour has a physiological significance or instead it is an experimental artefact is yet to be established.

New, as-yet biochemically uncharacterised, mononuclear molybdenum enzymes are now appearing in genomes of various organisms at an overwhelming rate and there is evidently much work to be done in elucidating the natural substrates of these enzymes and ultimately being able to express these proteins in sufficient quantities to enable full characterisation and then utilisation in novel biosensors.

Despite the many opportunities for determining the solution concentrations of substrates that present significant challenges for wet chemical methods, Mo enzymes biosensors are yet to reach commercial development, but this is a familiar tale. The phenomenal success of the glucose oxidase biosensor was driven by a need in the marketplace for such a device.[87] The technological issues for development of Mo-enzyme-based biosensors are no more challenging than those presented by the glucose biosensor but economics and market forces will decide which systems warrant intensive development into commercially viable biosensors.[88] At the moment the Mo-enzyme systems that probably warrant the most attention are those for the determination of sulfite (sulfite oxidase/dehydrogenase) and arsenite (arsenite oxidase) due to the potentially adverse health effects of each to humans and the difficulties surrounding their determination by wet chemical methods. This chapter has hopefully provided a perspective of the systems that are available for future development and an impetus for their further study.

References

1. R. Hille, *Met. Ions Biol. Syst.*, 2002, **39**, 187.
2. R. Hille, *Chem. Rev.*, 1996, **96**, 2757.
3. S. Leimkühler, A. L. Stockert, K. Igarashi, T. Nishino and R. Hille, *J. Biol. Chem.*, 2004, **279**, 40437.
4. A. E. G. Cass, G. Davis, G. D. Francis, H. A. O. Hill, W. J. Aston, I. J. Higgins, E. V. Plotkin, L. D. L. Scott and A. P. F. Turner, *Anal. Chem.*, 1984, **56**, 667.
5. F. A. Armstrong, *J. Chem. Soc., Dalton Trans.*, 2002, 661.
6. F. A. Armstrong, *Curr. Opin. Chem. Biol.*, 2005, **9**, 110.
7. P. V. Bernhardt, *Aust. J. Chem.*, 2006, **59**, 233.
8. C. Enroth, B. T. Eger, K. Okamoto, T. Nishino, T. Nishino and E. F. Pai, *Proc. Nat. Acad. Sci. USA*, 2000, **97**, 10723.
9. J. J. Truglio, K. Theis, S. Leimkühler, R. Rappa, K. V. Rajagopalan and C. Kisker, *Structure*, 2002, **10**, 115.
10. E. Watanabe, K. Ando, I. Karube, H. Matsuoka and S. Suzuki, *J. Food Sci.*, 1983, **48**, 496.
11. G. Volpe and M. Mascini, *Talanta*, 1996, **43**, 283.
12. Y. Yano, N. Kataho, M. Watanabe, T. Nakamura and Y. Asano, *Food Chem.*, 1995, **52**, 439.
13. J. Niu and J. Y. Lee, *Anal. Commun.*, 1999, **36**, 81.
14. P. Kotzian, P. Brazdilova, K. Kalcher and K. Vytras, *Anal. Lett.*, 2005, **38**, 1099.
15. F. Xu, L. Wang, M. Gao, L. Jin and J. Jin, *Talanta*, 2002, **57**, 365.
16. G. J. Moody, G. S. Sanghera and J. D. R. Thomas, *Analyst*, 1987, **112**, 65.
17. L. Mao, F. Xu, Q. Xu and L. Jin, *Anal. Biochem.*, 2001, **292**, 94.
18. E. Llaudet, N. P. Botting, J. A. Crayston and N. Dale, *Biosens. Bioelectr.*, 2003, **18**, 43.
19. H. De Groot, H. De Groot and T. Noll, *Biochem. J.*, 1985, **230**, 255.
20. E. M. D'Urso and P. R. Coulet, *Anal. Chim. Acta*, 1990, **239**, 1.
21. S. D. Haemmerli, A. A. Suleiman and G. G. Guilbault, *Anal. Biochem.*, 1990, **191**, 106.
22. E. M. D'Urso and P. R. Coulet, *Anal. Chim. Acta*, 1993, **281**, 535.

23. T. Yao, *Anal. Chim. Acta*, 1993, **281**, 323.
24. S. Hu and C. C. Liu, *Electroanalysis*, 1997, **9**, 1174.
25. J. H. T. Luong and K. B. Male, *Enzyme Microb. Technol.*, 1992, **14**, 125.
26. E. Gonzalez, F. Pariente, E. Lorenzo and L. Hernandez, *Anal. Chim. Acta*, 1991, **242**, 267.
27. G. Arai, S. Takahashi and I. Yasumori, *J. Electroanal. Chem.*, 1996, **410**, 173.
28. J. L. Fernandez, N. Mano, A. Heller and A. J. Bard, *Angew. Chem. Int. Ed.*, 2004, **43**, 6355.
29. L. Mao and K. Yamamoto, *Anal. Chim. Acta*, 2000, **415**, 143.
30. S. S. Hu and C. C. Liu, *Electroanalysis*, 1997, **9**, 372.
31. H. Okuma, H. Takahashi, S. Sekimukai, K. Kawahara and R. Akahoshi, *Anal. Chim. Acta*, 1991, **244**, 161.
32. K. F. Aguey-Zinsou, P. V. Bernhardt and S. Leimkuehler, *J. Am. Chem. Soc.*, 2003, **125**, 15352.
33. P. V. Bernhardt, M. J. Honeychurch and A. G. McEwan, *Electrochem. Commun.*, 2006, **8**, 257.
34. M. Resch, H. Dobbek and O. Meyer, *J. Biol. Inorg. Chem.*, 2005, **10**, 518.
35. H. Dobbek, L. Gremer, O. Meyer and R. Huber, *Proc. Natl. Acad. Sci. USA*, 1999, **96**, 8884.
36. H. Dobbek, L. Gremer, R. Kiefersauer, R. Huber and O. Meyer, *Proc. Natl. Acad. Sci. USA*, 2002, **99**, 15971.
37. A. P. F. Turner, W. J. Aston, I. J. Higgins, J. M. Bell, J. Colby, G. Davis and H. A. O. Hill, *Anal. Chim. Acta*, 1984, **163**, 161.
38. K. Fischer, G. G. Barbier, H.-J. Hecht, R. R. Mendel, W. H. Campbell and G. Schwarz, *Plant Cell*, 2005, **17**, 1167.
39. C. Kisker, H. Schindelin, A. Pacheco, W. A. Wehbi, R. M. Garrett, K. V. Rajagopalan, J. H. Enemark and D. C. Rees, *Cell*, 1997, **91**, 973.
40. E. Karakas, H. L. Wilson, T. N. Graf, S. Xiang, S. Jaramillo-Busquets, K. V. Rajagopalan and C. Kisker, *J. Biol. Chem.*, 2005, **280**, 33506.
41. N. Schrader, K. Fischer, K. Theis, R. R. Mendel, G. Schwarz and C. Kisker, *Structure*, 2003, **11**, 1251.
42. U. Kappler and S. Bailey, *J. Biol. Chem.*, 2005, **280**, 24999.
43. C. C. Moser, J. M. Keske, K. Warncke, R. S. Farid and P. L. Dutton, *Nature*, 1992, **355**, 796.
44. A. Pacheco, J. T. Hazzard, G. Tollin and J. H. Enemark, *J. Biol. Inorg. Chem.*, 1999, **4**, 390.
45. C. Feng, R. V. Kedia, J. T. Hazzard, J. K. Hurley, G. Tollin and J. H. Enemark, *Biochemistry*, 2002, **41**, 5816.
46. A. Isaac, J. Davis, C. Livingstone, A. J. Wain and R. G. Compton, *Trend. Anal. Chem.*, 2006, **25**, 589.
47. B. Timbo, K. M. Koehler, C. Wolyniak and K. C. Klontz, *J. Food Protect.*, 2004, **67**, 1806.

48. M. Situmorang, D. B. Hibbert, J. J. Gooding and D. Barnett, *Analyst*, 1999, **124**, 1775.
49. E. Dinckaya, M. K. Sezgintuerk, E. Akyilmaz and F. N. Ertas, *Food Chem.*, 2006, **101**, 1540.
50. S. B. Adeloju, J. N. Barisci and G. G. Wallace, *Anal. Chim. Acta*, 1996, **332**, 145.
51. S. B. Adeloju, A. Ohanessian and N. N. Duc, *Synth. Met.*, 2005, **153**, 17.
52. E. E. Ferapontova, T. Ruzgas and L. Gorton, *Anal. Chem.*, 2003, **75**, 4841.
53. R. Dronov, D. G. Kurth, H. Moehwald, R. Spricigo, S. Leimkuehler, U. Wollenberger, K. V. Rajagopalan, F. W. Scheller and F. Lisdat, *J. Am. Chem. Soc.*, 2008, **130**, 1122.
54. J. P. Hart, A. K. Abass and D. Cowell, *Biosens. Bioelectr.*, 2002, **17**, 389.
55. L. A. Coury Jr., B. N. Oliver, J. O. Egekeze, C. S. Sosnoff, J. C. Brumfield, R. P. Buck and R. W. Murray, *Anal. Chem.*, 1990, **62**, 452.
56. L. A. Coury Jr., R. W. Murray, J. L. Johnson and K. V. Rajagopalan, *J. Phys. Chem.*, 1991, **95**, 6034.
57. L. A. Coury Jr., L. Yang and R. W. Murray, *Anal. Chem.*, 1993, **65**, 242.
58. L. Yang, L. A. Coury Jr. and R. W. Murray, *J. Phys. Chem.*, 1993, **97**, 1694.
59. K.-F. Aguey-Zinsou, P. V. Bernhardt, U. Kappler and A. G. McEwan, *J. Am. Chem. Soc.*, 2003, **125**, 530.
60. S. J. Elliott, A. E. McElhaney, C. Feng, J. H. Enemark and F. A. Armstrong, *J. Am. Chem. Soc.*, 2002, **124**, 11612.
61. M. Abo, M. Dejima, F. Asano, A. Okubo and S. Yamazaki, *Tetrahedron: Asymmetry*, 2000, **11**, 823.
62. M. Abo, Y. Ogasawara, Y. Tanaka, A. Okubo and S. Yamazaki, *Biosens. Bioelectron.*, 2003, **18**, 735.
63. K. Heffron, C. Leger, R. A. Rothery, J. H. Weiner and F. A. Armstrong, *Biochemistry*, 2001, **40**, 3117.
64. K.-F. Aguey-Zinsou, P. V. Bernhardt, A. G. McEwan and J. P. Ridge, *J. Biol. Inorg. Chem.*, 2002, **7**, 879.
65. J. Dias, M. Than, A. Humm, G. P. Bourenkov, H. D. Bartunik, S. Bursakov, J. Calvete, J. Caldeira, C. Carneiro, J. J. Moura, I. Moura and M. J. Romao, *Structure*, 1999, **7**, 65.
66. P. Arnoux, M. Sabaty, J. Alric, B. Frangioni, B. Guigliarelli, J.-M. Adriano and D. Pignol, *Nat. Struct. Biol.*, 2003, **10**, 928.
67. M. G. Bertero, R. A. Rothery, M. Palak, C. Hou, D. Lim, F. Blasco, J. H. Weiner and N. C. J. Strynadka, *Nat. Struct. Biol.*, 2003, **10**, 681.
68. M. Jormakka, D. Richardson, B. Byrne and S. Iwata, *Structure*, 2004, **12**, 95.
69. M. G. Bertero, R. A. Rothery, N. Boroumand, M. Palak, F. Blasco, N. Ginet, J. H. Weiner and N. C. J. Strynadka, *J. Biol. Chem.*, 2005, **280**, 14836.

70. L. J. Anderson, D. J. Richardson and J. N. Butt, *Faraday Discuss.*, 2000, **116**, 155.
71. L. J. Anderson, D. J. Richardson and J. N. Butt, *Biochemistry*, 2001, **40**, 11294.
72. S. J. Elliott, K. R. Hoke, K. Heffron, M. Palak, R. A. Rothery, J. H. Weiner and F. A. Armstrong, *Biochemistry*, 2004, **43**, 799.
73. B. Frangioni, P. Arnoux, M. Sabaty, D. Pignol, P. Bertrand, B. Guigliarelli and C. Leger, *J. Am. Chem. Soc.*, 2004, **126**, 1328.
74. S. Cosnier, C. Innocent and Y. Jouanneau, *Anal. Chem.*, 1994, **66**, 3198.
75. S. Da Silva, D. Shan and S. Cosnier, *Sens. Actuators, B*, 2004, **B103**, 397.
76. B. Strehlitz, B. Gruendig, K. D. Vorlop, P. Bartholmes, H. Kotte and U. Stottmeister, *Fresenius J. Anal. Chem.*, 1994, **349**, 676.
77. D. Kirstein, L. Kirstein, F. Scheller, H. Borcherding, J. Ronnenberg, S. Diekmann and P. Steinrucke, *J. Electroanal. Chem.*, 1999, **474**, 43.
78. P. V. Bernhardt and J. M. Santini, *Biochemistry*, 2006, **45**, 2804.
79. K. B. Male, S. Hrapovic, J. M. Santini and J. H. T. Luong, *Anal. Chem.*, 2007, **79**, 7831.
80. K. R. Hoke, N. Cobb, F. A. Armstrong and R. Hille, *Biochemistry*, 2004, **43**, 1667.
81. J. Xu, J. J. Trimble, L. Steinberg and B. E. Logan, *Water Res.*, 2004, **38**, 673.
82. L. M. Steinberg, J. J. Trimble and B. E. Logan, *FEMS Microbiol. Lett.*, 2005, **247**, 153.
83. K. S. Bender, C. Shang, R. Chakraborty, S. M. Belchik, J. D. Coates and L. A. Achenbach, *J. Bacteriol.*, 2005, **187**, 5090.
84. A. F. W. M. Wolterink, E. Schiltz, P.-L. Hagedoorn, W. R. Hagen, S. W. M. Kengen and A. J. M. Stams, *J. Bacteriol.*, 2003, **185**, 3210.
85. C. Okeke Benedict, G. Ma, Q. Cheng, E. Losi Mark and T. Frankenberger William Jr., *J Microbiol. Methods*, 2007, **68**, 69.
86. R. Hille, *J. Biol. Inorg. Chem.*, 1996, **1**, 397.
87. J. D. Newman and A. P. F. Turner, *Biosens. Bioelectr.*, 2005, **20**, 2435.
88. P. T. Kissinger, *Biosens. Bioelectr.*, 2005, **20**, 2512.

CHAPTER 2
Scanning Probe Analyses at the Bioelectronic Interface

JASON J. DAVIS, BEN PETERS, YUKI HANYU AND WANG XI

Chemistry Research Laboratory, University of Oxford, Mansfield Road, Oxford OX1 3TA, UK

2.1 Introduction

In both understanding the electron transfer events that underpin many fundamental biological processes and developing new diagnostic techniques, the robust interfacing of man-made electronic configurations with biomolecules is an important and wide-reaching requirement. Despite significant progress in establishing communication between protein and electrode, the non-destructive, high-coverage and, ideally, highly homogeneous immobilisation of proteins or enzymes on planar surfaces remains experimentally challenging. Heterogeneous immobilisation processes will adversely impact the definition of thermodynamic and kinetic analyses and signal reproducibility. As surface chemical analysis and imaging methods have evolved so has our ability to utilise highly specific interfacial interactions to orientate biomolecules, control the rate of electron transfer to the surface (through distance modulation) and perform analysis at molecular levels of spatial resolution. In the latter case, the advent and development of high-resolution scanning probe microscopy (SPM) has been fundamental and enables the routine imaging of interfaces under a variety of controllable conditions. Such analyses help considerably in both the interpretation of voltammetric data and the refinement of electrode topography and chemistry. In appropriate configurations it is, additionally, possible to

---

Engineering the Bioelectronic Interface: Applications to Analyte Biosensing and Protein Detection
Edited by Jason Davis
© 2009 Royal Society of Chemistry
Published by the Royal Society of Chemistry, www.rsc.org

utilise imaging probes as electrodes and to (in effect) perform electronic and electroanalytical analyses in the near field; that is, at levels of the single molecule. The aim of this chapter is to introduce scanning probe technology, its application to biomolecular imaging and the bioelectronic interface, and its evolution beyond imaging to direct, single molecule, experimental analysis.

2.1.1 Scanning Probe Microscopy

Scanning probe microscopy (SPM) has become, over the last decade, a routinely used tool for the investigation of sub-micron surface structure. This is, in part, due to the increase in commercial availability of instrumentation designed for a range of applications; from large-scale, industrial parallel analyses of, for example, hard-disk surfaces to sub-nanometre resolution of atomic structure under UHV or electrolytic solution. SPM hardware can be combined with optical microscopes for the study of macroscopic biological systems such as living cells or the utilisation of near-field optical effects. Operational modes are highly variant and can be tuned to the sample or imaging analysis as required; probe-sample interactions can be minimal or a key part of, for example, force or surface potential mapping, conductance assessments or redox-tuned tunnelling analyses. The SPM has, therefore, been utilised as a powerful tool across the fields of science and engineering. In this section we will outline the modes of operation of the two most commonly used scanning probe techniques, Scanning Tunnelling Microscopy (STM) and Atomic Force Microscopy (AFM), and describe examples where they have been applied to the bioelectronic interface.

2.1.1.1 Scanning Tunnelling Microscopy

The world's first scanning probe microscope was developed in the IBM Laboratory, Zurich, in 1981.[1] The microscope was named the Scanning Tunnelling Microscope (STM), a reflection of its operating mechanism, and was novel in that it completely lacked the optical components found in traditional microscopes. Imaging of surface topological and electronic features was instead performed by monitoring the tunnelling current flowing between an (ideally) atomically sharp probe and conductive surface through a surface raster-scan (Figure 2.1). Alternatively, the tunnelling current can be fed into a feedback loop to actuate the probe (or underlying surface) and maintained constant. In either mode, spatial resolutions are routinely atomic in scale. Despite considerable advances in mechanical controls and software, operational principles remain the same today. Tunnelling microscopes are efficiently integrated into solution and electrochemical (as well as UHV) environments, directly leading to powerful imaging and spectroscopic applications at biological and bioelectrochemical interfaces.[2-5]

In a tunnelling configuration, the current I has an inversely exponential relationship with the probe-substrate separation d and is recorded pixel-by-pixel

Scanning Probe Analyses at the Bioelectronic Interface 27

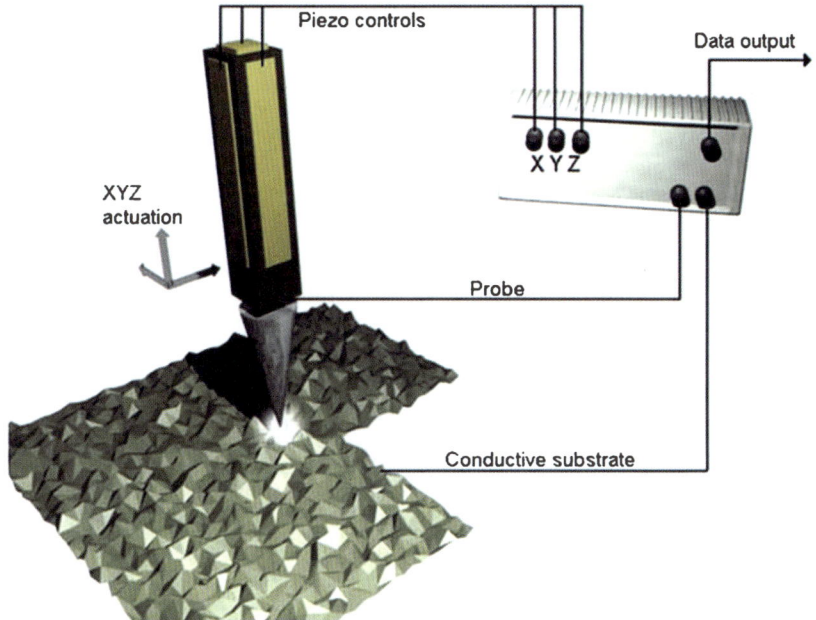

Figure 2.1 Schematic representation of an STM. An electrochemically etched wire is brought to within a few nanometres of a conductive substrate, and, by the application of a voltage bias between it and the underlying surface, a measurable tunnelling current flows. The variance in this as the probe/sample raster scan with respect to each other can be used to map topographic/electronic characteristics with atomic scales of spatial resolution. This is achieved by an XY piezo actuator controlling the lateral raster scan, a Z piezo and PID feedback electronics controlling the tip-sample separation (the action record of feedback loop also being exported through data output).

as the surface and probe scan relative to each other. With homogeneous materials, such as a metallic single crystal, this current map directly reflects surface topography.[6] In heterogeneous surfaces, image contrast is a convolution of topographic and electronic contributions. In addition to mapping the inversely exponential *I-d* relation, modern electronics are capable of measuring currents of the order of femtoamperes. This extreme sensitivity enables STM to reliably detect surface features as small as 0.1 Å in height. Lateral resolution is limited by the probe geometry, most notably its radius of curvature, and the sample mechanical and immobilisation characteristics. Compared to AFM (see below), STM has a higher lateral resolving power and, with care, can be operated under conditions where mechanical interactions with the sample are minimal (high tunnelling resistance; the tunnelling current can pass through an insulating gap and so does not require the probe to be in physical contact with the surface of interest). The requirement of measurable current flow does, however, preclude the imaging of large biological systems such as live cells;

typical tunnelling currents can only penetrate features of the order of a few nanometres thick. STM systems are, therefore, most effectively applied to atomic and molecular-scale analyses.

An STM has two image channels – current and topography. The latter is a spatially resolved record of actuation of the scanning probe z-piezo by the feedback electronics; the former is a record of the raw tunnelling current. There are two fundamental controllable parameters available to optimise the imaging conditions: the bias voltage applied between the tip and underlying surface and the tunnelling current set-point the software is asked to generate and maintain. For typical organic adsorbates these values would be 0.1–1 V of bias and 0.05–1 nA of current, under which conditions the probe will typically scan ~ 5–10 Å above the surface,[6] facilitating non-destructive imaging (this distance is tuned by varying the tip-substrate bias and tunnelling set-point, enabling tip-sample interactions to be investigated if desired). Since the applied bias is dropped entirely across the small tip-substrate gap, the electric field gradient is immense, typically exceeding $100 \, MV \, m^{-1}$. From a bioelectronic perspective, molecules confined to the tip-sample gap are potentially less solvated and may, accordingly, exhibit reorganisation energies which differ from those measured by bulk electroanalyses.[7]

Probe Considerations. The most commonly used STM probe materials include tungsten, iridium-platinum and gold, all of which display reasonable chemical inertness and mechanical pliability. Tungsten wire is typically electrochemically etched to a sharp probe in NaOH solution. Iridium-platinum probes are prepared by mechanical incision of the alloy wire or etched. Gold is inert, but expensive and soft; probes can be prepared by either mechanical incision or electrochemical etching in an ethanol/HCl mixture.[8] Resistance to ambient (or electrolytic) oxidation is important in maintaining a consistent density of electronic states, probe geometry and conductivity; a change in any of these parameters may profoundly affect imaging. Tungsten is low cost, but is relatively easily oxidised (leading to potential instability under electrochemical conditions). Platinum-iridium tips are mechanically hard and inert. Mechanical cutting tends to generate an irregular apex that can be difficult to reliably coat. Electrochemical etching in concentrated hydroxide, though, is effective.[9] Although time-consuming, there is a procedure that uses innocuous alternatives.[10–12]

Modes of Imaging. The STM may be operated in one of two modes: constant *current* and constant *height*. In the former, the feedback loop is configured to maintain the tunnelling current at a certain set-point through z-actuation. When the probe encounters a protrusion during a scan, there is an increase in tunnelling current due to the closer proximity of the probe to the surface (the chemical/atomic composition of the protrusion will also affect the tunnelling probability). This increase in current is detected and countered by the feedback loop which increases the probe-substrate separation until the set-point current is restored. Likewise, in the case of an indentation the probe-substrate separation is reduced. The record of z-movement during a surface scan is utilised in the construction of a height map. One limitation of this mode is the response time of the feedback electronics; the loop may fail if tunnelling current changes too rapidly. In allowing more time for the feedback loop to adjust its parameters to detected current modulation by scanning slowly, a better representation of the surface is generally acquired (though this must be balanced against the possibility of mechanical drift during the elongated scan duration).

In constant height mode a constant z-position is maintained while scanning laterally and the topographically or electronically induced current modulation is recorded (with no attempt to counteract this). In highly homogeneous samples there is a near-direct correlation between current and topology, and while this is not the case for biomolecules on conductive substrates, this imaging mode can demonstrably reveal sub-molecular structure[13] and be used for tracking and quantifying triggered molecular conductance change (see Section 2.2).

2.1.1.2 Atomic Force Microscopy

Though offering potentially exquisite imaging resolution (and studies which align nicely with redox analyses – see later), STM methods cannot be applied to bulk insulators and produce images which are potentially complex convolutions of topological and electronic properties. In 1986, Binnig and Quate developed another form of SPM, the atomic force microscope (AFM), an instrument based on the accurate measurement of minute force interactions between an appropriate scanning probe and an underlying surface.[14] Their AFM consisted of an atomically sharp probe mounted at one end of a cantilever, the deflection of which was detected by an STM probe. The most common modern AFM consists of a probe mounted on a microfabricated (smaller levers have higher resonant frequencies and are less susceptible to ambient vibration) cantilever, a piezoelectric scanner to control the (xyz) positions of either sample or probe and a laser optical system to detect any deflection of the cantilever (Figure 2.2). Cantilever deflection data, detected by a position-sensitive photodetector, is processed and fed into a feedback loop that operates an electro-mechanical positioning device in the z-direction, in a similar manner to STM, capable of displacements between 1 Å and 100 μm.[15] As with STM, slow scan rates allow the feedback loop to follow more accurately the surface topology. As a result, the topological image becomes more accurate and z-feedback error values decrease. The force interactions responsible for generating images are often not confined to those at the truly atomic scale (from which the microscope bears its name) and will typically include electrostatic and meniscus forces, among others.

In the AFM setup illustrated in Figure 2.2, three imaging modes are possible: contact mode, intermittent contact (IC) mode and frequency-modulated

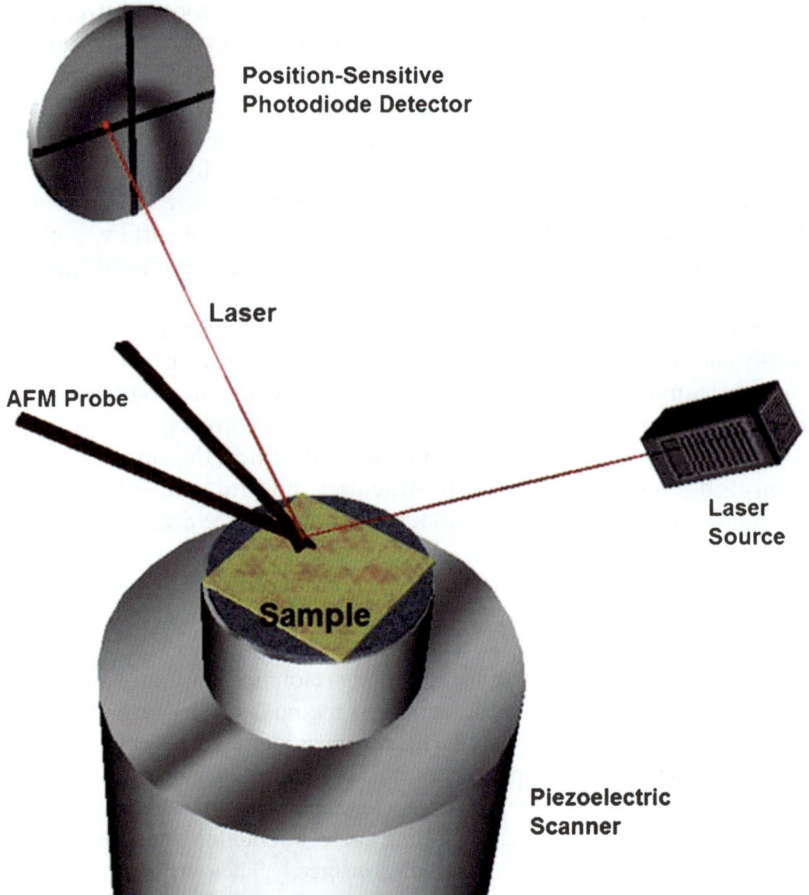

Figure 2.2 Schematic representation of an AFM. The deflection of a microfabricated cantilever in response to surface features is detected *via* a laser reflected from its rear. By setting this (deflection) to a pre-determined value, the force imparted by the tip on the surface can be controlled to sub-nanonewton levels of resolution in contact mode. In intermittent contact mode, the cantilever is driven at close to its natural frequency and the interaction of the tip with the surface measured by the change in amplitude of the cantilever response. As with the STM, relative tip and sample raster scanning is software controlled and piezo driven (either the tip or the sample are piezo mounted).

tapping mode. In contact mode, the feedback loop maintains cantilever *deflection* at a set level by moving the z-piezo in response to vertical movements of the cantilever. In IC mode the cantilever is driven close to its resonant frequency by applying an alternating current to the z-piezo, resulting in the tip oscillating in the z direction. Probe oscillation can also be driven magnetically, in which case it is called magnetic AC mode (or MAC mode, patented by Agilent Technologies); this can be particularly beneficial in fluid environments

Table 2.1 Imaging channels and corresponding output in different AFM imaging modes. STM channel information is given in brackets for comparison.

Imaging modes	z-feedback data	Error signal	Damping data
Contact mode	Topography	Deflection	Friction
Tapping mode	Topography	Amplitude	Phase
FM-Tapping mode	Topography	Frequency-shift	Dissipation
(STM)	(Topography)	(Current)	

where mechanical responses of the lever are simpler and more isolated than possible with acoustic excitation.[16] The amplitude of this is greatest when the cantilever is away from the surface and free from potentially damping interactions. When the tip is brought into contact with an underlying surface, the amplitude is decreased by a combination of a shift in the cantilever's power spectrum and energy dissipation to the surface. In normal imaging modes, the feedback loop serves to keep this amplitude constant. In addition to height and oscillation amplitude information, the phase of the cantilever motion can be measured with respect to the driving signal. Materials of differing mechanical characteristics introduce different phase lags, a phenomenon which can be utilised in building up mechanical maps of a surface. In frequency-modulated tapping mode a constant oscillation *frequency* is maintained. This mode operates in the attractive (or adhesive) region some distance from the surface.[17]

In all three of these modes the AFM acquires data through two main channels: error signal and displacement. The error signal is the discrepancy between the set-point and measured value (cantilever deflection, amplitude or frequency), while displacement is the movement in z-direction the feedback loop exerts in order to restore (ideally) zero error signal. Additionally, a third channel, damping, is any external force on the AFM probe to hinder its movement. The three channels are summarised in Table 2.1.

Imaging Considerations. The primary data input for an AFM is from the photodetector that monitors cantilever deflection *via* the reflected laser signal; a highly-reflective cantilever is, therefore, desirable for high-sensitivity, low-noise, imaging. The back side of cantilevers is usually coated with aluminium or gold (called the *reflective coating*) in order to maximise reflection. One drawback of adding a reflective coating of a differing material from the probe is a thermal non-equilibrium of the cantilever arising from incident laser heating and potentially generating up to $\pm 25\,\mathrm{pN}$ of thermal noise.[18] If imaging is performed in air or vacuum, thermal equilibrium is quickly reached upon irradiation due to low heat capacity of the AFM probe material. The thermal non-equilibrium becomes more unpredictable in fluid environments as equilibrium is relatively slowly attained. In such cases, the experimenter can wait until thermal equilibrium is reached, lower the temperature[19] or install a heating element to pre-condition the imaging environment.[20]

An additional fundamental limitation of AFM is that defined by the probe geometry.[21] In theory, the ideal probe geometry is an infinitely sharp needle: one of zero radius of curvature and infinitely high aspect ratio (impossible in practice and unlikely to be maintained in potentially blunting or contaminating environments). The effect of the finite tip size on image quality is called *tip convolution*. For example, a simple geometric model predicts that imaging a protein of 5 nm in diameter with an 8 nm probe would lead to a measured protein diameter of ~ 18 nm.[22] Commercial AFM probes aim to minimise the radius of curvature and maximise aspect ratio while maintaining mechanical strength (quoted radii of curvature are typically less than 8 nm). The use of carbon-nanotube-modified probes is the current closest approximation to the theoretically ideal "infinitely sharp needle",[23] and has been used to improve the quality of biomolecule imaging. For example, Bunch *et al.*[24] demonstrated a decrease in the measured width of a surface-immobilised DNA from ~ 10–20 nm as measured with a conventional silicon tip to 3.5 nm when imaged with a carbon-nanotube-modified probe.

Force Considerations. While the tip radius of curvature ultimately limits AFM lateral resolution, the cantilever spring constant, k, has a great bearing on measured feature heights, on image quality and on the force of interaction between tip and sample. For higher sensitivity and minimal sample deformation, a cantilever with lowest possible force constant (typically 0.1–1 N m^{-1}) is used; this is particularly important in contact mode imaging of soft (biological) samples.[25] For IC mode, a stiff cantilever (force constants typically ~ 40 N m^{-1}) with a high natural resonant frequency (100–500 kHz) is used. The danger of probe-induced surface damage/perturbation during imaging can be reduced in this mode, particularly when operated in a region away from the surface or so-called "attractive regime". The presence of meniscus forces and films of contamination in ambient environments can, however, prohibit operation in this regime.[26] Operation in the "repulsive regime" brings with it potentially considerable vertical compression of the sample under analysis but maintains minimal lateral force interaction. Equivalent consideration applies to the adhesive forces present in typically ambient contact mode operation (these can be tens of nanonewtons in magnitude). Operation under fluid, though more experimentally demanding, facilitates a dramatic reduction in probe-sample interaction force.[27] Compressional forces exerted on biological samples are, additionally, directly modulated by the imaging feedback conditions (see below).

The monitored cantilever deflection inherent in any AFM configuration can be utilised in both force sensing and controlled nanomanipulation. A "force spectrum" is a plot of cantilever deflection against tip-sample distance (see Section 2.1.2.2). In the retracted position, when the tip is far from the surface, no force operates between the tip and sample and the spectrum lies at its background level. As the probe gradually approaches the surface and ultimately meets it at the point of contact, atomic forces come into effect and the cantilever deflects. As the cantilever moves away from the surface, for a distance beyond the point of contact, an adhesive force acts between it and the underlying surface, the origin of which is largely van der Waals and water meniscus based. As the cantilever retracts further through the adhesive domain, the restoration force being built up as it deflects ultimately overcomes the adhesive force and the tip jumps to the completely retracted position. In studies relating to biorecognition, bioaffinity and mechanical properties of nano-objects, the point of scrutiny is this exact moment of the probe's breakage from adhesive domain and facilitates powerful application in spatially resolved force mapping or molecular unfolding (see Section 2.1.2).

Lateral Forces. As a probe scans over a sample, it experiences a lateral shear force, or "twisting force", due to frictional interactions. The torsional deflection generated can be recorded to produce a friction image of a surface, such as that shown in Figure 2.3. Though the origin of friction force is somewhat complex, and typically involves a multitude of contributing interactions (van der Waals, hydrophobic, electrostatic and meniscus adhesion),[28] this constitutes a powerful means of generating "chemical contrast" across a surface (normal modes of AFM operation, contact or IC, cannot, to a first approximation, determine the chemical identity of an underlying surface),[29] particularly when the probe is chemically modified (chemical force

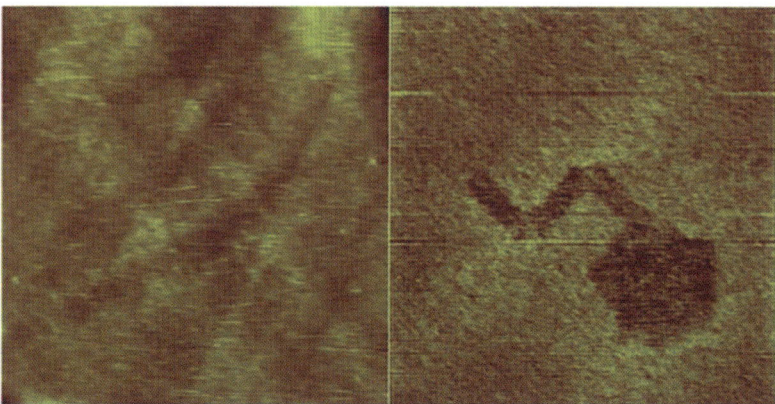

Figure 2.3 An example of pattern nanofabricated by catalytic nanolithography. As a metal nanoparticle modified probe scans over a vinyl-terminated SAM in the presence of 4-iodobenzoic acid (in this example) a localised catalytic Heck reaction occurs which couples the solution phase reagent to the surface reagent (introducing local carboxylic acid terminal functionality). Unlike replacement nanolithography, also known as "nanografting", this technique does not impinge on the structural integrity of the host SAM. The surface functional group changes are too small to be detected topographically (left) but reliably resolved using chemical force microscopy (right). Both images are 1 μm by 1 μm in size. The spatial resolution of such catalytic nanolithography can reach 10 nm.[39]

microscopy, CFM). CFM has been used to distinguish hydrophilic from hydrophobic regions,[30] interactions between specific functional groups such as polar moieties, and acid-base interactions with nanometre scales of resolution.[31] In biological analyses, CFM can be employed to detect specific intermolecular interactions such as protein-substrate interactions and DNA pairing events.[32,33]

2.1.1.3 Nanolithography

The discussed mechanical interactions between a scanning probe and its underlying surface can be utilised in inducing highly spatially resolved changes in the latter. Collectively known as scanning probe lithography (SPL), a multitude of methods now exist of achieving nano-scale manipulation of chemical, electronic and magnetic properties. Capable of greater spatial resolution than "more industrial" microfabrication methods such as focused ion beam (and at much reduced cost and increased environmental and sample flexibility), force controls can also be sufficiently refined so as to facilitate the controlled manipulation of single atoms and molecules.[34,35] More recently, meniscus forces between an AFM probe and an underlying surface have been used in molecular and nanoparticle delivery.[36,37] Even more refined has been the recent demonstration of functionally active probes in inducing local chemical reactions.[38-40] These near-field abilities can also be applied to nanometre-scale biological engineering, such as the fabrication of protein or DNA arrays,[41] the manipulation of single protein molecules[42] and DNA strands,[43] and the controlled immobilisation of cells and viruses for further analysis.[44]

2.1.1.4 Dynamic Imaging and Temporal Resolution

The spatial resolution of an SPM scan depends, not only on the physical properties of the probe and sample, but also on the scan rate. Faster scan rates tend to reduce resolution as the control system requires a finite time to respond to changes in surface topography. Typically, a scan rate of 1–2 Hz is used for scans under a micron in size, and so an image takes 4–9 minutes to acquire. On this timescale, biomolecules are imaged in one frame (statically) with any motion being blurred out. Many biomolecules partake in dynamic processes and, while a typical commercial system may be limited in its frame rate, efforts have been made to capture this using SPM as a "video device" rather than a "stills camera".[45] Providing issues such as mechanical drift can be minimised, time-lapse imaging of biological systems can be fruitful. An environmental change may be induced to trigger, for example, the opening and closing of the membrane Nuclear Core Complex in response to the introduction of calcium.[45] A time-lapse AFM has been used to watch the growth of amyloid fibrils over a number of hours.[46] Many biological processes of potential interest do, however, operate on timescales much shorter than traditional SPM data acquisition permits. To this end, AFMs and STMs capable of tens of frames per second have been constructed and used to image biomolecules at work, notably the motor proteins myosin and dynein.[47] Recently an AFM with the capability of imaging at up to 1200 frames per second has been reported[48] and used to image collagen fibrils lying on a substrate. The development of a "video-STM" and its application to a real-time imaging of gold surface reconstruction has also been reported.[49]

2.1.1.5 Conductive-probe AFM

The ability to reversibly and rapidly engage (and potentially pre-image) molecules with high spatial control can facilitate high-throughput electronic analyses. In a typical conductive-probe AFM (CP-AFM) configuration, a metal-molecule-metal (mMm) junction is created with the tip and substrate acting as electrodes across which a bias voltage is applied. Conductive tips are typically fabricated by coating conventional probes with a metal (gold or platinum are common) or with doped diamond. The CP-AFM, while similar to STM, has an advantage in that the tip position is independent of the bias across or current through the molecule of interest. This means that topographic and conductance imaging can be carried out simultaneously without the convolution effects of electronic structure seen in STM, and current-voltage spectroscopy data (see Section 2.2.4) can be obtained with tip feedback engaged and at a quantifiable force. A disadvantage of the technique is that it is currently difficult to generate probes capable of operation under electrolytic solution (requiring that the surface is robustly insulated everywhere but the tip apex) in a manner that retains the cantilever's bending properties (see also Section 2.2). This remains an active area of research (potentially enabling conductance analyses with electrochemical gating and quantifiable probe-molecule mechanical coupling) and has yielded promising developments in recent years.[50–52]

2.1.2 SPM Applications at the Biomolecular Interface

The SPM functionalities outlined above lend themselves directly to fields such as microengineering, nanofabrication and materials science, and enable surface characteristics such as stiffness, wear, electrochemical and morphological properties to be probed at a sub-micron level.[53–56] Here we shall concentrate on the application of SPM to analyses at the bioelectrochemical interface. The interfacial nature, spatial resolution and environmental flexibility available within standard hardware configurations can offer considerable insight into biological surfaces with imaging modes providing information on surface coverage, distribution and homogeneity. They support analyses carried out by, for example, surface plasmon resonance (SPR), ellipsometry, FTIR spectroscopy, electrochemistry, Raman spectroscopy, *etc.*, and enable a direct visualisation of molecular-scale events such as the docking of a protein to surface-tethered DNA and the dissociation protein dimers on electrodes.[24,57] Modified probes can be utilised to detect molecular recognition events and quantify the stability of a protein fold, and imaging can be carried out in an electrochemical environment where immobilised biomolecules are able to simultaneously turn over substrate and/or communicate with underlying electrodes. ECSTM is perhaps the most powerful imaging mode in the latter context (see Section 2.2).

2.1.2.1 General Bioimaging Considerations

As discussed, AFM modes are suitable for imaging large biological specimens such as whole cells with sub-cellular features routinely identified.[58] Imaging can be used to assess the structure and stability of cells in the imaging environment[59] or to assess drug impact, for example by combining with force spectroscopy.[60] The spatial capabilities of a modern scanning probe system really come to the fore, however, in molecular imaging applications. Within this context, AFM imaging has been used widely to visualise proteins and DNA on surfaces, and to monitor biological interactions such as motor-protein-microtubule,[61] inter-protein[62] or protein-DNA binding *in situ*.[63] When imaging biomolecules the choice of modality is dependent on a number of factors, including the natural environment of the biomolecule, the strength of its surface adsorption, the physical dimensions and its surface coverage. The low intrinsic conductance and mechanical vulnerability of immobilised biomolecules, together with the considerable experimental demands of bio-STM (see Section 2.2.3), make AFM the most common imaging mode in biological analyses. Even with force-quantified AFM, however, care must be taken to ensure the imaging probe interacts, as much as possible, in a non-perturbative manner with the surface. Weakly bound molecules of interest may be displaced in contact-mode imaging, severely restricting data quality or volume. In such circumstances, intermittent-contact mode may provide a better imaging solution. The imaging environment is also of critical importance; in some cases, imaging under controlled pH and ionic strength is necessary if native biomolecular structure is to be retained during the experiment. Operation under fluid, additionally, greatly reduces potentially perturbative adhesion force, the so-called "snap to contact" that is largely unavoidable in ambient AFM imaging.

AFM analyses of DNA have been extensive and progressively more refined and reliable. Bunch *et al.*[24] have, for example, been able to non-destructively resolve DNA-DNA polymerase complexes using carbon-nanotube-modified tips in non-contact mode. The visualisation of DNA nanostructures has also come to the fore recently as a complementary technique to more traditional methods such as gel electrophoresis.[64] Such structures can be engineered in a modular fashion and sub-units, repeating patterns or individual structures have been visualised by AFM.[65] These patterns have been used as a base for deposition of proteins[66,67] and nanoparticles[68,69] and have potential future application in sensing and electronics.

As with DNA, the molecularly resolved imaging of proteins/enzymes on surfaces has become, in recent years, a matter of routine. The purple membrane of the bacteria *Halobacterium halobium* has attracted much interest as a photoactive model ion-pumping membrane protein[70] and has been the subject of some of the most highly resolved AFM analyses.[71,72] Müller et al.,[73] for example, used AFM to image the native purple membrane, revealing the trimeric crystalline structure expected from X-ray crystallographic data. Obtaining images in contact mode under a buffered solution they were both able to resolve the structural sub-units of the protein and to demonstrate the sensitivity of two-dimensional molecular packing on crystallisation conditions. In related experiments, Scheuring et al.[74] imaged the surface layer of *Corynebacterium glutamicum*, a bacterium, adsorbed on mica by AFM and were able to resolve PS2 proteins within the membrane. The stacking and hydrophobic/hydrophilic characteristics of the native layer were demonstrated by dissecting the top layer with the probe to reveal an alternatively patterned layer below. As with other membrane proteins, such as bacteriorhodopsin, this work demonstrated the densely packed, crystalline nature of proteins in some membranes and also combines high-resolution imaging with "nano-dissection", a method closely related to AFM-based lithographic techniques. Protein structural characteristics may also be revealed in controlled force imaging. In a study of electron transport through ferritin, which may be prepared with its mineral core empty (apoferritin) or filled with hydrous-ferric oxide (holoferritin), Davis et al. used topographic and phase imaging to resolve compression of the protein matrix and the relative rigidity of the core (Figure 2.4).[75]

Chemically assembled (as opposed to crystallised or physisorbed) molecular monolayers of cytochromes, blue copper proteins and enzymes have also been imaged at high resolution under fluid by both AFM and STM.[76–79] These electroactive species play a vital role in electron transport chains, and can be robustly interfaced with electrodes through a variety of chemical or engineered methods, some of which have been utilised in molecular electronic assessments (see below). The molecularly resolved electrochemical characteristics of yeast and horse-heart cytochromes *c* have, for example, been resolved in electrochemical STM experiments.[79,80]

2.1.2.2 Force Spectroscopy

The force spectroscopy of biomolecules by AFM involves the tethering or physisorption of the molecule of interest to a substrate or AFM probe prior to tip withdrawal from a point of contact whilst the lever deflection is tracked, resulting in a deflection-distance curve. During this process the molecule is stretched and, depending on its structure, extends, unfolds or breaks.[81,82] This can be detected by stepwise jumps in the deflection-distance curve corresponding to the unfolding of specific peptide chains or domains, information useful in determining the structure and stability of the protein fold in conjunction with X-ray crystallographic data *etc*. The relative stability of domains can be assayed and a direct measurement of chain length is possible.

In the membrane protein imaging work referred to above, force spectroscopic methods were utilised to extract single molecules from their supporting

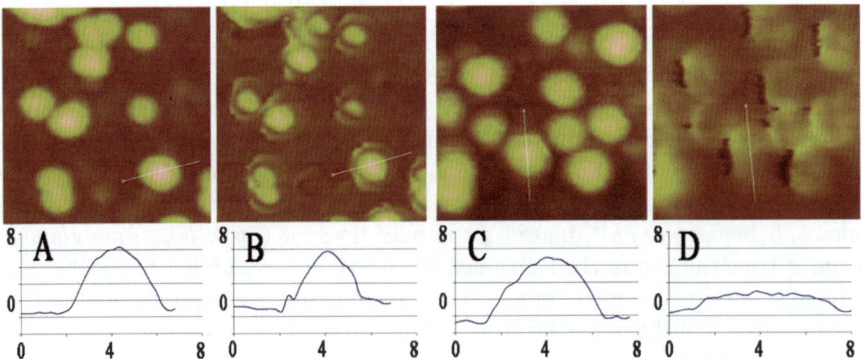

Figure 2.4 IC-AFM images of the holo (A&B) and apo (C&D) forms of the iron-storage protein ferritin on HOPG, with molecular cross sections (scales in nm). The holo form has a hard mineral centre, while this is missing from the apo form. Figures A and C show both forms at low imaging force, with molecular heights similar to the 6 nm expected from X-ray crystallography structures. Figures 2.4B and 2.4D show the effect of increasing imaging force. The protein shell is compressed but the mineral core remains for holo-ferritin in Figure 2.4B, while Figure 2.4D demonstrates that the shell collapses at high force for apo-ferritin. These visualisations can be correlated with CP-AFM IV spectroscopy to explain the changing electronic behaviour. Figure reproduced with permission of Institute of Physics, *Nanotechnology*.[75]

membrane. Müller *et al.* demonstrated, for example, the extraction of bacteriorhodopsin molecules with a force of ~240 pN and that the stretching distance (~2.8 nm) correlated with the size of the membrane hydrophobic region. Though force spectroscopy can reveal much about the unfolding of proteins by expansive forces, the compression of a biomolecule is harder to analyse by equivalent means. For this, other methods must be employed such as imaging individual molecules or performing current-voltage spectroscopy at higher loads. This will be discussed later. Closely related to protein removal/unfolding assessment is molecular recognition by AFM force spectroscopy. In this methodology a modified tip can be engaged and retracted from a surface supporting complementary molecules of interest. If the two molecules "recognise" each other with an attractive force, an additional adhesion can be detected in the force spectra[83–85] or, alternatively, by friction imaging.[86] In addition to probing the fundamental interactions of biomolecules on a molecule-by-molecule basis, molecular recognition analyses have obvious potential immunoassay relevance and can detect the presence of antibodies or misfolded proteins.[87]

2.1.3 Summary

Scanning probe microscopes are able to image biomolecules on both conductive and insulating surfaces under a range of imaging modes that can be chosen to suit the application and experimental information required.

Molecular and sub-molecular imaging, force, recognition or conductance/tunnelling spectroscopic analyses are possible under UHV, ambient, buffered and electrolytic conditions. The ability to operate under electrolyte, on a variety of surfaces, with the potential of both *in situ* electroanalysis and direct quantitative assessment of molecule-electrode electronic coupling makes applications to bioelectronic interfaces both irresistible and potentially powerful.

2.2 Bioelectronic Analyses

2.2.1 Electrode Surface Considerations

In the electrochemical analysis of biomolecules, electrodes are typically polished mechanically or electrochemically (or both), and often modified with a monolayer designed to promote or modulate electron transfer as required. The cleaning and polishing processes are surface roughening but improve voltammetric responses on both gold[88] and graphitic interfaces.[89,90] For reliable high-resolution (molecular) imaging, however, surface roughness should be considerably lower than molecular height. Atomically flat electrodes can be prepared by a number of methods. For non-electronic investigations cleaved planes of mica are of considerable value. Highly oriented pyrolytic graphite (HOPG) can be similarly cleaved to reveal fresh conductive atomic faces, microns in size. Gold deposited on a support (commonly mica or glass) can be flame annealed to form Au(111) atomic terraces hundreds of nanometres across. Alternatively mica can be used as a template onto which gold is evaporated; peeling off the mica allows one to utilise the underlying, very flat, gold (commonly called template-stripped gold or TSG).[91] Though the surface roughness of macro-scale electrodes typically exceeds that where molecular-scale imaging is possible, analyses have been able to confirm, for example, the general phenomenon of increased metalloprotein electrochemical response with increasing surface roughness,[88] and the preferential aggregation of protein at electrode surface discontinuities.[92]

In the following sections we outline several case studies in which SPM methods have been used to assess the surface assembly, distribution, homogeneity and robustness of biomolecular immobilisation (*i.e.* ability of the layer to withstand washing, repeated imaging, variant applied probe loads, *etc.*) on electrode surfaces. We also outline the application of EC-STM to the direct, molecular-scale, analysis of biomolecular redox activity.

2.2.2 AFM Imaging Case Studies

2.2.2.1 Electrode Confined Metalloproteins

The interactions of proteins and enzymes with electrode surfaces have been investigated for several decades.[93–96] Such studies not only shed light on the fundamental thermodynamic and kinetic characteristics utilised by life-sustaining processes such as respiration, but also facilitate hugely beneficial applications to sensing.[97,98] The direct imaging of these interfaces enables both improved designs of surface and associated chemistry and a more robust interpretation of thermodynamic and (especially) kinetic data.[92] With appropriate care, very high (molecular and sub-molecular) *in situ* imaging resolution is possible for surface-confined metalloproteins (Figure 2.5A). The type I (blue) copper protein azurin (Figure 2.5B), an electron transporter in the respiratory chain of denitrifying bacteria, has been studied widely, and imaged on a variety of surfaces using AFM[92] and STM.[99] It is electroactive on both bare and modified gold surfaces and can be controllably chemisorbed through either a natural disulfide or engineered surface cysteines.[100,101] The robustness of this mode of immobilisation facilitates repeated imaging in non-contact, fluid contact or tunnelling configurations. This ability to control the interfacing of these molecules with the underlying electrode has been utilised in engineering resulting electroanalytical responses.[95,102]

The metalloprotein yeast iso-1-cytochrome *c* (YCC), also an electron shuttle, has been shown to exhibit fast interfacial electron transfer at gold.[96] Interestingly, AFM imaging resolved molecular clusters at the surface and led to a refined analysis of the redox interactions between this cytochrome and an enzyme partner. This is an example of how a scanning probe study can be used to compliment and extend traditional electrochemical studies, as well as to assess the properties of adsorbed molecular layers.

2.2.2.2 Peptide Aptamers

Peptide aptamers are scaffold proteins that are used to host a peptide sequence of interest. Since this sequence can be engineered to interact with a particular target protein species, these have considerable application in protein detection and protein-based therapeutics. The ideal scaffold will be biologically inert, and interact specifically with targets. By modifying the scaffold to contain a solution-exposed cysteine residue (or other tags) the molecules can be controllably immobilised on electrode surfaces and their recognition characteristics analysed by Surface Plasmon Resonance (SPR),[98] or electroanalysis. Scanning probe imaging can prove particularly beneficial if those confined molecules are not easily probed by either optical or redox electrochemical means. Figure 2.6 shows the resolved homogeneity of a chemisorbed molecular film of the STM aptamer scaffold protein on an evaporated film gold electrode and provides evidence that the protein is not conglomerating or denaturing. Indirectly, such imaging confirms the inherent stability and molecular coverage (required in subsequent formation constant calculations) of these molecular layers (on repeated imaging and washing) and provides data highly complementary to that generated by SPR or other spectroscopic tools.

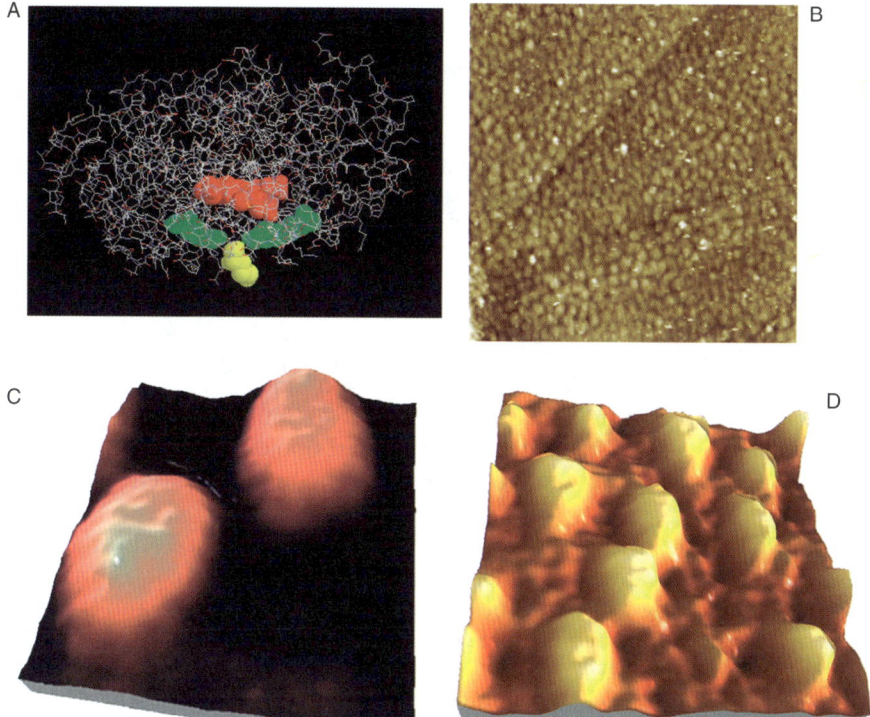

Figure 2.5 A: By engineering a single cysteine residue into the surface of a cytochrome P450cam enzyme, Davis *et al.* were able to demonstrate not only the robust surface assembly and *direct* (by virtue of orientationally controlled proximity of the heme to the underlying surface) voltammetry of the enzyme but also a molecularly resolved tunnelling profile of the monolayer (B; image 250 × 250 nm). Figure reproduced with permission of the Royal Society of Chemistry, *Faraday Discuss.*[76] C: *In situ* tunnelling image of two cytochromes immobilised on a single crystal gold electrode surface. The resolution of this image (8 × 8 nm in size) represents probably the maximum accessible through this technology. Figure reproduced with permission of the Royal Society of Chemistry, *J. Mater. Chem.*[126] D: *In situ* tunnelling image of an azurin mutant directly immobilised on gold. Individual molecules, which are simultaneously electrochemically addressable, are clearly resolved.

2.2.3 The Direct Imaging of Electrochemistry and Enzyme Activity

2.2.3.1 *Direct Biomolecule Imaging on Electrodes*

STM may be used to image biomolecular monolayers in the same manner as AFM, and is capable of molecular and, in some circumstances, sub-molecular resolution under electrolytic media.[4,13,103] In one case study, Davis and Hill demonstrated the fabrication of homogeneous enzyme monolayers on

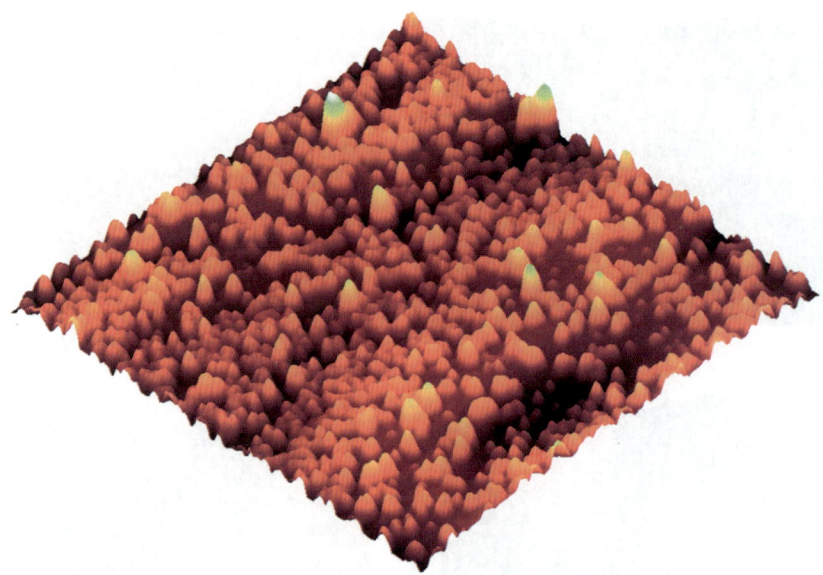

Figure 2.6 IC-mode AFM image of an STM-Cys$^+$ aptamer adlayer chemically immobilised on a SAM coated (OEG-COOH/OEG-OH 1 : 100) gold chip. The individually adsorbed aptamer species are clearly visible and within the receptive monolayer (image size $750 \times 750\,\text{nm}^2$). Repeated imaging (after surface rinsing) demonstrates the robustness of protein-surface chemical coupling and also indicates that the monolayer is robustly coupled to the gold electrode. Subsequent SPR analyses confirm orientated chemisorption.

atomically flat gold electrodes by controlled chemisorption.[76] Significant within this work was the ability to utilise this controlled and orientated immobilisation of a large and delicate enzyme to bring the heme prosthetic group into coupling distance with the electrode (the engineered molecule was electrochemically addressable under conditions where the wild-type is not). In related work by the same group, single molecule resolution tunnelling imaging of two redox-active metalloproteins was demonstrated under solution at reduced temperatures; it is known that the Young's modulus of a protein fold increases significantly on cooling[19] and this may have benefits in reducing tip-induced protein deformation and maximising image resolution.[77]

From its early development, the tunnelling imaging of biomolecules generated questions about the mechanisms of charge flux method across the interelectrode (tunnelling) gap.[104–106] Given that proteins are generally considered to be bulk "insulators", electronic tunnelling across a typical protein diameter (30–50 Å) should, in theory, be extremely inefficient.[107] Though the mechanisms of electron movement across these structures during imaging remain somewhat debatable, and may only have a passing resemblance to native or biological tunnelling, in some cases at least, the evolved and optimised designs of these redox biomolecules are resolvable. This is particularly clear in cases where natural redox sites lie directly on the tip-substrate electron transport pathway and is directly accessible in electroanalytical experiments. This phenomenon is exploited in EC-STM imaging described next.

2.2.3.2 Metalloprotein-electrode Imaging under Potential Control

Since an STM image is a convolution of topographic and electronic features, a change in the electron transport efficiency through a molecular adsorbate will manifest itself as a change in the apparent STM feature height (increased height implying increased conductance). An additional feature of some STM configurations is the ability to set up a four-terminal system under bipotentiostatic control such that the tunnelling ("imaging") tip is also a working electrode (Figure 2.7).

In this manner, the potential of the substrate on which the imaged molecular species resides can be controlled relative to a reference electrode (imaging/conductance analyses can be carried out as the electrode absolute potentials are swept through the redox site half-wave potential). This effectively enables imaging/current flow analyses to be carried out under conditions where electrochemical current (redox electron transfer between an immobilised molecule and its underlying supporting electrode) is switched on or off. Several studies have now noted a surge in conductance when this "redox current" is switched on (*i.e.* the molecular conductance is tuneable through potential).[2,108,109] EC-STM, effectively, facilitates electroanalysis at the single molecule level.

Cu-azurin chemisorbed on bare gold electrodes has been shown by Alessandrini and coworkers to have an apparent height modulated by applied potential,[108] and to be maximised at a value related to the formal (bulk) electrochemical half-wave potential. Work within our group has confirmed this phenomenon and demonstrated it further for cytochromes *c*, adsorbed both directly to gold and electrostatically to a monolayer-modified surface (Figure 2.8).[79] Such investigations show that it is possible to tune and measure biomolecular conductance at a single molecule level, and also to elucidate the effect of decoupling the protein from the electrode; in the cytochrome *c* height modulation experiments the effect of introducing a spacing monolayer is to reduce current increase at the resonance potential, indicating a weaker electronic coupling of the protein to its underlying electrode (and thus a reduced effect of "tuning" the potential of the latter). As with the protein directly adsorbed on bare gold electrodes, the single molecule conductance varies systematically with surface potential, being maximised at values close to the electrochemical half wave (Figure 2.8B); the molecule-to-molecule variance of this conductance and redox tuning analysis is shown in the histogram of Figure 2.8C. (The nature of electronic transport may be obtained from the precise position of the conductance maximum.[7]) An ability to carry out STM on redox active molecules under conditions where electroanalyses can be performed facilitates a direct visualisation, not only of the thermodynamics of single

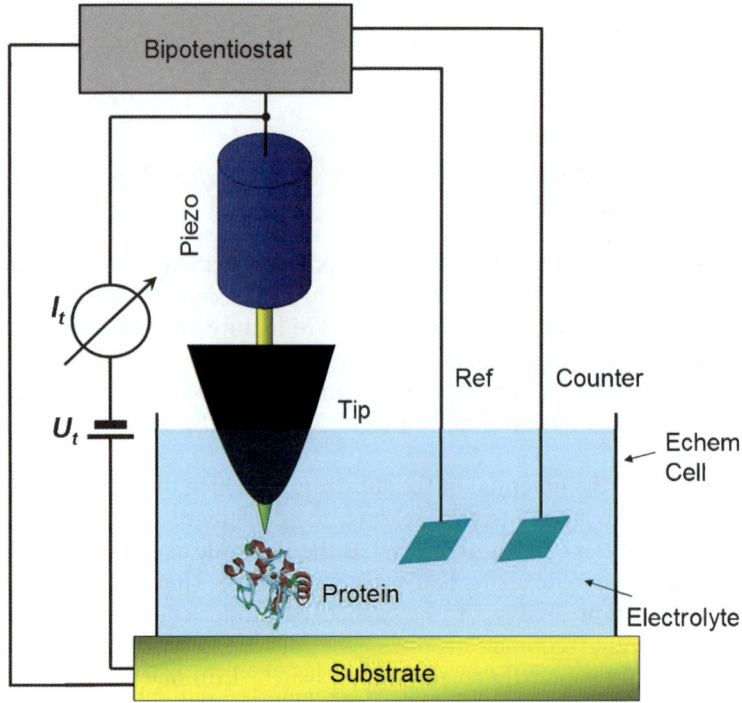

Figure 2.7 Schematic representation of the four-terminal EC-STM configuration – by controlling substrate or tip potential, the redox state of electrochemically active adsorbates can be tuned, and the effect investigated *via* height modulation or current-voltage spectroscopy. In such experiments the tip must be effectively insulated everywhere bar its very apex in order to reduce ionic contributions to current (which may swamp any tunnelling currents).

molecule electron transfer (the surface potential at which the electrochemical conductance channel becomes available), but also of the kinetics of its coupling to an underlying surface by *in situ* imaging of adsorption, and the dispersion of molecular properties within any given surface-confined population (something fundamental to the way in which bulk electroanalyses are interpreted). For example, the spread in the potential at which an individual metalloprotein conductance maximum is observed (Figure 2.8C) is a direct measure of the variations in electrochemical properties on the molecular scale that contribute to dispersion as observed by, for example, the width of voltammetric wave in a bulk measurement.

2.2.3.3 Scanning Electrochemical Microscopy

The Scanning Electrochemical Microscope (SECM) is a member of the scanning probe family of microscopes in which a microelectrode, typically between

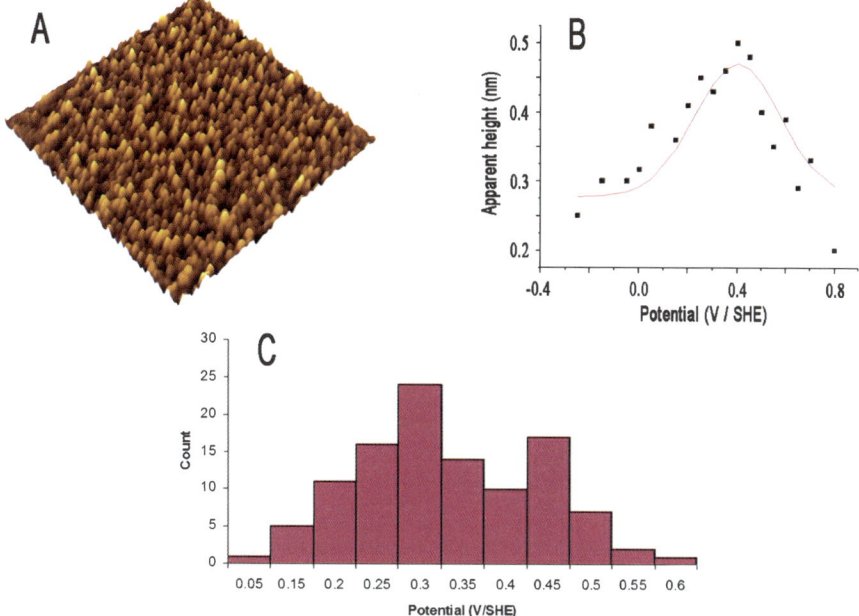

Figure 2.8 A: *In situ* electrochemical STM image of a densely packed monolayer of yeast iso-1-cytochrome *c* directly chemisorbed *via* a surface cysteine onto a gold electrode surface under buffered electrolyte. Tunnelling current 0.1 nA, tip-substrate bias 0.1 V, substrate potential 0.18 V *vs.* Ag/AgCl. $200 \times 200\,\text{nm}^2$ scan. B: variation of the height/conductance of one YCC molecule with substrate potential as measured by STM. The peak height is the state of maximum conductance, and the enhancement in magnitude between on and off resonance conductance is here ~ 20 times. C: Histogram of the potential at which height peaks for each individual molecule. The modal peak potential is close to the half-wave potential of the adsorbed YCC layer (as determined in bulk electroanalysis), suggesting that the cause of the resonance is the redox centre.

1 and 25 μm[110] in diameter, is scanned across a surface, whilst collecting a current.[110] Unlike STM, which operates through the generation and measurement of a tunnelling current and must therefore be in very close proximity to an underlying surface, SECM probes act as "bulk" electrodes and measure faradaic (electrochemical) currents as a function of spatial position. Though of somewhat limited (typically several microns) spatial resolution, SECM methods enable a direct probe of electrode characteristics and, in some cases, the redox properties of surface-confined molecules.[111,112]

In combining the redox current collection characteristics of an SECM probe with the deflection-based imaging characteristics of an AFM probe, SECM can straddle the boundary between high-resolution topographical imaging, molecular electronics and nanoscale sensing. Appropriately generated and modified (the surface insulated except for at the very apex) probes of this kind can act as nanoelectrodes and be used to simultaneously image electrochemical activity and topography across a surface. The probe insulation (required to minimise the loss of redox current within capacitative noise) required in these studies must be inert, offer complete coverage of the conductive surface and be mechanically flexible enough to enable cantilever bending. A number of different configurations has been employed to satisfy these demands, including the use of electrochemically etched platinum coated in an anodic electrodeposition paint,[51,113] polymer-coated carbon nanotubes[114,115] and a layered tip-cantilever assembly with a platinum film insulated by SiO_2.[52,116] In a subtle variation of this approach, the beneficial effects of locating the redox current (conductive) collecting region of the probe away from its apex (and thereby avoiding potential damaging or fouling during imaging) have been noted and utilised in the direct visualisation of glucose oxidase enzyme activity within micron-sized pits.[117] In this, and a similar study where surface confined horseradish peroxidase activity was mapped,[111] the power of a localised probe to screen both electrode and biological activity is demonstrable.

2.2.4 Spectroscopic Assessment Electrode-biomolecule Electronic Coupling

We have seen thus far that scanning probe imaging methods, as applied to bioelectronic interfaces, facilitate a characterisation of electrode layer surface topography, chemical variance, biomolecule immobilisation and homogeneity, and single molecule redox characteristics. The spatially refined scanning of a sharp conductive probe above an electrode surface, additionally, enables conductance and redox characteristics to be probed spectroscopically; of particular interest in this context are current-voltage and current-potential spectroscopic methods. Closely related to STM height modulation, these techniques offer further insight into the electrochemical coupling between an adsorbed molecule and its underlying electrode surface and related gating of molecular properties.

2.2.4.1 CP-AFM–Current-Voltage Spectroscopy

In bringing a metal-coated and voltage biased AFM probe into stable force feedback with a pre-assembled layer of protein or enzyme on an electrode surface, the layer/molecular conductance can be probed under controllable environmental conditions. In carrying out such experiments across a range of compressional loads it is possible to demonstrate the effects of protein structure and deformation on electron flux and, seemingly, the mechanisms by which electrons flow across a redox active structure.[78,80,118] A comparative analysis of holo and apo forms of the iron storage protein ferritin, for example, has demonstrated the role of the metal oxide mineral core in both facilitating current flow and maintaining a gross, uncompressed structure when pressurised (observations additionally confirmed by AFM imaging at calibrated load).[75]

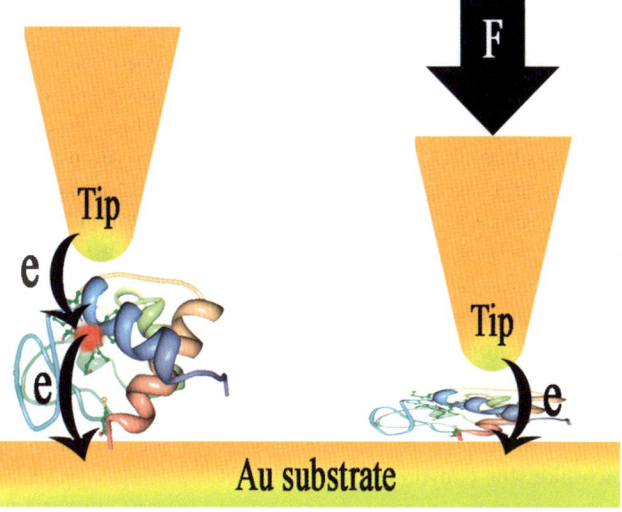

Figure 2.9 Schematic representation of a potential change in charge transfer mechanism across a proximal probe junction confined metalloprotein. If the tip electrode-molecule coupling is weak, the protein is structurally unperturbed and presents a large tunnelling distance to electrons attempting to span the interelectrode gap. Direct tunnelling is then inefficient, making resonant tunnelling through the redox active centre an efficient alternative pathway. On compressing the molecule with the probe electrode, the tunnelling distance is reduced; the exponential relationship between current and distance facilitates a considerable increase in direct tunnelling. Any contributions from resonant or two-step transport are lost within this dominant current.

With redox addressable metalloproteins it has been reported that redox centres lying along the tunnelling pathway play a significant role in current flow only under specific conditions – that is when probe electrode contact is robust but not grossly perturbative. Figure 2.9 depicts a possible cause of an observed switch in tunnelling from resonant tunnelling at low loads to direct tunnelling at high loads. The protein matrix is compressed at large forces leading to a shorter tunnelling gap and efficient direct tunnelling. Further to this it is likely that the electronic structure is highly perturbed from the natural configuration, possibly moving redox active electronic states away from electrode states. The possible relevance of these observations to "bulk" electroanalytical assessments of immobilised metalloprotein is intriguing and an objective of ongoing experimentation.[75,78,80]

2.2.4.2 In situ EC-STM gating – Current-Potential and Current-Voltage Spectroscopy

In the EC-STM height modulation experimental configuration outlined in Section 2.2.3.2 (Figure 2.7), surface-confined biomolecular species are imaged at a constant tip-substrate bias and changes in current flux are expressed through change in measured height as a function of substrate (working electrode) potential. The molecularly confined redox centre may be oxidised or reduced in the process. It is possible to measure this potential-tuned change in current directly by holding the tip stationary over the molecule, turning the microscope feedback off (so that the tip doesn't move in response to the changing current) and sweeping the substrate potential at fixed tip-substrate bias. In this mode the measured dependence of tunnelling current on potential is called current-potential, or *IE*, spectroscopy. The resulting current has been shown to peak near the midpoint electrochemical potential of the molecule (as with the height modulation configuration). While this protocol has been largely applied to organic compounds,[119,120] it has also been demonstrated on electrode immobilised proteins.[2] *IE* spectroscopy, like electrochemical imaging, gives a measure of enhanced current through a biomolecule as redox-based electron transfer between it and the contact electrodes becomes energetically viable. Studies like this can quantify the electronic coupling of a single active centre to its supporting electrode (larger on-resonance currents are expected when the electrode-molecule electronic coupling is greater) and provide direct access (through the width of the current-potential curve) to reorganisation energies at the single-molecule level.[121]

An alternative means of probing molecule-electrode communication and electrochemical characteristics is to hold the tip over the molecule and sweep the bias applied between it and the underlying planar electrode (usually over the order of ±1–2 V). If feedback in the z direction is turned off, again to stop the tip moving in response to changing current, the resulting current through the junction is measured *versus* applied bias (current-voltage, or *IV*, spectroscopy). In electrochemically controlled *IV* spectroscopic analyses (in which all potentials are fixed to a real reference value) spectroscopic features specific to the presence of redox centres can be revealed and directly correlated with electrochemical potentials. These features are due to the interaction of the tunnelling electrons with accessible molecular orbitals; the molecule may be reduced and subsequently oxidised,[122] or electron transport may proceed through the orbitals with no molecular relaxation.[123]

Our own group has used STM current-voltage spectroscopy to show that molecular junctions formed on electrochemically addressable copper-azurin monolayers exhibit both current switching[124] and negative differential resistance (NDR),[77] phenomena not observed when the copper centre is replaced with redox-inactive zinc atom. YCC has been investigated in a similar manner, with NDR and switching observed and gated with applied potential.[125] In these spectra, the current surge observed when the electrodes are able to exchange electrons with the redox site (and a redox current is "switched on") can be several orders of magnitude (Figure 2.10). IV spectroscopic analyses, when compared to redox-inactive controls, confirm that the active centre plays a vital role in resonant transport across electron-transporting metalloproteins, enable, in principle, direct access to single-molecule electrode potentials and a

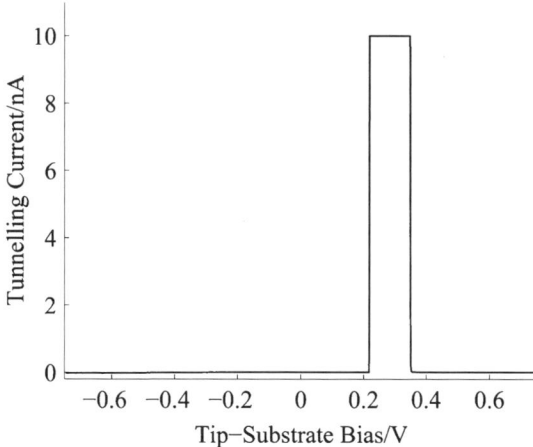

Figure 2.10 Example of an electrochemically gated current-voltage curve (within an ECSTM configuration) obtained on a yeast cytochrome c monolayer, substrate potential 0.1 V $vs.$ Ag/AgCl. The tip is held stationary over a region densely packed with protein, the current feedback turned off (so that the tip does not respond to the changing current) and the tip-substrate bias swept. The sudden increase in current at positive voltage bias is attributed to resonant tunnelling through the protein redox centre (the preamplifier saturating at 10 nA). By altering the substrate potential the relative energies of the electrodes and the active centre can be tuned, and the bias position of the resonance systematically controlled (confirming the electrochemical origin of this phenomenon). Investigations such as these essentially resolve single molecule electrochemical activity (the surface potential at which a redox site is accessible to an electron and the efficiency of redox site-electrode electronic coupling).[80]

qualitative assessment (at least) of electrode-protein electron transfer rate constant.

2.3 Summary

An understanding of biomolecular electron transport has become crucially important, particularly across the biomedical sciences, but also in the disparate but highly interconnected fields of physics, chemistry, biology and bioengineering. The ability of a scanning probe system to image and interrogate surfaces at nanometre-resolution in a biologically relevant environment provides a powerful means of not only aiding the construction of these interfaces but also interpreting their behaviour. Though basic imaging systems are conceptually simple, an entire range of derived force, friction, chemical imaging, electroanalytical, conductance, lithographic, nanomanipulation and fast-scanning techniques now exist. Electrochemical/electronic modes (SECM, STM,

EC-STM, CP-AFM) are particularly suited to analysis at the bioelectronic interface and enable not only a direct visualisation of molecular coverage, distribution and robustness of immobilisation, but also an assessment of local activity, conductance or, in the case of EC-STM (or current-voltage spectroscopy), the resolution of single molecule electrochemical activity. These innovations have transformed SPM from a powerful imaging tool to a highly versatile means of probing molecular properties, an ability that is redefining our interpretation of bulk bioelectronic/bioelectrochemical assays.

References

1. G. Binnig and H. Rohrer (International Business Machines Corp., USA). Application: EP, 1981, p. 37.
2. A. Alessandrini, M. Salerno, S. Frabboni and P. Facci, *Appl. Phys. Lett.*, 2005, **86**, 133902.
3. T. Albrecht, W.-W. Li, W. Haehnel, P. Hildebrandt and J. Ulstrup, *Bioelectrochemistry*, 2002, **69**, 193.
4. E. P. Friis, J. E. T. Andersen, Y. I. Kharkats, A. M. Kuznetsov, R. J. Nichols, J.-D. Zhang and J. Ulstrup, *Proc. Natl. Acad. Sci. USA*, 1999, **96**, 1379.
5. J. Zhang, M. Grubb, A. G. Hansen, A. M. Kuznetsov, A. Boisen, H. Wackerbarth and J. Ulstrup, *J. Phys. Condens. Matter.*, 2003, **15**, S1873.
6. J. Frommer, *Angew. Chem. Int. Ed. Engl.*, 1992, **31**, 1298.
7. E. P. Friis, Y. I. Kharkats, A. M. Kuznetsov and J. Ulstrup, *J. Phys. Chem. A*, 1998, **102**, 7851.
8. B. Ren, G. Picardi and B. Pettinger, *Rev. Sci. Instrum.*, 2004, **75**, 837.
9. R. M. Penner, M. J. Heben and N. S. Lewis, *Anal. Chem.*, 1989, **61**, 1630.
10. A. J. Nam, A. Teren, T. A. Lusby and A. J. Melmed, *J. Vac. Sci. Technol. B*, 1995, **13**, 1556.
11. A. G. Guell, I. Diez–Perez, P. Gorostiza and F. Sanz, *Anal. Chem.*, 2004, **76**, 5218.
12. J. Lindahl, T. Takanen and L. Montelius, *J. Vac. Sci. Technol. B*, 1998, **16**, 3077.
13. P. B. Lukins and T. Oates, *Biochim. Biophys. Acta, Bioenerg.*, 1998, **1409**, 1.
14. G. Binnig, C. F. Quate and C. Gerber, *Phys. Rev. Lett.*, 1986, **56**, 930.
15. L. A. Bottomley, *Anal. Chem.*, 1998, **70**, 425R.
16. W. H. Han, S. M. Lindsay and T. W. Jing, *Appl. Phys. Lett.*, 1996, **69**, 4111.
17. C. W. Yang, I. S. Hwang, Y. F. Chen, C. S. Chang and D. P. Tsai, *Nanotechnology*, 2007, **18**.
18. B. T. Marshall, K. K. Sarangapani, J. H. Wu, M. B. Lawrence, R. P. McEver and C. Zhu, *Biophys. J.*, 2006, **90**, 681.

19. Y. Y. Zhang, S. T. Sheng and Z. F. Shao, *Biophys. J.*, 1996, **71**, 2168.
20. B. D. Sattin and M. C. Goh, *Rev. Sci. Instrum.*, 2004, **75**, 4778.
21. N. Gadegaard, *Biotech. Histochem.*, 2006, **81**, 87.
22. U. D. Schwarz, H. Haefke, P. Reimann and H. J. Guntherodt, *J. Microsc.*, 1994, **173**, 183.
23. J. H. Hafner, C. L. Cheung, A. T. Woolley and C. M. Lieber, *Prog. Biophys. Mol. Biol.*, 2001, **77**, 73.
24. J. S. Bunch, T. N. Rhodin and P. L. McEuen, *Nanotechnology*, 2004, **15**, S76.
25. J. H. Wu, Y. Fang, D. Yang and C. Zhu, *J. Biomech. Eng., Trans. Asme*, 2005, **127**, 1208.
26. Z. Liu, Z. Li, H. Zhou, G. Wei, Y. Song and L. Wang, *Micron*, 2005, **36**, 525.
27. G. Binnig, *Ultramicroscopy*, 1992, **42–44**, 7.
28. Y. Kaibara, K. Sugata, M. Tachiki, H. Umezawa and H. Kawarada, *13th European Conference on Diamond, Diamond-Like Materials, Carbon Nanotubes, Nitrides and Silicon Carbide*, 2003, **12**, 560.
29. E. Tocha, H. Schonherr and G. J. Vancso, *Langmuir*, 2006, **22**, 2340.
30. R. D. Piner and C. A. Mirkin, *Langmuir*, 1997, **13**, 6864.
31. Nanocraft, *Nanocraft*, 2007.
32. L. T. Mazzola, C. W. Frank, S. P. A. Fodor, C. Mosher, R. Lartius and E. Henderson, *Biophys. J.*, 1999, **76**, 2922.
33. A. Noy, *Surf. Interface Anal.*, 2006, **38**, 1429.
34. S. Decossas, F. Mazen, T. Baron, G. Bremond and A. Souifi, *Nanotechnology*, 2003, **14**, 1272.
35. S. W. Hla, L. Bartels, G. Meyer and K. H. Rieder, *Phys. Rev. Lett.*, 2000, **85**, 2777.
36. B. Mokaberi, Y. Jaehong, M. Wang and A. A. G. Requicha, in *Robotics and Automation, 2007 IEEE International Conference on*, 2007, pp. 1406.
37. A. A. G. Requicha, *Proc. IEEE*, 2003, **91**, 1922.
38. C. Blackledge, D. A. Engebretson and J. D. McDonald, *Langmuir*, 2000, **16**, 8317.
39. J. J. Davis, C. B. Bagshaw, K. L. Busuttil, Y. Hanyu and K. S. Coleman, *J. Am. Chem. Soc.*, 2006, **128**, 14135.
40. W. T. Muller, D. L. Klein, T. Lee, J. Clarke, P. L. McEuen and P. G. Schultz, *Science*, 1995, **268**, 272.
41. K. B. Lee, S. J. Park, C. A. Mirkin, J. C. Smith and M. Mrksich, *Science*, 2002, **295**, 1702.
42. *Progr. Biophys. Mol. Biol.*, 1996, **65**, 22.
43. T. Tano, M. Tomyo, H. Tabata and T. Kawai, *Jpn. J. Appl. Phys., Part. 1*, 1998, **37**, 3838.
44. K. Salaita, Y. H. Wang and C. A. Mirkin, *Nat. Nanotechnol.*, 2007, **2**, 145.
45. D. Stoffler, K. N. Goldie, B. Feja and U. Aebi, *J. Mol. Biol.*, 1999, **287**, 741.

46. C. Goldsbury, J. Kistler, U. Aebi, T. Arvinte and G. J. S. Cooper, *J. Mol. Biol.*, 1999, **285**, 33.
47. T. Ando, T. Uchihashi, N. Kodera, A. Miyagi, R. Nakakita, H. Yamashita and M. Sakashita, *Jpn. J. Appl. Phys.*, 2006, **45**, 1897.
48. L. M. Picco, L. Bozec, A. Ulcinas, D. J. Engledew, M. Antognozzi, M. A. Horton and M. J. Miles, *Nanotechnology*, 2007, **18**, 044030.
49. M. Labayen and O. M. Magnussen, *Surf. Sci.*, 2004, **573**, 128.
50. A. V. Patil, R. Vlijm and T. Oosterkamp, *NSTI Nanotechnology Conference and Trade Show, Boston, MA, United States*, 2006.
51. J. V. Macpherson and P. R. Unwin, *Anal. Chem.*, 2000, **72**, 276.
52. P. L. T. M. Frederix, M. R. Gullo, T. Akiyama, A. Tonin, N. F. d. Rooij, U. Staufer and A. Engel, *Nanotechnology*, 2005, **16**, 997.
53. K.-H. Chung, C.-E. Jang and D.-E. Kim, *J. Micromech. Microeng.*, 2007, **17**, 1877.
54. L. Harding, W. P. King, X. Dai, D. Q. M. Craig and M. Reading, *Pharm. Res.*, 2007, **24**, 2048.
55. B. Deng, X. Yan, Q. Wei and W. Gao, *Mater. Charact.*, 2007, **58**, 854.
56. R. Hiesgen, D. Eberhardt, E. Aleksandrova and K. A. Friedrich, *J. Appl. Electrochem.*, 2007, **37**, 1495.
57. C. M. Halliwell, J. A. Davies, J. C. Gallop and P. W. Josephs-Franks, *Bioelectrochemistry*, 2004, **63**, 225.
58. D. Anselmetti, N. Hansmeier, J. Kalinowski, J. Martini, T. Merkle, R. Palmisano, R. Ros, K. Schmied, A. Sischka and K. Toensing, *Anal. Bioanal. Chem.*, 2007, **387**, 83.
59. J. J. Davis, H. A. O. Hill and T. Powell, *Cell Biol. Internat.*, 2001, **25**, 1271.
60. C. Rotsch and M. Radmacher, *Biophys J.*, 2000, **78**, 520.
61. D. Turner, C. Chang, K. Fang, P. Cuomo and D. Murphy, *Anal. Biochem.*, 1996, **242**, 20.
62. M. B. Viani, L. I. Pietrasanta, J. B. Thompson, A. Chand, I. C. Gebeshuber, J. H. Kindt, M. Richter, H. G. Hansma and P. K. Hansma, *Nat. Struct. Biol.*, 2000, **7**, 644.
63. L. Hamon, D. Pastré, P. Dupaigne, C. L. Breton, E. L. Cam and O. Piétrement, *Nucleic Acids Res.*, 2007, **35**, e58.
64. C. Niemeyer, M. Adler, B. Pignataro, S. Lenhert, S. Gao, L. Chi, H. Fuchs and D. Blohm, *Nucleic Acids Res.*, 1999, **27**, 4553.
65. P. W. K. Rothemund, *Nature*, 2006, **440**, 297.
66. J. D. Cohen, J. P. Sadowski and P. B. Dervan, *Angew. Chem. Int. Ed.*, 2007, **46**, 7956.
67. A. Kuzuya, K. Numajiri and M. Komiyama, *Angew. Chem. Int. Ed.*, 2008, **47**.
68. J. Sharma, R. Chhabra, Y. Liu, Y. Ke and H. Yan, *Angew. Chem. Int. Ed.*, 2006, **45**, 730.
69. S. Xiao, F. Liu, A. E. Rosen, J. F. Hainfeld, N. C. Seeman, K. Musier-Forsyth and R. A. Kiehl, *J. Nanopart. Res.*, 2002, **4**, 313.
70. J. K. Lanyi, *Biochim. Biophys. Acta, Bioenerg.*, 2000, **1460**, 1.

71. R. R. Birge, *Annu. Rev. Phys. Chem.*, 1990, **41**, 683.
72. P. C. Pandey, *Anal. Chim. Acta*, 2006, **568**, 47.
73. D. J. Muller, J. B. Heymann, F. Oesterhelt, E. Pebay-Peyroula, R. Neutze and E. M. Landau, *Biochim. Biophys. Acta, Bioenerg.*, 2000, **1460**, 119.
74. S. Scheuring, H. Stahlberg, M. Chami, C. Houssin, J.-L. Rigaud and A. Engel, *Mol. Microbiol.*, 2002, **44**, 675.
75. D. N. Axford and J. J. Davis, *Nanotechnology*, 2007, **18**, 1.
76. J. J. Davis, D. Djuricic, K. K. W. Lo, E. N. K. Wallace, L.-L. Wong and H. A. O. Hill, *Farad. Discuss.*, 2000, **116**, 15.
77. J. J. Davis, C. L. Wrathmell, J. Zhao and J. Fletcher, *J. Mol. Recognit.*, 2004, **17**, 167.
78. D. Axford, J. J. Davis, N. Wang, D. Wang, T. Zhang, J. Zhao and B. Peters, *J. Phys. Chem. B*, 2007, **111**, 9062.
79. J. Davis, B. Peters, W. Xi and D. Axford, *Current Nanoscience*, 2008, **1**, 62.
80. J. J. Davis, B. Peters and X. Wang, *J. Phys. Condens. Matter*, 2008, **20**, 374123.
81. J. M. Fernandez and H. B. Li, *Science*, 2004, **303**, 1674.
82. K. C. Neuman and A. Nagy, *Nat. Methods*, 2008, **5**, 491.
83. S. Allen, X. Chen, J. Davies, M. C. Davies, A. C. Dawkes, J. C. Edwards, C. J. Roberts, S. J. B. Tendler and P. M. Williams, *Appl. Phys. A*, 1998, **66**, S255.
84. P. Hinterdorfer, F. Kienberger, G. Kada, H. Mueller and P. Hinterdorfer, *J. Mol. Biol.*, 2005, **347**, 597.
85. S. Lin, Y.-M. Wang, L.-S. Huang, C.-W. Lin, S.-M. Hsu and C.-K. Lee, *Biosens. Bioelectron.*, 2007, **22**, 1012.
86. M. Ludwig, W. Dettmann and H. E. Gaub, *Biophys. J.*, 1997, **72**, 445.
87. C. McAllister, M. A. Karymov, Y. Kawano, A. Y. Lushnikov, A. Mikheikin, V. N. Uversky and Y. L. Lyubchenko, *J. Mol. Biol.*, 2005, **354**, 1028.
88. M. C. Leopold and E. F. Bowden, *Langmuir*, 2002, **18**, 2239.
89. R. Schlögl and H. P. Boehm, *Carbon*, 1983, **21**, 345.
90. F. A. Armstrong, P. A. Cox, H. A. O. Hill, B. N. Oliver and A. A. Williams, *J. Chem. Soc., Chem. Commun.*, 1985, **1**, 1236.
91. M. Hegner, P. Wagner and G. Semenza, *Surf. Sci.*, 1993, **291**, 39.
92. J. J. Davis, H. A. O. Hill and A. M. Bond, *Coord. Chem. Rev.*, 2000, **411**, 200.
93. S. R. Betso, M. H. Klapper and L. B. Anderson, *J. Am. Chem. Soc*, 1972, **94**, 8197.
94. M. J. Eddowes and H. A. O. Hill, *J. Chem. Soc. Chem. Commun.*, 1977, 771b.
95. L. Andolfi, D. Bruce, S. Cannistraro, G. W. Canters, J. J. Davis, H. A. O. Hill, J. Crozier, M. P. Verbeet, C. L. Wrathmell and Y. Astier, *J. Electroanal. Chem.*, 2005, **565**, 21.
96. H. A. Heering, F. G. M. Wiertz, C. Dekker and S. d. Vries, *J. Am. Chem. Soc.*, 2004, **126**, 11103.

97. J. Tkac, J. W. Whittaker and R. Tautgirdas, *Biosens. Bioelectron.*, 2007, **22**, 1820.
98. J. J. Davis, J. Tkac, S. Laurenson and P. K. Ferrigno, *Anal. Chem.*, 2007, **79**, 1089.
99. J. Zhang, A. M. Kuznetsov and J. Ulstrup, *J. Electroanal. Chem.*, 2003, **541**, 133.
100. E. P. Friis, J. E. T. Andersen, L. L. Madsen, N. Bonander, P. Moller and J. Ulstrup, *Electrochim. Acta*, 1997, **42**, 2889.
101. B. Bonanni, A. R. Bizzarri and S. Canistraro, *J. Phys. Chem. B*, 2006, **110**, 14574.
102. J. J. Davis and H. A. O. Hill, *Chem. Commun.*, 2002, 393.
103. J. J. Davis, H. A. O. Hill, A. Kurz and C. Jacob, *Phys. Chem. Comm.*, 1998, **1**, 12.
104. B. Barris, U. Knipping, S. M. Lindsay, L. Nagahara and T. Thundat, *Biopolymers*, 1988, **27**, 1691.
105. J. E. T. Andersen, G. J. Leggett, C. J. Roberts, P. M. Williams, M. C. Davies, D. E. Jackson and S. J. B. Tendler, *Langmuir*, 1993, **9**, 2356.
106. W. Han, E. N. Durantini, T. A. Moore, A. L. Moore, D. Gust, P. Rez, G. Leatherman, G. R. Seely, N. Tao and S. M. Lindsay, *J. Phys. Chem. B*, 1997, **101**, 10719.
107. C. C. Page, C. C. Moser, X. Chen and P. L. Dutton, *Nature*, 1999, **402**, 47.
108. A. Alessandrini, M. Gerunda, G. W. Canters, M. P. Verbeet and P. Facci, *Chem. Phys. Lett.*, 2003, 625.
109. Q. Chi, O. Farver and J. Ulstrup, *Proc. Natl. Acad. Sci. USA*, 2005, **102**, 16203.
110. A. J. Bard, F.-R. F. Fan, J. Kwak and O. Lev, *Anal. Chem.*, 1989, **61**, 132.
111. C. Kranz, A. Kueng, A. Lugstein, E. Bertagnolli and B. Mizaikoff, *Ultramicroscopy*, 2004, **100**, 127.
112. D. P. Burt and P. R. Unwin, *Electrochem. Commun.*, 2008, **10**, 934.
113. J. V. Macpherson and P. R. Unwin, *Anal. Chem.*, 2001, **73**, 550.
114. H. Dai, J. H. Hafner, A. G. Rinzler, D. T. Colbert and R. E. Smalley, *Nature*, 1996, **384**, 147.
115. C. T. Gibson, S. Carnally and C. J. Roberts, *Ultramicroscopy*, 2007, **107**, 1118.
116. M. R. Gullo, P. L. T. M. Frederix, T. Aklyama, A. Engel, N. F. deRooij and U. Staufer, *Anal. Chem.*, 2006, **78**, 5436.
117. A. Kueng, C. Kranz, A. Lugstein, E. Bertagnolli and B. Mizaikoff, *Angew. Chem. Int. Ed.*, 2003, **42**, 3237.
118. J. Zhao, J. J. Davis, M. S. P. Sansom and A. Hung, *J. Am. Chem. Soc.*, 2004, **126**, 5601.
119. T. Albrecht, A. Guckian, J. Ulstrup and J. G. Vos, *Nano Lett.*, 2005, **5**, 1451.
120. T. Albrecht, K. Moth-Poulsen, J. B. Christensen, J. Hjelm, T. Bjørnholm and J. Ulstrup, *J. Am. Chem. Soc.*, 2006, **128**, 6574.
121. N. J. Tao, *Phys. Rev. Lett.*, 1996, **76**, 4066.

122. J. Zhang, Q. Chi, A. M. Kuznetsov, A. G. Hansen, H. Wackerbarth, H. E. M. Christensen, J. E. T. Andersen and J. Ulstrup, *J. Phys. Chem. B.*, 2002, **106**, 1131.
123. W. Schmickler and C. Widrig, *J. Electroanal. Chem.*, 1992, **336**, 213.
124. J. J. Davis, D. A. Morgan, C. L. Wrathmell and A. Zhao, *IEE Proceedings – Nanobiotechnology*, 2004, **151**, 37.
125. B. Peters and J. J. Davis, *manuscript in preparation*.
126. J. J. Davis, D. A. Morgan, C. L. Wrathmell, D. N. Axford, J. Zhao and N. Wang, *J. Mater. Chem.*, 2005, **15**, 2160.

CHAPTER 3
Electrical Interfacing of Redox Enzymes with Electrodes by Surface Reconstitution of Bioelectrocatalytic Nanostructures

ITAMAR WILLNER, RAN TEL-VERED
AND BILHA WILLNER

Institute of Chemistry, The Hebrew University of Jerusalem, Jerusalem, 91904, Israel

3.1 Introduction

Electrical communication between the redox centres of enzymes and electrodes is an essential function for the development of enzyme electrodes for biosensors or biofuel cell elements.[1–3] Redox proteins usually lack direct electrical communication between their redox sites and the electrode support. This might be attributed to the fact that the redox centres are deeply embedded in the protein matrices, and may be theoretically explained by the Marcus theory.[4] The electron transfer rate constant between a donor and acceptor pair, k_{ET}, is given by eqn (3.1), where d is the distance separating the donor-acceptor pair, d_o is the Van der Waals distance between the components and $\Delta G°$ and λ are the free energy change and the reorganisation energy accompanying the electron transfer (ET) process, respectively. As the dimensions of redox proteins are within the range of 50–150 Å, and their redox sites are deeply embedded in the protein structure, there exists a steric barrier for ET communication between the redox sites and the electrodes that prevents the direct electrochemical activation of the redox enzymes.[5]

$$k_{ET} \alpha \exp[-\beta(d-d_o)] \exp[-(\Delta G° + \lambda)^2/(4RT\lambda)] \qquad (3.1)$$

Different methods were suggested to overcome this kinetic barrier for ET between the enzyme redox centres and the electrodes, by designing intermediary, short-distance, mediated ET reactions that communicate the redox centres of enzymes with the electrode. Diffusional redox mediators that attain close proximity to the redox centres of the proteins, exchange electrons with the redox sites and diffuse out of the protein shells to exchange electrons with the electrodes, were used to establish the mediated electrical communication between the proteins and electrodes.[6] Integrated electrically contacted redox enzyme-modified electrodes were designed by the chemical tethering of redox relay groups to the protein. These shorten the ET distances between the enzyme redox centres and the electrodes[7] and, thus, "wire" the charge transport between the enzyme redox centres and the electrode. Also, enzymes were immobilised in redox-active polymers containing tethered electroactive groups that mediate the ET between the redox enzymes and the electrodes.[8]

The electrical contacting of redox proteins and electrodes is not only a fundamental challenge in bioelectrochemistry, but it also has immense practical implications. Two important applications of electrically contacted redox enzymes are envisaged in Figure 3.1. These include the use of the enzyme electrodes as amperometric biosensors,[9] which results in an amperometric response that relates to the concentration of the substrate, Figure 3.1(A), and the development of biofuel cells, where the biocatalytic oxidation of a fuel substrate at the anode, and the biocatalysed reduction of an oxidiser, e.g. oxygen at the cathode, lead to the generation of electrical power,[10] Figure 3.1(B).

Although substantial progress has been reported in the electrical contacting of enzymes with electrodes, the electron transfer turnover rates are far lower than the ET effectiveness in the native systems that include O_2 as acceptor or cofactor-mediated ET. The reasons for the inefficient electrical contacting of redox enzymes with electrodes by artificial means rests on the fact that the relay units tethered to the proteins are randomly positioned on structurally non-optimised sites, and the fact that the immobilisation of the redox enzymes on the electrodes occurs with random orientations of the redox sites with respect to the electrode. For optimal ET communication, the desirable configuration of the enzyme would include an assembly where all redox protein units exist in an identical orientation, in which the redox centres are as close as possible to the electrode, and the charge transfer relay, R, is positioned between the protein redox centre and the electrode, to shorten the ET distances, Figure 3.2(A). Such "wiring" of the redox proteins would require the nano-engineering of the

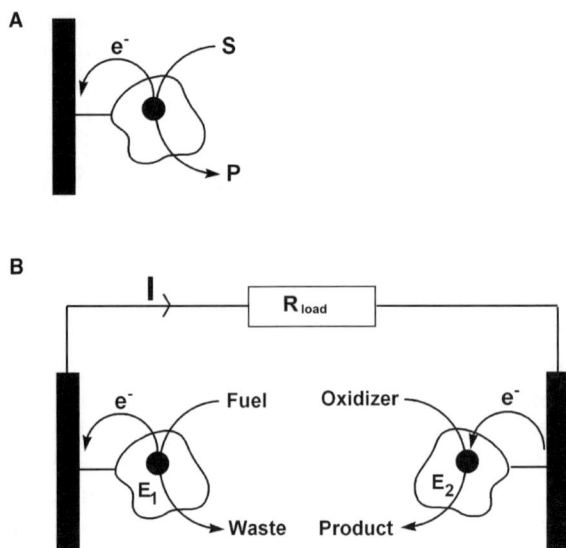

Figure 3.1 (A) Integrated enzyme electrodes for bioelectronic applications – bioelectrocatalysed oxidation of a substrate. (B) Schematic configuration of a biofuel cell composed of enzymes, E_1 and E_2, electrically contacted to the anode and the cathode, respectively.

proteins on the surfaces by chemical means. The extensive progress in the chemical modification of surfaces with monolayers, or thin films, of biomolecules paved the way to improving the wiring process by the hierarchical, oriented construction of wired redox protein assemblies on surfaces. This has been accomplished by the reconstitution of the protein on pre-designed molecular wires assembled on surfaces, Figure 3.2(B). According to this paradigm, the redox cofactor is extracted from the biocatalyst to yield the hollow-enzyme or apo-protein. The assembly of a relay-cofactor wire on the electrode, followed by the reconstitution of the apo-enzyme on the cofactor site, results in the structurally aligned enzyme with respect to the electrode, with the tailored positioning of the relay between the enzyme redox centre and the electrode. A further extension of this paradigm may be applied for redox enzymes that involve a diffusional ET mediating cofactor that yields an intermediary, structurally oriented, affinity complex between the cofactor and the redox protein. A relay-cofactor wire is then assembled on the electrode, and the structurally aligned affinity complex, which is formed with the redox protein, is cross-linked to yield a rigidified electrically contacted enzyme electrode, Figure 3.2(C).

The present article reviews the advances in the electrical contacting of redox proteins by the reconstitution process, and discusses the use of the engineered enzyme-modified electrodes as amperometric biosensors or biofuel cells.

Electrical Interfacing of Redox Enzymes

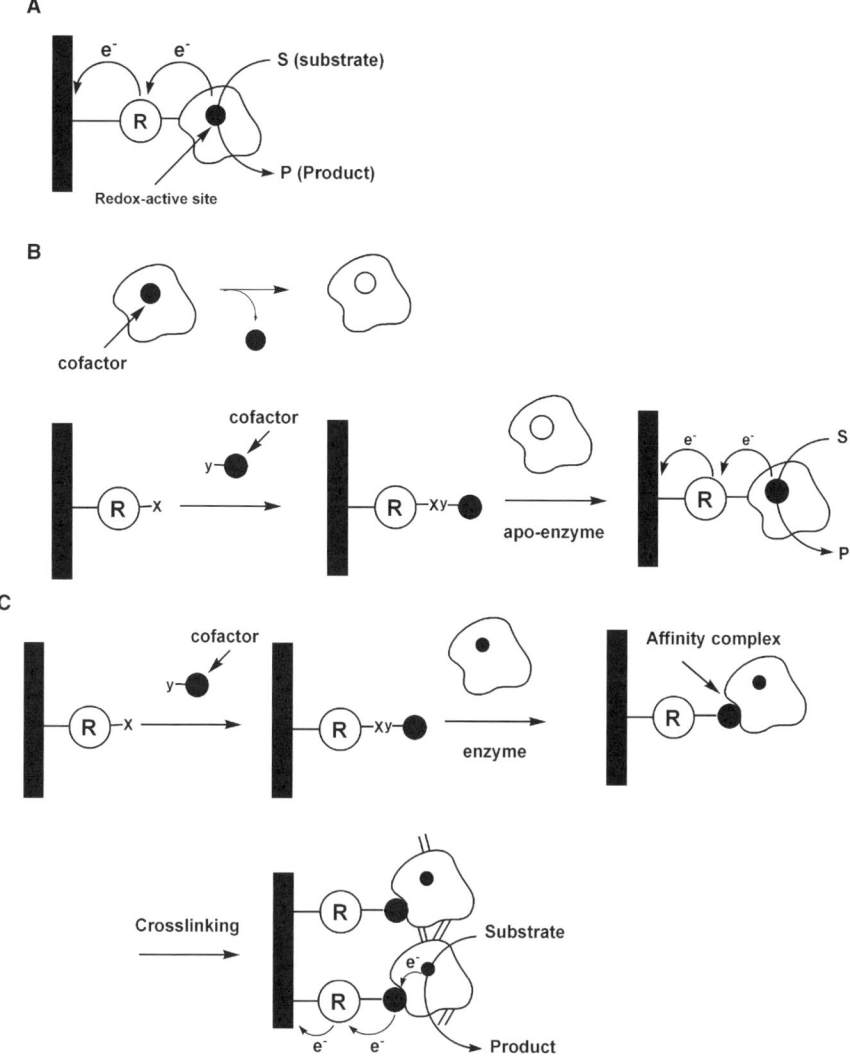

Figure 3.2 (A) A charge-transfer-relay-assisted bioelectrocatalysed oxidation of a substrate at an integrated enzyme electrode. (B) Reconstitution of an apo-enzyme on a relay-cofactor monolayer for the alignment and electrical wiring of a redox enzyme. (C) Electrical contacting of a redox enzyme by the assembly of a relay-cofactor monolayer on the electrode, and the surface cross-linking of the cofactor-enzyme affinity complex.

3.2 Reconstituted Enzyme Electrodes in Monolayer Configurations

The electrical contacting of the flavoenzyme glucose oxidase, GOx, was achieved by the reconstitution of apo-GOx on charge transporting wires consisting of a relay unit, as charge carrying element, and a flavin adenine dinucleotide cofactor as a reconstitution site. In one configuration,[11] Figure 3.3(A), pyrroloquinoline quinone, PQQ (**1**), was assembled on an Au electrode, and N^6-(2-aminoethyl-flavinadenine dinucleotide), amino-FAD (**2**), was covalently linked to the PQQ sites. The reconstitution of apo-glucose oxidase on the surface-bound FAD site led to a structurally aligned enzyme monolayer, with a surface coverage of $1.5 \times 10^{-12}\,\text{mol}\,\text{cm}^{-2}$. The PQQ electron relay units mediated ET between the FAD sites and the electrode, while activating the bioelectrocatalysed oxidation of glucose. The electron transfer turnover rate was estimated to be *ca.* $900\,\text{s}^{-1}$, a value that is similar to the exchange rate between the enzyme redox centre and its native electron acceptor, oxygen.[12] This efficient electrical communication between the enzyme redox centre and the electrode led to an oxygen-insensitive amperometric glucose-sensing electrode. Similarly, the effective ET communication between the biocatalyst and the electrode generated an amperometric glucose biosensor device that was insensitive to common glucose-sensing interferants, such as ascorbic acid or uric acid. The second route for the assembly of the electrically contacted GOx electrode is depicted in Figure 3.3(B). This method substitutes the synthetic FAD cofactor (**2**) with the natural FAD cofactor by applying "click chemistry" principles to construct the enzyme electrode.[13] The pyrroloquinoline quinone monolayer-functionalised electrode was reacted with 3-aminophenyl-boronic acid (**3**), and the native FAD cofactor (**4**) was linked by the vicinal hydroxyl groups to the boronic acid ligand. The reconstitution of apo-glucose oxidase on the FAD cofactor sites led to an electrically contacted enzyme electrode. Figure 3.3(C) depicts the cyclic voltammograms observed upon the bioelectrocatalysed oxidation of variable concentrations of glucose by the reconstituted enzyme electrode, and the derived calibration curve. Knowing the surface coverage of the enzyme on the electrode, $2 \times 10^{-12}\,\text{mol}\,\text{cm}^{-2}$, and the resulting saturation current, the turnover rate of electrons between the redox centre and the electrode was estimated to be $700\,\text{s}^{-1}$. This high turnover rate yielded a glucose-sensing electrode configuration that was insensitive to O_2, or other common interferences, such as ascorbic or uric acids.

Recent advances in supramolecular chemistry demonstrated the ability to design "molecular wires" on electrodes, on which threaded redox-active rings are "stopped" into semi-rotaxane configurations. The threaded "rings" were, then, electrochemically shuttled along the "wire",[14] and their vectorial motion along the "wire" enabled the ET communication between the redox site of the protein and the electrode. A monolayer consisting of a "molecular shuttle" in a semi-rotaxane configuration was used to electrically contact glucose oxidase with the electrode,[15] Figure 3.4(A). The bis-bipyridinium cyclophane (**5**) was threaded on "molecular wires" that included the bis-iminobenzene π-donor sites, which were assembled in a monolayer configuration on an Au electrode. The supramolecular π-donor-acceptor complexes between (**5**) and the bis-iminobenzene components were stopped with the amino-FAD cofactor (**2**), and apo-glucose oxidase was reconstituted on the FAD sites. The resulting integrated enzyme electrode revealed electrical communication between the enzyme

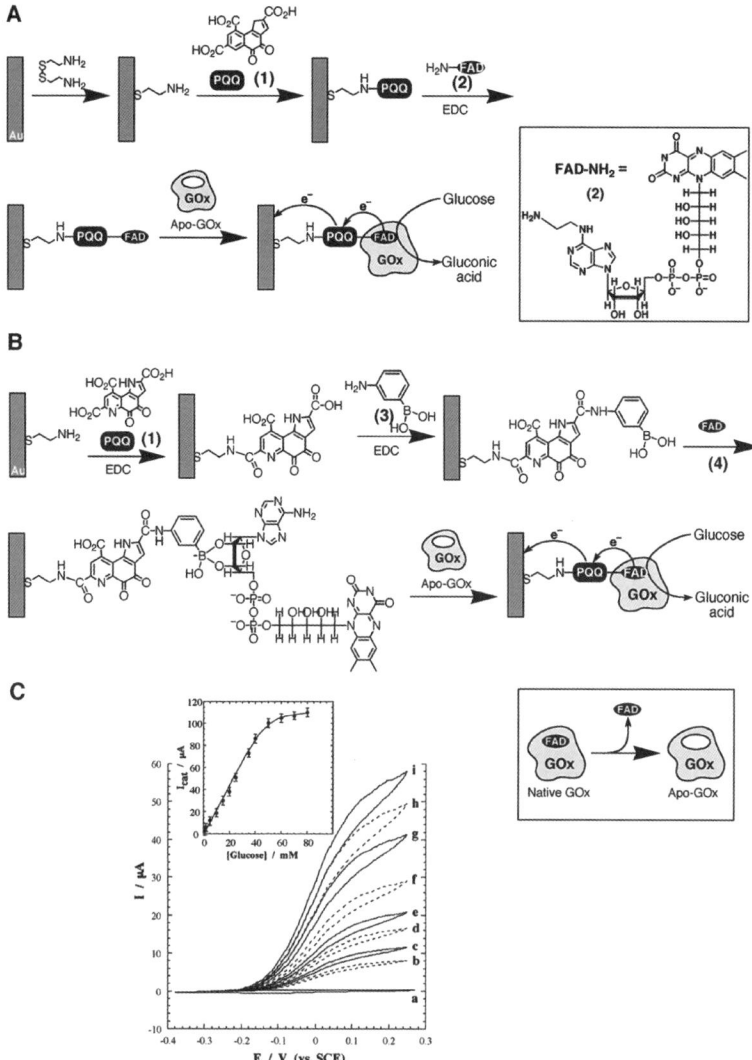

Figure 3.3 (A) Surface reconstitution of apo-glucose oxidase on a PQQ/FAD monolayer associated with an Au-electrode. (Reprinted in part with permission from reference 11. Copyright 1996 American Chemical Society.) (B) Assembly of the PQQ/FAD monolayer on an Au electrode *via* a boronic acid bridge, and the reconstitution of apo-GOx on the FAD cofactor sites. (C) Cyclic voltammograms corresponding to the bioelectrocatalysed oxidation of variable concentrations of glucose by the reconstituted glucose oxidase-functionalised electrode according to (B). Glucose concentrations correspond to (a) 0 mM, (b) 5 mM, (c) 10 mM, (d) 15 mM, (e) 20 mM, (f) 25 mM, (g) 35 mM, (h) 40 mM, (i) 50 mM. Inset: calibration curve correspond to the transduced electrocatalytic currents at different concentrations of glucose. (Reprinted with permission from reference 13. Copyright 2002 American Chemical Society.)

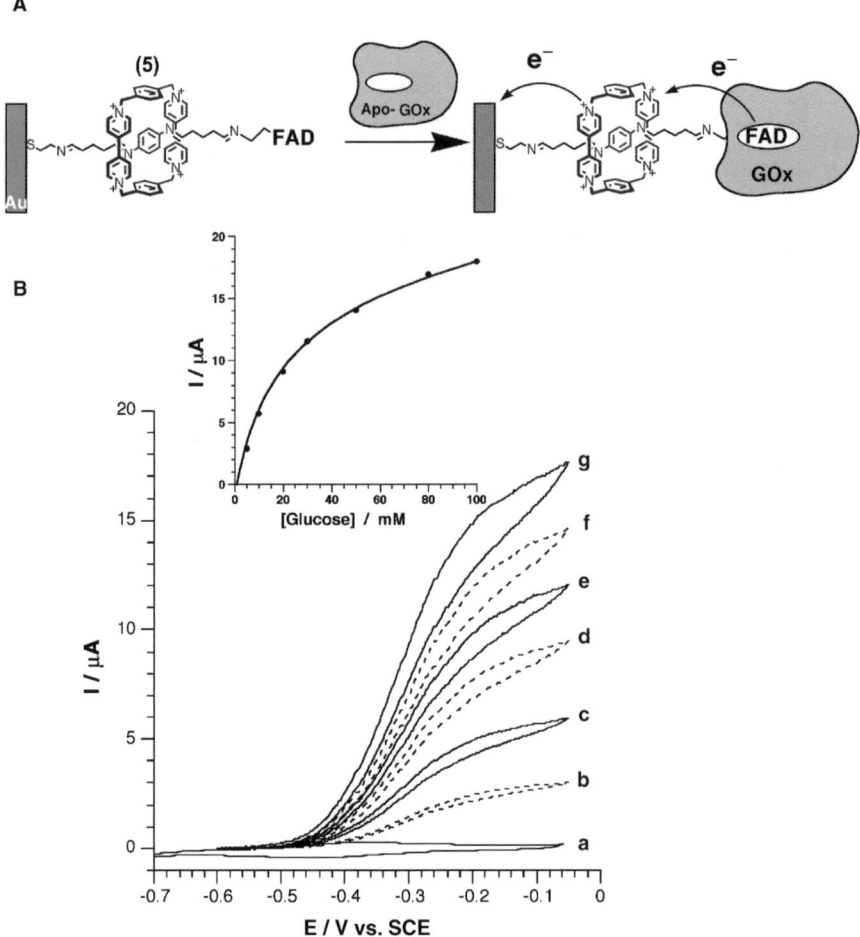

Figure 3.4 (A) The reconstitution of apo-glucose oxidase on an FAD cofactor that "stoppers" the cyclophane (**5**) on the molecular wire. The redox enzyme is contacted with the electrode by means of the electrochemically shuttled redox unit along the wire. (B) Cyclic voltammograms corresponding to the bioelectrocatalysed oxidation of different concentrations of glucose by the GOx-reconstituted electrode in the rotaxane structure: (a) 0 mM, (b) 5 mM, (c) 10 mM, (d) 20 mM, (e) 30 mM, (f) 50 mM, (g) 80 mM. Inset: calibration curve derived from the cyclic voltammograms at -0.1 V vs. SCE. (Reproduced with permission from reference 15. *Angew. Chem. Int. Ed.*, 2004, **43**, 3292–3300. Copyright Wiley-VCH Verlag GmbH & Co. KGaA.)

and the electrode. The redox potential of the threaded cyclophane is -0.43 V vs. SCE, ca. 100 mV more positive than that of the FAD cofactor associated with the enzyme. Upon the biocatalysed oxidation of glucose, the cyclophane unit oxidises the cofactor site, thereby transforming the cyclophane into the respective radical-cation, which lacks π-acceptor properties. The potential applied on the electrode, −0.43 V vs. SCE, attracts the reduced cyclophane, and the electrically shuttled cyclophane is being oxidised at the electrode to the tetracationic cyclophane. The latter product is dynamically shuttled to the bis-iminobenzene site, where a stable π-donor-acceptor complex is formed. Thus, the biocatalytic transformation and the concomitant electrochemical oxidation of the redox-relay, threaded on the wire, dynamically shuttle the electron relay unit between the electrode and the π-donor site, leading to the "wiring" of the enzyme with the electrode. In fact, the rates for the translocation of the reduced cyclophane to the electrode and the relocation of the oxidised cyclophane and the π-donor sites were estimated to be $320\,s^{-1}$ and $9.2\,s^{-1}$, respectively.[14] Figure 3.4(B) depicts the electrocatalytic anodic currents observed upon the bioelectrocatalysed oxidation of glucose, and the respective derived calibration curve. The biocatalysed oxidation of glucose occurs at an onset potential of −0.4 V vs. SCE. This is probably the lowest potential ever observed for the bioelectrocatalysed oxidation of glucose in an electrically wired assembly. The bioelectrocatalysed oxidation of glucose at such low potentials has immense significance in designing amperometric glucose biosensors, and the development of biofuel cells. Electrochemical glucose sensing is interfered by oxidisable compounds, such as ascorbate or uric acid, yet these electrochemical processes are prohibited at this negative potential and, thus, the amperometric analysis of glucose is not perturbed. Furthermore, the oxidation of glucose at such a negative potential, close to the thermodynamic potential of the redox centre of glucose oxidase, is invaluable for the design of future biofuel cells with high power outputs, a topic that will be addressed in a separate section in this chapter.

3.3 Electrical Wiring of Redox Proteins with Electrodes by their Reconstitution on Cofactor-Functionalised Metallic Nanoparticles (NPs) or Carbon Nanotubes (CNTs)

Metallic nanoparticles (NPs) such as Au, Ag or Pt NPs or carbon nanotubes (CNTs) with appropriate graphitic folding exhibit ballistic conductance. As the diameters of the NPs or folded CNTs are comparable with the dimensions of proteins, one may envisage that the conjugation of redox proteins with these nano-elements could yield hybrid systems with unique electronic properties. Specifically, one may substitute the molecular relay units with NPs or CNTs as nano-relay units that act as charge carriers, and electrically wire the redox centres of the enzymes with the respective electrodes.

Au NPs (1.4 nm in diameter) were used as charge relays for the electrical contacting of glucose oxidase with electrodes.[16] The amino-FAD cofactor (**2**) was covalently tethered to the Au NPs functionalised with a single N-hydroxysuccinimide functionality, Figure 3.5(A). Reconstitution of apo-glucose

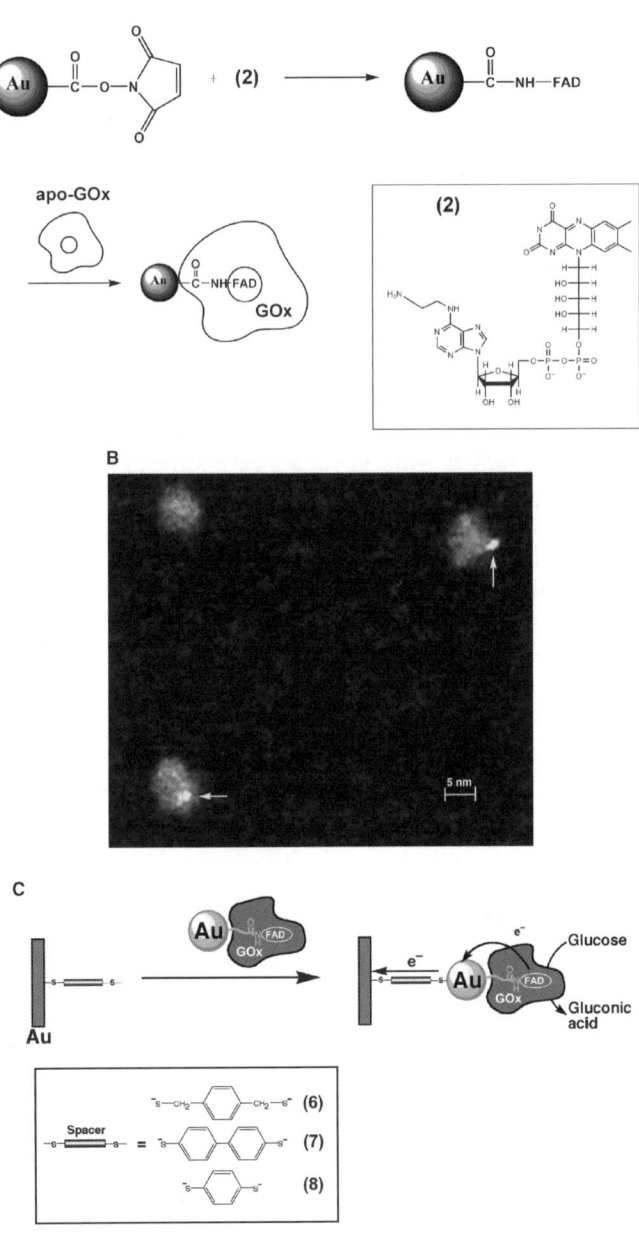

Figure 3.5 (A) The assembly of electrically contacted glucose oxidase by the reconstitution of apo-GOx on an FAD-functionalised Au nanoparticle (1.4 nm). (B) A STEM image of GOx reconstituted with the Au-FAD. Arrows show individual Au nanoparticles. (C) The assembly of an Au NP (1.4 nm) electrically contacted glucose oxidase electrode by the reconstitution of apo-GOx on the FAD-functionalised Au NP, and the immobilisation of the enzyme/nanoparticle hybrid on an electrode surface using different dithiol cross-linkers. (D) Cyclic voltammograms corresponding to the bioelectrocatalysed oxidation of variable concentrations of glucose by the

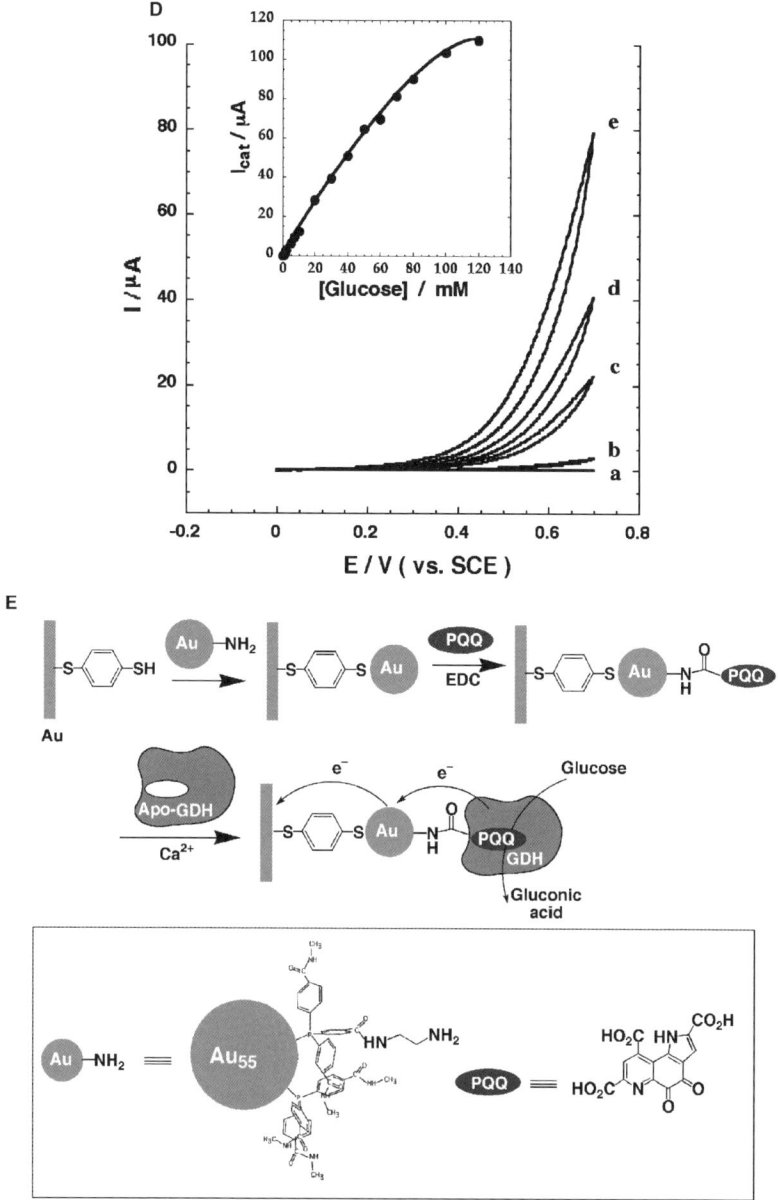

electrically contacted Au NP-reconstituted GOx-modified electrode. Glucose concentrations corresponding to (a) 0 mM, (b) 1 mM, (c) 2 mM, (d) 5 mM and (e) 10 mM. Inset: calibration curve corresponding to the electrocatalytic currents at different glucose concentrations. (E) Assembly of GDH reconstituted on the PQQ-functionalised Au NPs associated with an Au electrode. (Parts (B) and (D) reproduced from reference 16, reprinted with permission from AAAS. Part (E) reprinted with permission from reference 18. Copyright 2005 American Chemical Society.)

oxidase on the FAD-functionalised Au NPs yielded an Au NP-enzyme hybrid nanostructure. Figure 3.5(B) shows the STEM image of the GOx-Au NP conjugate, where the incorporation of the Au NPs into the protein matrices is clearly visible. The resulting Au NP-enzyme conjugates were, then, assembled on an Au electrode, using different dithiol cross-linkers as bridging units, *e.g.* p-xylene dithiol (**6**), 4,4'-dimercaptobiphenyl (**7**) and 1,4-benzene dithiol (**8**), Figure 3.5(C). The resulting enzyme electrode revealed unprecedented electrical contact efficiency between the redox centre of the enzyme and the electrode, and with benzene dithiol (**8**) as linker, a turnover rate of $5000\,s^{-1}$ was observed, Figure 3.5(D). This value should be compared to the ET exchange rate between the enzyme redox centre and its native electron acceptor, O_2, *ca.* $900\,s^{-1}$. The efficient electrical contact between the enzyme redox centre not only leads to the effective bioelectrocatalysed oxidation of glucose and to the sensitive analysis of glucose, but also has indispensable significance reflected by the specificity of the electrode. The resulting Au NPs electrically contacted enzyme electrode was found to be insensitive to common glucose-sensing interferants, or to atmospheric O_2. In this system, the Au NPs act as nano-relay units that mediate the ET between the redox centre of GOx and the electrode. In fact, the molecular dithiol bridges, which link the Au NPs to the electrode, were found to be the rate-limiting components in controlling the ET between the enzyme and the electrode. The effectiveness of mediated ET was found to follow the order (**8**) > (**7**) > (**6**). The effects of the "molecular bridges" on the ET efficiency are attributed to the degree of electron delocalisation in the "molecular wires". The bridge (**8**) consists of a fully conjugated wire, whereas the biphenyl linker (**7**) includes a perturbed π conjugation, due to the steric tilting of the two benzene rings. The least-effective charge-mediator is (**6**), and this is explained by the two sp^3 carbon units that "break" the delocalisation path of the linker. In fact, theoretical studies[17] on the charge transport through the three molecular wires (**6**), (**7**) and (**8**) confirmed the experimentally observed results. A similar approach was applied to electrically activate the pyrroloquinoline quinone (PQQ)-dependent glucose dehydrogenase, GDH.[18] Au NPs (1.4 nm) that included a single amine functionality were linked to a gold electrode, and the PQQ cofactor was covalently attached to the particles, Figure 3.5(E). The apo-GDH was then reconstituted on the PQQ cofactor, and the resulting Au NP/GDH conjugate revealed electrical communication with the electrode. From the saturation current generated by the electrode, and knowing the surface coverage of the enzyme, the ET turnover rate between the enzyme and the electrode was estimated to be $11\,800\,s^{-1}$.

The electronic coupling between the reconstituted enzyme redox centres and the Au NPs was used to follow the "bio-pumping" of electrons into Au NPs, which resulted in their charging. The changes in the dielectric properties of the NPs, as a result of their charging, altered the localised surface plasmon absorbance of the NPs, and changed the capacitance at the particle–solution interface. These physical changes enabled the probing of the charging process, by following the changes in the effects of the coupling of the surface plasmon of an Au-coated surface with the localised plasmon of the NPs, or by probing the capacitance changes of the particles, as a result of their charging.[19] Au NPs, 1.4 nm, were linked to an Au electrode by a long chain, 1,9-nonanedithiol monolayer, exhibiting low dielectric properties. The long-chain alkane dithiol monolayer acted as an electron tunnelling barrier that allowed the charging of the NPs. The amino flavin adenine dinucleotide (**2**) was covalently tethered to the particles, and apo-GOx was reconstituted on the cofactor sites, Figure 3.6(A). The changes in the surface plasmon resonance, as a result of the coupling of the localised plasmon of the particles, which are being charged by the biocatalysed transformation, with the surface plasmon wave, are depicted in Figure 3.6(B). The changes in the minimum reflectivity angles, at different glucose concentrations, and the respective calibration curve (curve (a) and inset) are shown in Figure 3.6(C). Similar changes in the surface plasmon spectra could be stimulated by the application of a potential on the electrode, which resulted in the charging of the NPs by the electrode, in analogy to their charging by the biocatalysed process. Similarly, the capacitance changes at the NP/GOx-functionalised electrode, as a result of the charging of the NPs by the biocatalysed process, are depicted in Figure 3.6(D). A calibration curve that corresponded to the capacitance changes at variable concentrations of glucose was derived, Figure 3.6(D), inset. The surface plasmon resonance changes, as well as the capacitance changes, were observed only with the reconstituted enzyme on the NPs (*cf.* Figure 3.6(C), curves (b) and (c)), implying that the reconstituted protein exhibits electronic coupling with the NPs, and this allows their electrical charging. Also, the bridging of the NP to the electrode with shorter alkane dithiol chains enabled the leakage of electrons from the NPs to the electrode and, thus, the charging of the NPs became inefficient. By following the effects of the external potential on the charging of the NPs, and using the derived value of capacitance changes, as a result of the biocatalysed oxidation of glucose, it was estimated that at a concentration of 1×10^{-2} M of glucose, and using the nonane dithiol as tunnelling barrier, the bioelectrocatalytic oxidation of glucose resulted in the average charging of a single particle with 10 electrons.[19]

Carbon nanotubes (CNTs) provide an additional conducting nano-object. The "armchair" or "zigzag" folding motifs of the graphitic layer yield CNTs that reveal metallic conductivity.[20] The integration of biomolecules with CNTs attracted substantial research efforts,[21] and biomolecule-CNT hybrid systems found different applications in designing electrical or optical sensors,[22] nanocircuitry[23] or nanoscale devices.[24] The diameter of CNTs (*ca.* 1.8 nm), and their conductivity properties, suggested their potential use as electrical nano-connectors for wiring redox proteins with electrodes.[25] Carbon nanotubes were subjected to oxidative cleavage under harsh acidic conditions, a process that led to the shortening of the CNTs, and to the substitution of the edges of the CNTs with carboxylic acid functionalities. The resulting CNTs were then separated into several fractions that included CNTs with narrow distributions of lengths, which corresponded to 25–30 nm, 40–50 nm, 80–100 nm and 200–230 nm. The assembly of the CNTs-bridged glucose oxidase, GOx-functionalised electrode is depicted in Figure 3.7(A).[26] The carboxylic acid-modified CNTs were covalently anchored to a

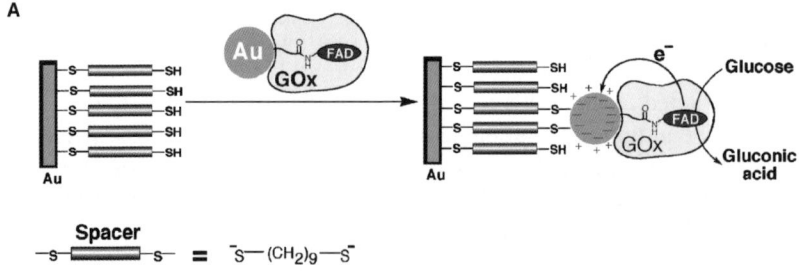

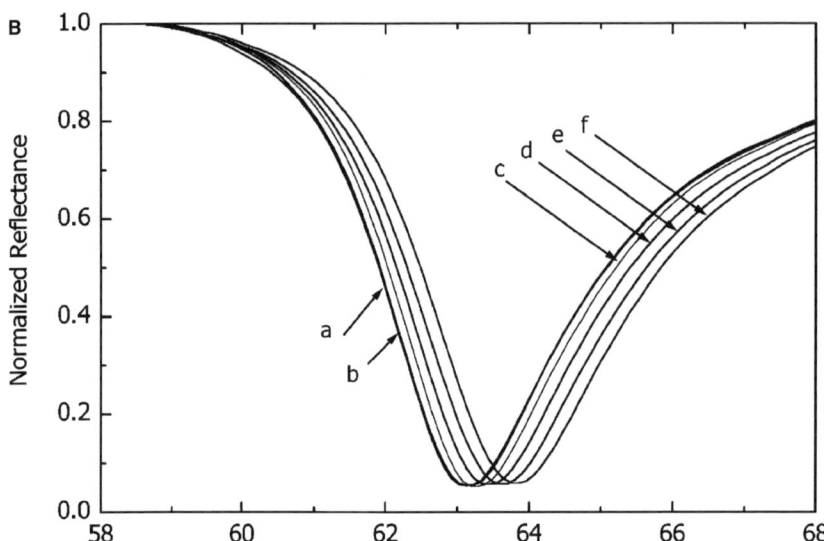

Figure 3.6 (A) Assembly of the electrically contacted reconstituted GOx/Au-NP systems on Au-electrodes using a 1,9 nonanedithiol bridging unit. (B) SPR spectra corresponding to the GOx-functionalised-Au-NP arrays bridged to the Au-electrodes upon the addition of different concentrations of glucose: (a) 0 mM, (b) 0.3 mM, (c) 1.6 mM, (d) 8 mM, (e) 40 mM, (f) 100 mM. (C) Shifts of the plasmon angles induced upon addition of different concentrations of glucose to: (a) the aligned GOx reconstituted Au-NPs bridged to the Au-electrode, (b) the non-aligned GOx randomly covalently bound to the Au-NPs bridged to the Au-electrode, (c) the Au-electrode surface modified with 1,9-nonanedithiol monolayer. Inset: the relative changes of the plasmon angle upon the addition of various glucose concentrations to the aligned GOx/Au-NP system bridged to the Au-electrode by 1,9-nonanedithiol. (D) Time-dependent capacitance changes of the GOx/Au-NP system upon the addition of 100 mM glucose to the aligned GOx/Au-NP system. Inset: effect of various concentrations of glucose on the capacitance values of the aligned GOx/Au-NP system. (Reprinted with permission from reference 19. Copyright 2004 American Chemical Society.)

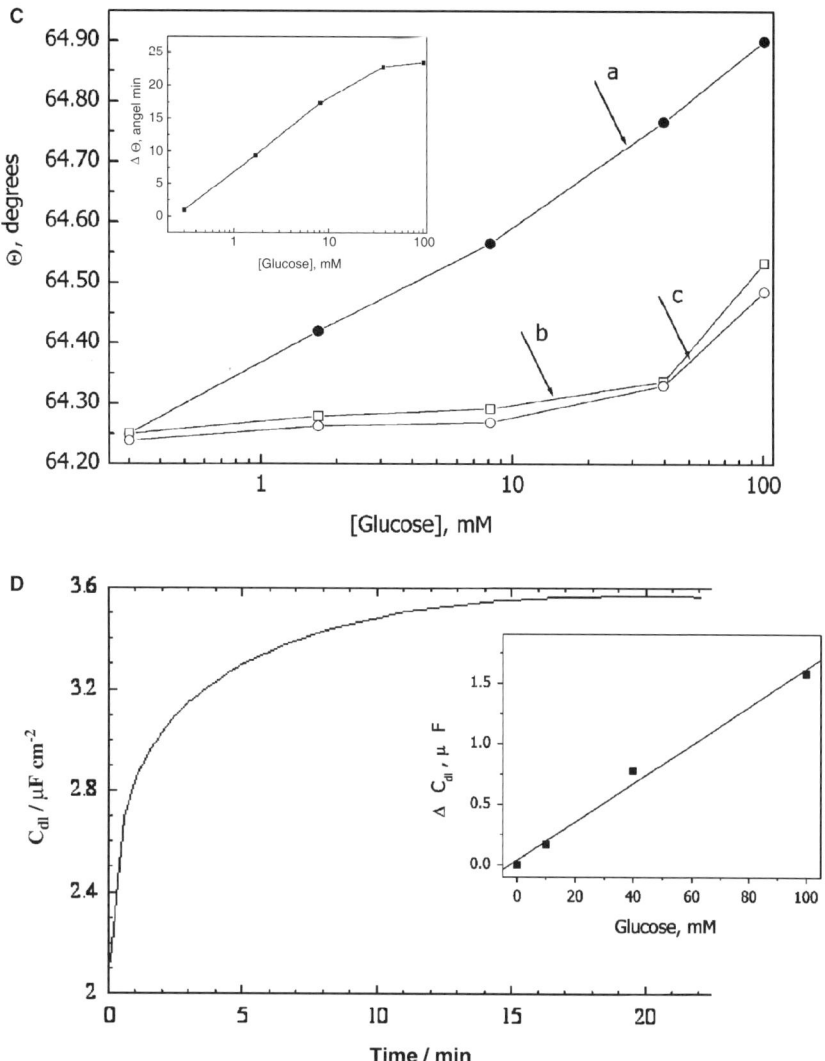

Figure 3.6 Continued.

cystamine-functionalised electrode, and the amino-FAD cofactor (**2**) was linked to the carboxylic acid functionalities at the ends of the CNTs. The reconstitution of apo-GOx on the FAD units yielded, then, the integrated enzyme-CNT electrode. AFM and TEM imaging confirmed the formation of the reconstituted enzyme at the ends of the CNTs. The reconstitution of the standing CNTs on the Au surface generated nanostructures with a height corresponding to 5 nm, consistent with the formation of reconstituted protein units on the ends of the CNTs, Figure 3.7(B). The AFM and TEM images of solution-suspended CNTs, which were modified at their two ends with the FAD cofactor (**2**), and, subsequently, were reconstituted with apo-GOx, confirmed the tethering of the proteins to the ends of the CNTs,

Figures 3.7(C) and (D), respectively. The GOx reconstituted onto the CNTs revealed electrical contacting with the electrode, and with 25–30 nm-long connecting CNTs, the turnover rate of electrons was estimated to be $4100\,s^{-1}$, Figure 3.7(E). The electrical contacting efficiency was, however, controlled by the length of the CNTs, and as the tubes were longer, the turnover rate was lower. This was attributed to defects introduced into the side walls of the CNTs upon the scission of the long non-substituted CNTs. As the number of defects increased with the length of the CNTs, the barriers for charge transport along the CNTs were higher for the enzyme systems bridged to the electrode with longer CNTs.

3.4 Reconstitution of apo-Enzymes in Thin Films of Redox Polymers

The incorporation of redox enzymes in redox-active polymers is a common practice for electrical contacting enzymes with electrode surfaces, and this method was frequently applied to develop amperometric biosensors[27] or biofuel cells.[28] The enzyme-polymer hybrid systems reveal the advantage of relatively high loading of the biocatalysts in the polymer matrices. These systems suffer, however, from the disadvantage that the enzymes are randomly distributed in the polymers, and thus, the proteins lack steric orientation for optimal electrical coupling between the redox centres and the ET mediator units associated with the polymers. Thus, the reconstitution of redox proteins in electroactive polymer films may combine the benefits of high enzyme loading and optimised steric orientation of the biocatalytic redox centres in respect to the ET-mediating polymer films.

This approach was materialised by the reconstitution of redox enzymes in polyaniline redox-active films. A composite of polyaniline (PAn) and polyacrylic acid (PAA) was electropolymerised onto electrodes, and the resulting PAn/PAA films revealed quasi-reversible redox activity at neutral pH values, due to the doping of the PAn with the negatively charged PAA chains, Figure 3.8(A). The amino-FAD cofactor (**2**) was, then, covalently linked to the PAA chains, and apo-glucose oxidase was reconstituted onto the FAD cofactor sites.[29] The resulting enzyme–polymer composite revealed electrical communication with the

Figure 3.7 (A) Assembly of the CNT-modified electrically contacted GOx electrode. (B) AFM image of GOx reconstituted on the CNTs associated with the Au surface. (C) AFM image of CNTs reconstituted at their ends with GOx units. (D) HRTEM image of a CNT modified at its ends with GOx units (the enzymes were stained with uranyl acetate). (E) Cyclic voltammograms corresponding to the bioelectrocatalysed oxidation of different concentrations of glucose by the GOx-reconstituted CNT-functionalised electrode: (a) 0 mM, (b) 20 mM, (c) 60 mM, (d) 160 mM. Inset: derived calibration curve corresponding to the amperometric responses of the reconstituted electrode at 0.45 V *vs.* SCE in the presence of different concentrations of glucose. (Reproduced with permission from reference 26. *Angew. Chem. Int. Ed.*, 2004, **43**, 2113–2117. Copyright Wiley-VCH Verlag GmbH & Co. KGaA.)

Electrical Interfacing of Redox Enzymes

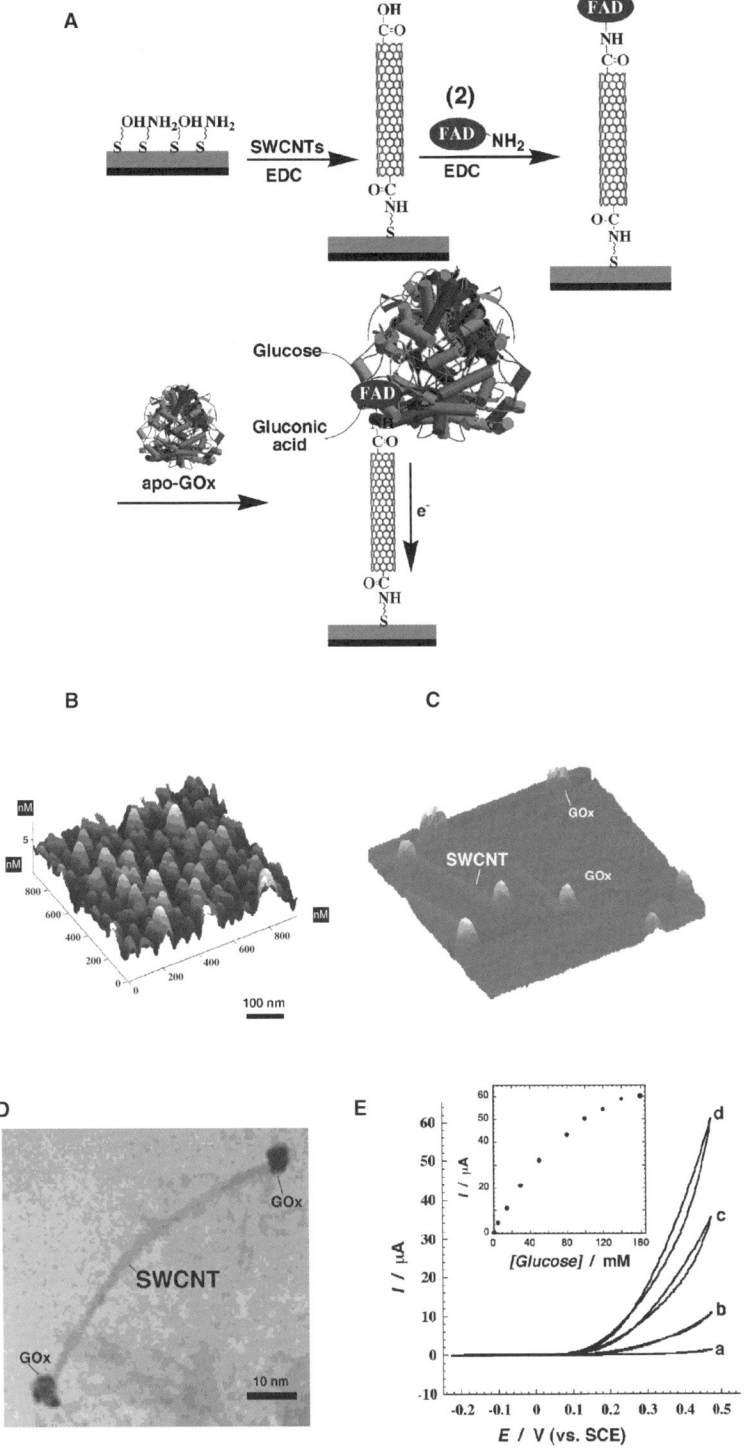

electrode, Figure 3.8(B). The bioelectrocatalytic activation of the reconstituted enzyme was attributed to the alignment of the enzyme in an appropriate orientation, with respect to the electroactive PAn/PAA film that facilitated mediated ET between the cofactor site embedded in the enzyme and the electrode. In a control experiment in which the enzyme glucose oxidase was randomly linked to the PAA chains, a 10^2-fold lower bioelectrocatalytic activity of the enzyme, as compared to the reconstituted GOx, was observed. This result demonstrated the significance of steric orientation of the biocatalyst in the polymer film for optimal ET communications between the enzyme and the electrode. The coverage of the film with GOx was found to be $3 \times 10^{-12}\,\text{mol cm}^{-2}$, and from the resulting saturation value of the electrocatalytic anodic current, $i = 0.3\,\text{mA cm}^{-2}$, the ET turnover rate between the enzyme redox site and the electrode was estimated to be $1000\,\text{s}^{-1}$. This value is comparable to the ET exchange between the redox site of GOx and its native electron acceptor (O_2), $900\,\text{s}^{-1}$. A similar approach was adapted to electrically activate the pyrroloquinoline quinone-dependent enzyme glucose dehydrogenase, GDH.[30] The PAA chains in the PAn/PAA composite associated with the electrode were modified with 1,4-diaminobutane, and PQQ was covalently linked to the resulting amine functionalities. The apo-GDH was, then reconstituted on the cofactor sites, Figure 3.8(C). The resulting enzyme/polymer composite revealed electrical contact with the electrode, and the PAn polymer mediated the bioelectrocatalysed oxidation of glucose.

3.5 Design of Electrically Contacted Enzyme Electrodes by the Crossing of Surface-confined Cofactor-enzyme Affinity Complexes

The previous chapters demonstrated the electrical contacting of enzymes through the reconstitution of apo-enzymes on the respective cofactor-modified electrodes. This paradigm can be applied for enzymes that include protein-embedded cofactors that enable the generation of a hollow apo-enzyme and, in turn, allow the firm, non-detachable reconstitution of the apo-enzyme on the cofactor-modified electrode. Numerous redox enzymes use, however, diffusional cofactors that mediate electron transfer between the enzyme redox centres and their surroundings by the formation of affinity complexes. For these enzymes, a "reconstitution-like" method to establish electrical communication between the enzymes and the electrodes was developed, Figure 3.2(C). According to this approach, the diffusional cofactor (or a synthetic analogue of the native cofactor) is attached to the electrode surface. Formation of the affinity complex between the enzyme and the cofactor is accompanied by the chemical cross-linking of the protein units on the electrode, a process that yields a non-detachable enzyme array. The formation of the cofactor-enzyme affinity complex aligns the active site of the biocatalyst into an optimal steric orientation for ET communication with the cofactor. Thus, the cross-linked enzyme array may exhibit electrical contact with the electrode surface, provided that the cofactor sites reveal direct electrical communication with the electrode, or

Electrical Interfacing of Redox Enzymes

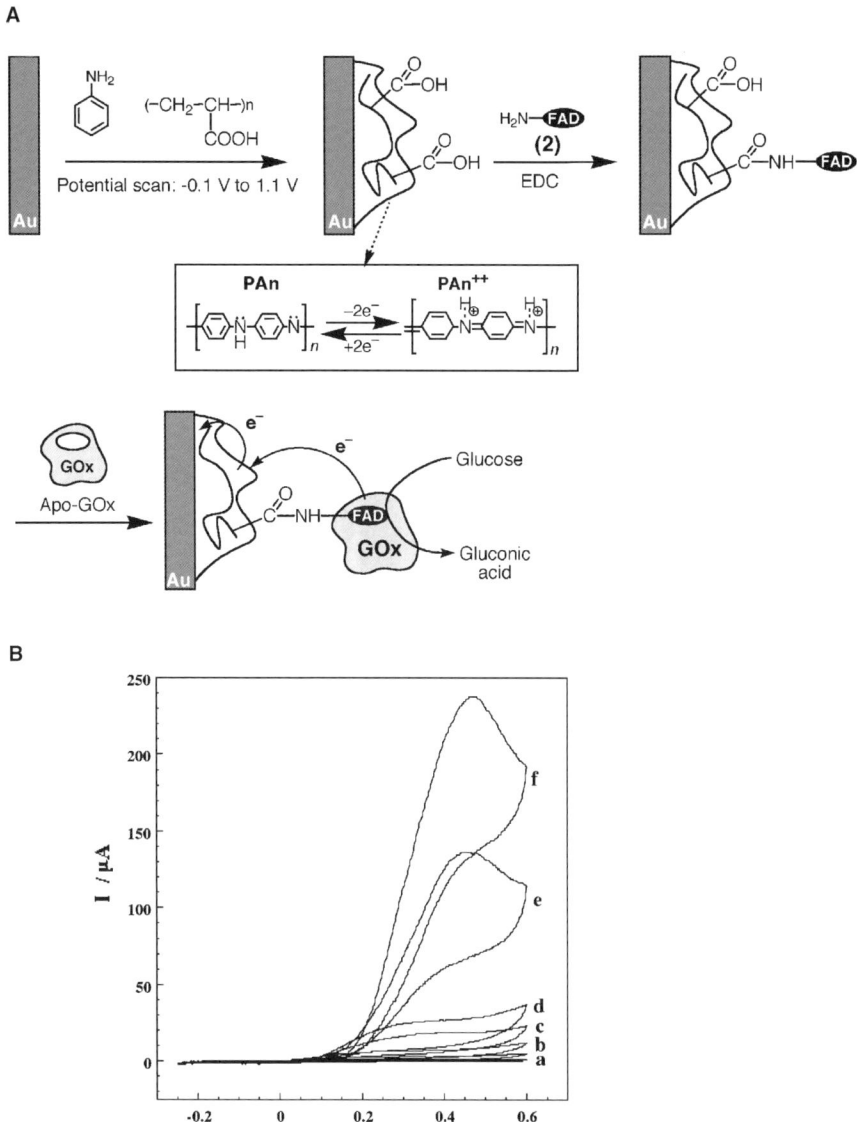

Figure 3.8 (A) Assembly of an electrically contacted polyaniline-reconstituted glucose oxidase electrode. (B) Cyclic voltammograms corresponding to the bioelectrocatalysed oxidation of variable concentrations of glucose by the integrated, electrically contacted polyaniline-reconstituted glucose oxidase electrode. Glucose concentrations correspond to (a) 0 mM, (b) 5 mM, (c) 10 mM, (d) 20 mM, (e) 35 mM, (f) 50 mM. (Parts A and B reprinted with permission from reference 29. Copyright 2002 American Chemical Society.) (C) Assembly of the electrically contacted polyaniline/PQQ-reconstituted glucose dehydrogenase electrode, GDH. (Reproduced from reference 30 by permission of The Royal Society of Chemistry.)

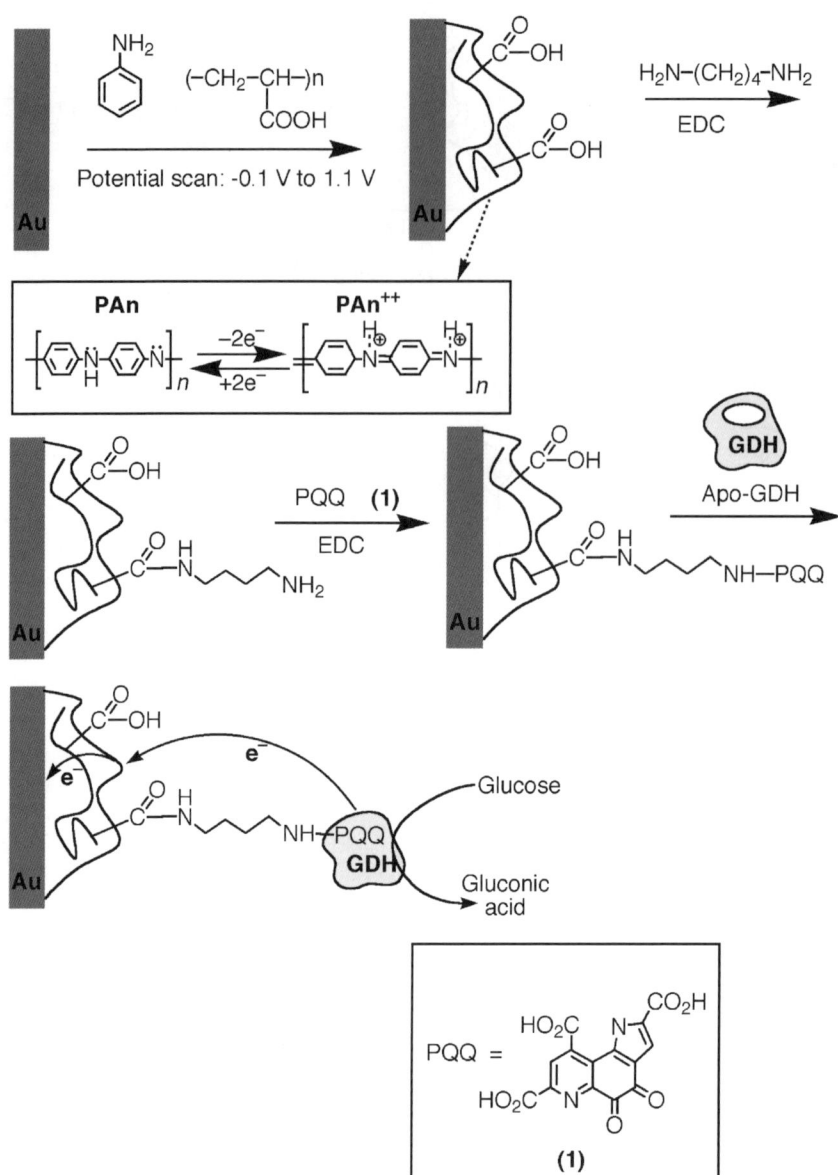

Figure 3.8 Continued.

mediated ET in the presence of an electron relay. Among the diffusional cofactors that activate redox proteins, the nicotinamide adenine dinucleotide (phosphate), $NAD(P)^+$, or hemoproteins, such as cytochrome c, should be mentioned. The labile affinity complexes between these cofactors and the redox proteins were used to construct integrated, electrically contacted enzyme electrodes.

Although the NAD(P)$^+$/NAD(P)H cofactor couple is a common redox mediator in biological transformations, the ET exchange of the couple with electrodes is hindered due to kinetic barriers, and due to unfavoured side reactions.[31] The origin of the kinetic barriers for ET between the electrode and the NAD(P)$^+$/NAD(P)H cofactors is attributed to the fact that the redox processes of these cofactors involve 2e$^-$ and H$^+$ transformations. The oxidation of the NAD(P)H cofactors was, however, electrocatalysed and mediated by different organic relay compounds or transition metal complexes.[32–34] One of the effective relay units for the electrocatalysed oxidation of the NAD(P)H cofactors is pyrroloquinoline quinone, PQQ, (**1**).[35] It enabled the tailoring of integrated, electrically contacted, monolayer-functionalised electrodes of NAD(P)$^+$-dependent enzymes.[13] A monolayer of pyrroloquinoline quinone, PQQ, was assembled on Au electrodes, and 3-aminophenyl boronic acid (**3**) was covalently linked to the PQQ monolayer. The formation of the boronate complex between the boronic acid ligand and the ribose site associated with nicotinamide adenine dinucleotide phosphate, NADP$^+$ (**9**), resulted in the attachment of the cofactor to the electrode. The subsequent generation of the labile affinity complex between the NADP$^+$ cofactor and malate dehydrogenase, MalD, followed by cross-linking of the surface-confined cofactor-enzyme complex with glutaric dialdehyde, yielded an integrated, rigid, electrically contacted enzyme electrode that activated the bioelectrocatalysed oxidation of malate to pyruvate, Figure 3.9(A). Figure 3.9(B) shows the cyclic voltammograms observed upon the oxidation of variable concentrations of malate, and the respective calibration curve. The turnover rate of electrons between the enzyme and the electrode in the integrated system was estimated to be 190 s^{-1}.

A similar approach was applied to electrically activate the NAD$^+$-dependent enzyme, lactate dehydrogenase, LDH, Figure 3.10.[13] The nicotinamide adenine dinucleotide, NAD$^+$ (**10**), was linked to the PQQ/boronic acid monolayer-functionalised electrode. The availability of two ribose units in the cofactor structure enables, however, two binding modes of the cofactor to the boronic acid ligand (I or II in Figure 3.10). The formation of the affinity complex between the NAD$^+$ cofactor and LDH, followed by the cross-linking of the labile enzyme-cofactor complexes with glutaric dialdehyde, yielded a rigid electrically contacted enzyme film that mediated the bioelectrocatalysed oxidation of lactate to pyruvate. Detailed chronoamperometric experiments revealed that, indeed, the NAD$^+$ cofactor binds to the boronic acid ligand by the two modes, I and II, and that the binding mode II is *ca.* 10-fold more efficient as compared to I in mediating the bioelectrocatalysed oxidation of lactate.[13]

Other electron relays, and particularly organic redox-active dyes, act as electrocatalysts for the oxidation of NADH.[36] Indeed, integrated electrically contacted NAD(P)$^+$-dependent enzyme electrodes were constructed by the immobilisation of the biocatalysts on carbon nanotubes (CNTs), and the deposition of the resulting nanostructures on glassy carbon electrodes.[37] The redox-active dye Nile blue, NB (**11**), was adsorbed onto the CNTs by π–π

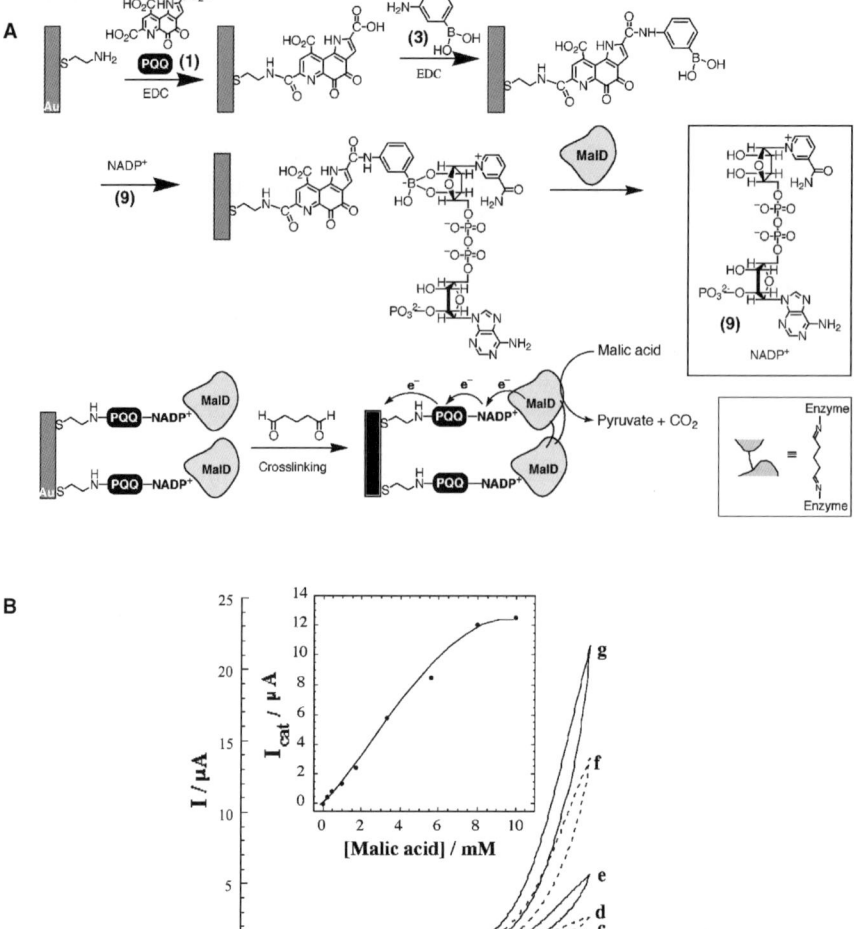

Figure 3.9 (A) Assembly of an integrated malate dehydrogenase, MalD, electrode for the bioelectrocatalysed oxidation of malate by the surface cross-linking of an affinity complex formed between MalD and a boronate-linked PQQ/$NADP^+$ monolayer. (B) Cyclic voltammograms corresponding to the bioelectrocatalysed oxidation of different concentrations of malate by the enzyme. Malate concentrations are (a) 0 mM, (b) 0.25 mM, (c) 0.5 mM, (d) 1 mM, (e) 1.7 mM, (f) 3.3 mM and (g) 5.6 mM. Inset: calibration curve corresponding to the electrocatalytic anodic currents as a function of malate concentration. (Reprinted with permission from reference 13. Copyright 2002 American Chemical Society.)

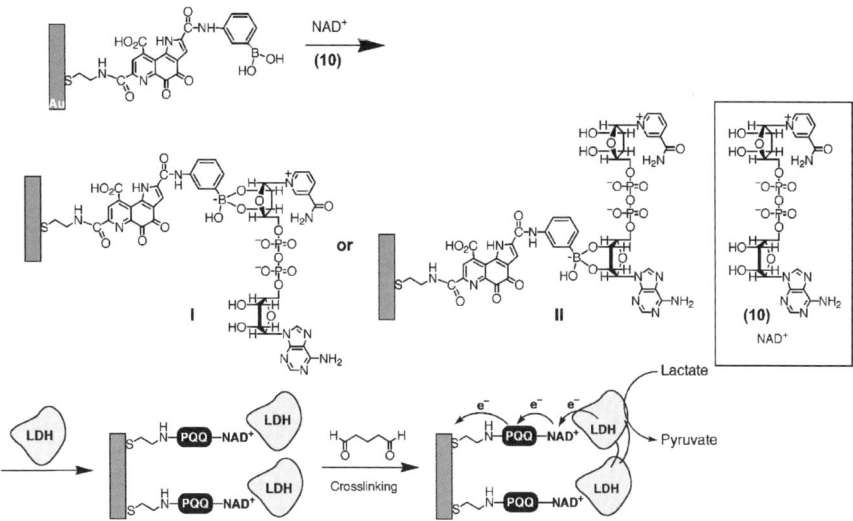

Figure 3.10 Assembly of an integrated lactate dehydrogenase, LDH, electrode for the bioelectrocatalysed oxidation of lactate by the surface cross-linking of an affinity complex formed between LDH and different structures of a boronate-linked PQQ-NAD$^+$ monolayer. (Reprinted with permission from reference 13. Copyright 2002 American Chemical Society.)

interactions, and it revealed a quasi-reversible 2-electron redox process. The modified electrodes were deposited onto a glassy carbon electrode, and 4-carboxyphenyl boronic acid (12) was covalently linked to the NB units. Subsequently, the boronic acid ligand acted as a robust anchor site for the association of the NAD(P)$^+$ cofactors,[37] Figure 3.11, (A) and (B). The NADP$^+$-dependent glucose dehydrogenase, GDH, was then interacted with the cofactor units to form affinity complexes that were cross-linked with glutaric dialdehyde to yield the integrated, electrically contacted, enzyme electrode, Figure 3.11(A). Similarly, the NAD$^+$ cofactor was linked to the boronic acid ligand, the affinity complexes between the enzyme alcohol dehydrogenase, AlcDH, and the cofactor sites were generated and these were cross-linked with glutaric dialdehyde. Two binding modes, I and II, of the two different ribose units to the boronic acid ligand were identified by chronoamperometry. The cofactor consisting of the boronate complex with the ribose adjacent to adenine group was found to be *ca.* 6-fold more efficient in electrically contacting AlcDH with the electrode, as compared to the NAD$^+$ coordinated to the boronic acid ligand by the ribose unit close to the nicotinamide site. Figures 3.11(C) and (D) depict the voltammetric responses of the GDH and of the AlcDH integrated enzyme electrodes, in the presence of variable concentrations of the respective substrates, and the respective calibration curves of the systems.

Other diffusional cofactors can be similarly used to form integrated enzyme electrodes by the cross-linking of cofactor-enzyme affinity complexes. This was demonstrated with the cytochrome *c*, Cyt. *c*, cofactor.[38] The cofactor was

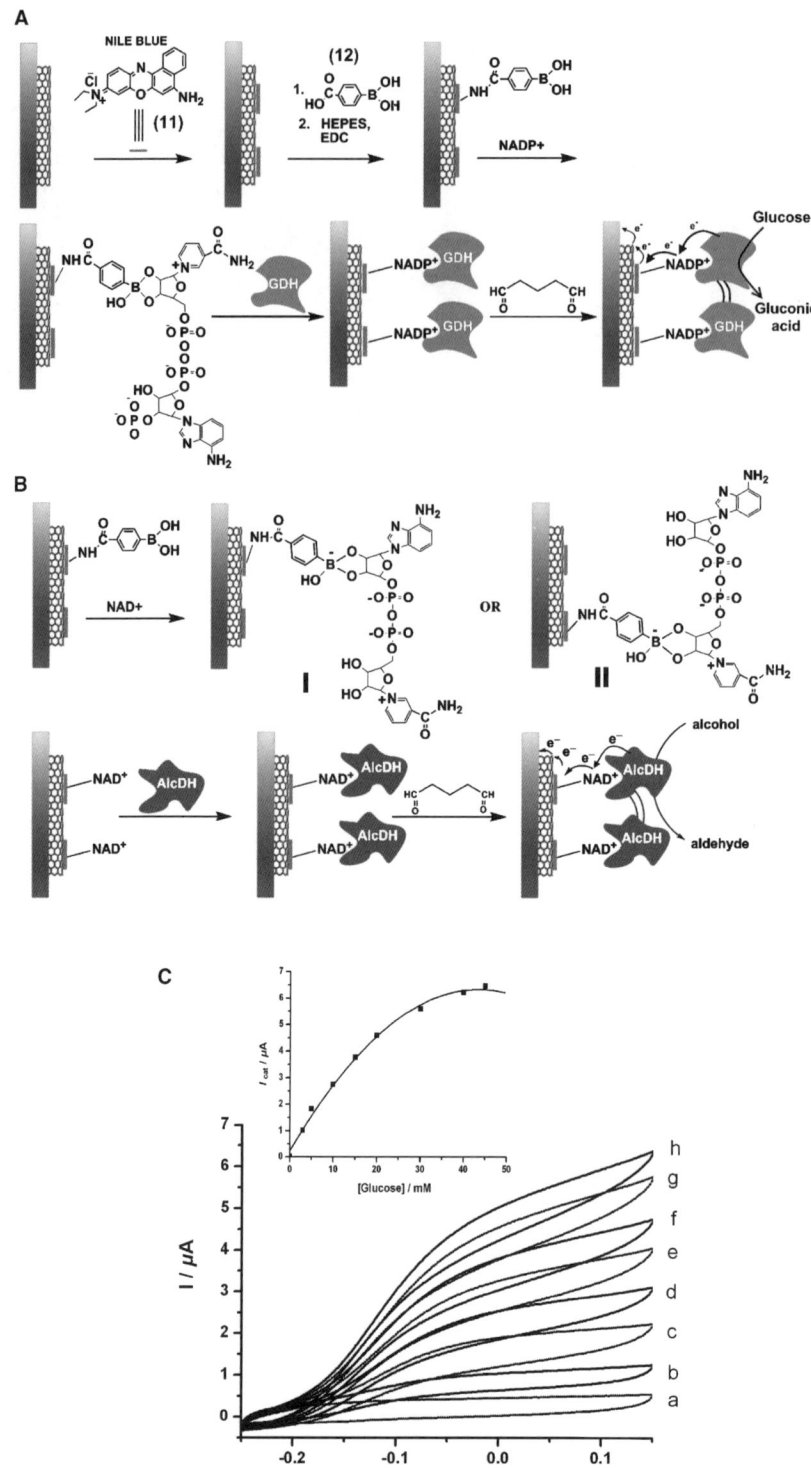

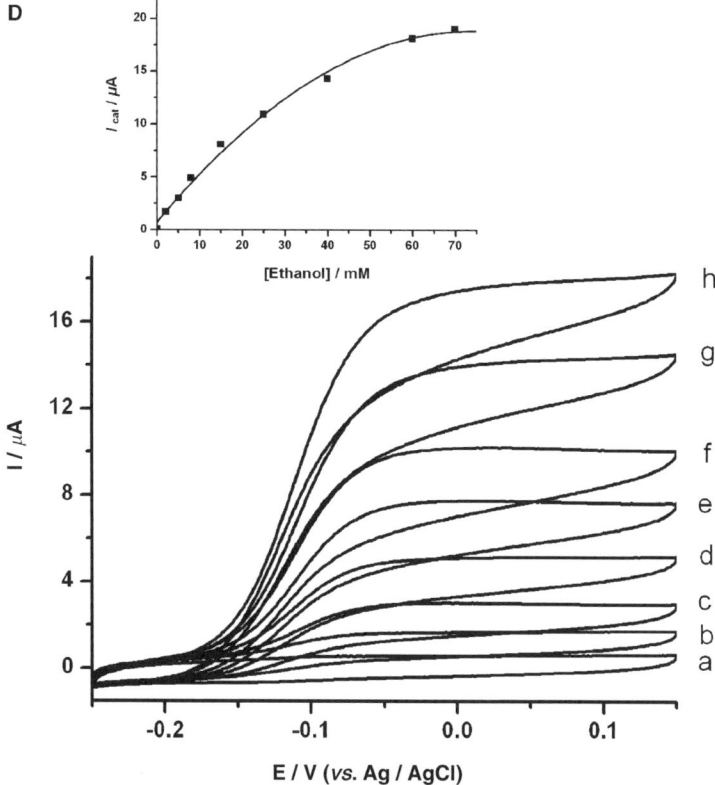

Figure 3.11 (A) Assembly of the electrically contacted GDH-functionalised electrode by the surface cross-linking of an affinity complex of GDH with Nile blue-$NADP^+$ associated with CNTs. (B) Assembly of an integrated alcohol dehydrogenase, AlcDH, electrode by the surface cross-linking of an affinity complex between AlcDH and Nile blue-NAD^+ linked to CNTs. (Note that two modes of binding of NAD^+ to the Nile blue units by the boronic acid ligand are possible, structure I or II). (C) Cyclic voltammograms corresponding to the integrated GDH associated with $NADP^+$-(**12**)-(**11**)-functionalised SWCNT electrode in the presence of different concentrations of glucose: a) 0, b) 3, c) 5, d) 10, e) 15, f) 20, g) 30, h) 40 mM. Data were recorded in 0.1 M phosphate buffer (pH 7.0) under Ar at a potential scan rate of $5\,mV\,s^{-1}$. Inset: calibration curve corresponding to the electrocatalytic currents measured at $E = 0.1\,V$ vs. SCE at variable concentrations of glucose. (D) Cyclic voltammograms corresponding to the integrated AlcDH associated with a NAD^+-(**12**)-(**11**)-functionalised SWCNT electrode in the presence of different concentrations of ethanol: a) 0, b) 2, c) 5, d) 8, e) 15, f) 25, g) 40, h) 60 mM. Data were recorded in 0.1 M phosphate buffer (pH 7.0) under Ar at a potential scan rate of $5\,mV\,s^{-1}$. Inset: calibration curve corresponding to the electrocatalytic currents measured at $E = 0\,V$ vs. SCE at variable concentrations of ethanol. (Reproduced with permission from reference 37. Copyright Wiley-VCH Verlag GmbH & Co. KGaA.)

covalently tethered to an Au electrode. The resulting modified electrode was interacted with cytochrome oxidase, COx, to form an affinity complex with the appropriate orientation of Cyt. *c* and COx for inter-protein electron transfer, Figure 3.12(A). The cross-linking of the affinity complex yielded an integrated electrode that revealed bioelectrocatalytic activity towards the reduction of O_2 to H_2O. Furthermore, synthetic analogues of natural cofactors may be used to assemble integrated bioelectrocatalytically active electrodes. A four-helix bundle *de-novo* protein (**13**) was covalently linked to a maleimide monolayer-functionalised Au electrode. The Fe(III)-protoporphyrin IX (**14**) was incorporated into the *de-novo* protein, and two heme sites were linked to the protein bundle through ligation to histidine residues, Figure 3.12(B).[39] The two heme sites were found to exhibit slightly different redox potentials. While the heme site close to the electrode surface revealed a redox potential of -0.43 V *vs.* SCE, the heme unit linked at the remote position relative to the electrode exhibited a redox potential of -0.36 V *vs.* SCE. This potential gradient resulted in a vectorial electron transfer cascade from the electrode to the remote heme site. In the presence of nitrate reductase, NR, an affinity complex between the enzyme and the *de-novo* hemoprotein was formed, and the subsequent cross-linking of the affinity complex resulted in an integrated, electrically contacted enzyme electrode that electrocatalysed the reduction of NO_3^- to nitrite (NO_2^-). Figure 3.12(C) depicts the electrocatalytic cathodic currents generated by the functionalised electrode, in the presence of different concentrations of NO_3^-. The resulting calibration curve is shown in Figure 3.12(C), inset.

A different "reconstitution-like" approach to electrically wire redox proteins with electrodes involved the application of inhibitors tethered to conductive molecular wires, as functional units to electrically communicate the enzyme redox centre with the electrode surface. The inhibitor unit acts as a targeting "arrowhead" that binds to the active site of the protein, while the molecular wire unit facilitates the charge transport between the active centre and the electrode. Charge transport through molecular wires is a subject of extensive theoretical[17] and experimental[27,40] research. Oligophenylacetylenes were used as stiff linear conjugated molecular wires for charge transport and, hence, a monolayer consisting of the thiolated diethylaniline oligophenylacetylene (**15**) was used to electrically contact amine oxidase with an Au electrode.[41] The enzyme includes a topaquinone/Cu^{2+} redox centre, and it catalyses the oxidation of amines to aldehydes. The diethylamine group tethered to the thiolated wire (**15**) linked to the electrode binds to the active site of the enzyme. The association of the wire "headgroup" to the enzyme aligned the redox protein on the electrode surface, and the "molecular wire" was found to electrically couple the redox protein with the conducting surface.

(**15**)

Electrical Interfacing of Redox Enzymes

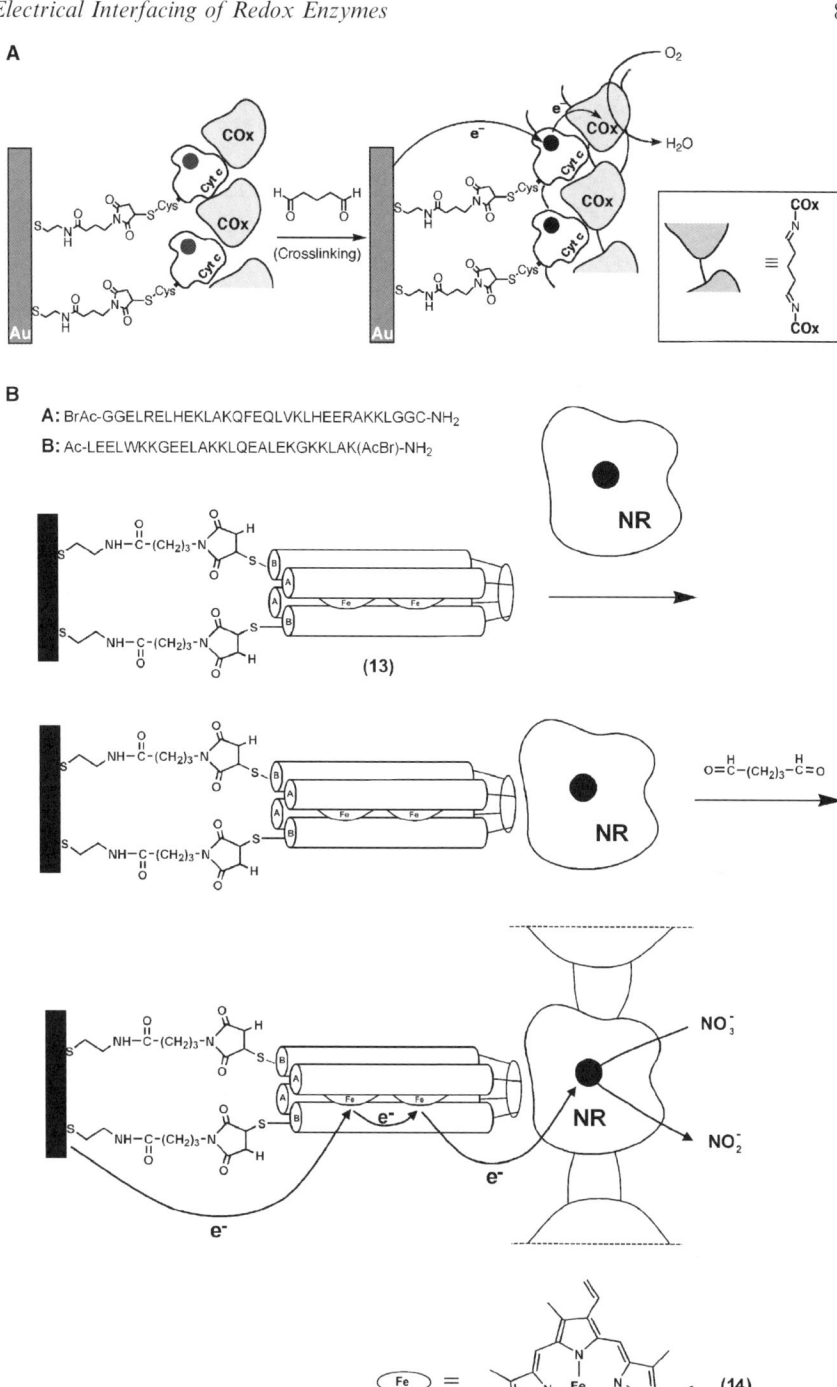

A: BrAc-GGELRELHEKLAKQFEQLVKLHEERAKKLGGC-NH$_2$
B: Ac-LEELWKKGEELAKKLQEALEKGKKLAK(AcBr)-NH$_2$

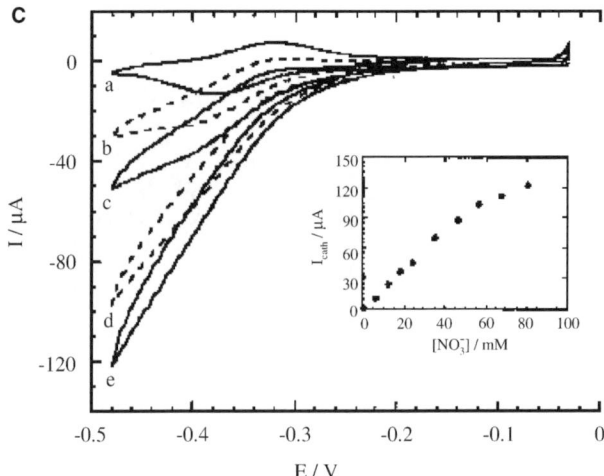

Figure 3.12 (A) Assembly of the integrated bioelectrocatalytic Cyt c/COx-electrode. (Reproduced from reference 38 by permission of The Royal Society of Chemistry.) (B) Assembly of the integrated nitrate reductase/Fe(III)-protoporphyrin IX-reconstituted *de-novo* protein-layered Au electrode. (C) Cyclic voltammograms of the integrated nitrate reductase/Fe(III)-protoporphyrin IX-reconstituted *de-novo* protein-layered Au electrode at different concentrations of nitrate: (a) 0, (b) 12, (c) 24, (d) 46 and (e) 68 mM. Inset: calibration curve for the amperometric responses (at $E = -0.48$ V *vs.* SCE) of the electrode at different NO_3^- concentrations. Data were recorded in 0.1 M phosphate buffer (pH 7.0) at a potential scan rate of 5 mV s^{-1}. (Reprinted with permission from reference 39. Copyright 1999 American Chemical Society.)

3.6 Reconstituted Enzyme Electrodes for Biofuel Cell Applications

Besides using the electrically contacted enzyme electrodes as amperometric biosensors, the functionalised electrodes find growing interest in the development of biofuel cell elements.[2,28] The biofuel cell consists of two electrodes, Figure 3.1(B), where the anode stimulates the biocatalysed oxidation of a fuel-substrate (such as glucose, alcohol or α-hydroxy acids), and the cathode drives the biocatalysed reduction of an oxidiser. While O_2 is the most abundant oxidiser, other compounds, such as H_2O_2, were employed as oxidisers in biofuel cell elements.[42] The power output of the biofuel cell is given by eqn (3.2), where ΔV_{ac} is the potential difference between the anode and the cathode, and I_{ac} is the current passing through the external circuit. For maximum power output, the anodic oxidation of the fuel, and the concomitant reduction of the oxidiser, should proceed as close as possible to the thermodynamic redox potentials of the fuel and oxidiser, respectively. The current generated by the biofuel cell is controlled by the turnover rates of the electrons at the anode/cathode, or by the effectiveness of the electrical contacting between the biocatalysts associated with the electrode.

Different potential applications of biofuel cells were suggested. Besides the possibility to generate electrical energy from biomass substances, implantable biofuel cells might use biological fluids, *e.g.* glucose in blood, to generate electrical power for the activation of mechanical units, such as pacemakers or prosthetic elements. Alternatively, the generation of electrical power from biomass or plants could provide new means to operate computers, or to power imaging and communication instrumentation in remote, isolated regions.

$$P = \Delta V_{ac} \cdot I_{ac} \qquad (3.2)$$

After the first report on the construction of a non-compartmentalised biofuel cell,[43] substantial research efforts were directed to improve the power output of the biofuel cells, by tailoring electrically contacted enzyme electrodes.[2,28,3b] In most of the reported systems, the biocatalysed oxidation of glucose by glucose oxidase (GOx), or the oxidation of glucose by glucose dehydrogenase (GDH), were used as the anodic processes. The biocatalysts that were applied for the reduction of O_2 to water usually included one of the copper proteins, laccase or bilirubin oxidase. For example, glucose oxidase and bilirubin oxidase, electrically contacted with electrodes by means of redox polymers, were used as biocatalytic materials for the development of a glucose/O_2 miniaturised biofuel cell element.[44] Glucose oxidase, GOx, was immobilised in a polymer consisting of tris-dialkylated-N,N'-bisimidazole $Os^{2+/3+}$ complex tethered to a polyvinylpyridinium hydrogel, as an electrical wiring matrix. This electrode acted as an anode for the bioelectrocatalysed oxidation of glucose at -0.36 V *vs.* Ag/AgCl. The cathode of the biofuel cell consisted of bilirubin oxidase, immobilised in a redox-active polymer composed of the $Os^{2+/3+}$ bis-(4,4'-dichloro-2,2'-bipyridine) complex, that was tethered to a polyacrylamide-polyvinyl imidazole polymer, through ligation of the imidazole to the $Os^{2+/3+}$ centre. The electrically contacted enzyme reduced O_2 at $+0.36$ V *vs.* Ag/AgCl. The integrated biofuel cell was found to operate with a power output of 4.3 µW.

The effective electrical contacting of redox enzymes with electrodes by means of the reconstitution process, and the demonstration that the bioelectrocatalysed oxidation of glucose by the surface-reconstituted glucose oxidase is insensitive to O_2, paved the way to construct the first integrated, non-compartmentalised biofuel cell element.[43] This cell, Figure 3.13(A), consisted of an anode composed of apo-glucose oxidase that was reconstituted on a pyrroloquinoline quinone (PQQ)/FAD monolayer, linked to the electrode (*cf.* the detailed construct of the electrically contacted enzyme electrode as shown in Figure 3.3(A)). The cathode was composed of an integrated, electrically contacted Cyt. *c*/COx electrode (*cf.* Figure 3.12(A)) that was assembled on an Au electrode. Figure 3.13(B) depicts the performance of the monolayer-functionalised GOx and Cyt. *c*/COx electrodes as a biofuel cell. Figure 3.13(B) shows the cell voltage and currents at different external resistances. Figure 3.13(B), inset, shows the respective power output of the resulting biofuel cell. An optimal power output of 4 µW was observed with an external load of 0.9 kΩ. Although the resulting power is quite low, the result demonstrated the

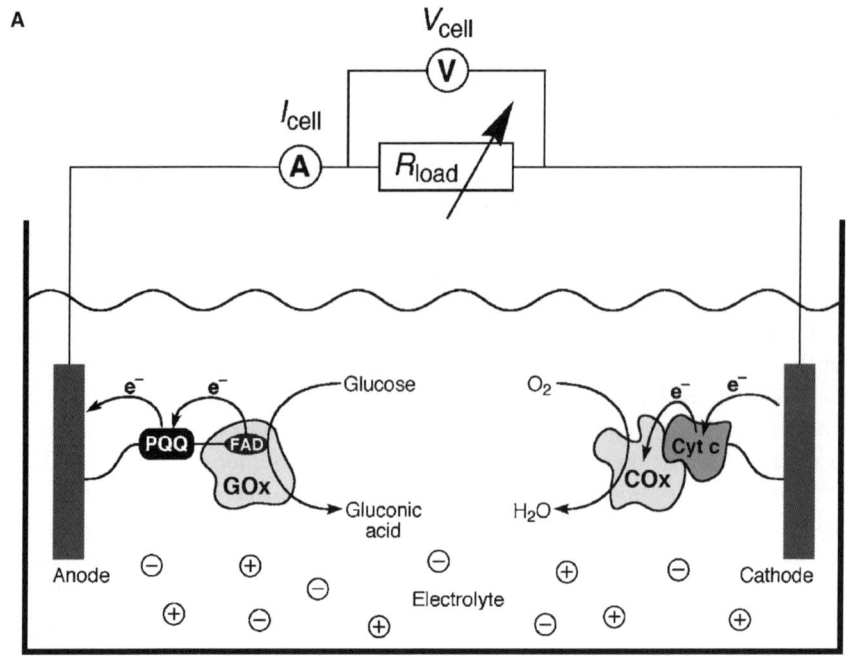

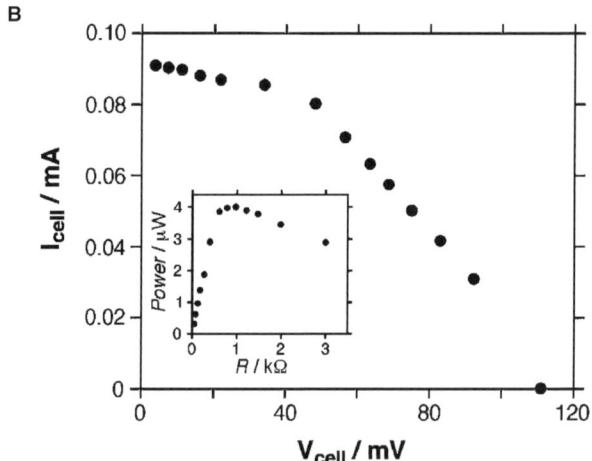

Figure 3.13 (A) Schematic configuration of a non-compartmentalised biofuel cell employing glucose and O_2 as fuel and oxidiser, and using PQQ/FAD/GOx and Cyt c/COx-functionalised Au electrodes as biocatalytic anode and cathode, respectively. (Adapted with permission from reference 9. Copyright Elsevier.) (B) Current–voltage behaviour of the biofuel cell at different external loads. Inset: the electrical power extracted from the biofuel cell at different external loads. (Reproduced with permission from reference 43. Copyright Elsevier.)

feasibility to assemble biofuel cells, and the system paved the way to search for improved systems. In fact, the detailed electrochemical characterisation of the biofuel cell, consisting of the reconstituted GOx and Cyt. c/COx electrodes, enabled the tracing of the origins for the low power output, and the results indicated directions to improve the power output: (i) The biofuel cell configuration depicted in Figure 3.13(A) consists of monolayer-functionalised bioelectrocatalytic electrodes. Means to increase the content of the biocatalysts, which communicate with the electrodes, are anticipated to increase the current generated by the system. The immobilisation of the enzymes in redox polymer films[8] (rather than monolayer structures), the fabrication of electrically contacted enzyme multilayers[7e] or the roughening of the electrode could all provide methods to enhance the content of biocatalysts associated with electrodes. (ii) While the GOx reconstituted enzyme electrodes revealed effective electrical contacting, and a turnover rate of electrons of $ca.$ $600\,s^{-1}$ was identified, the Cyt. c-mediated bioelectrocatalysed reduction of O_2 was found to be kinetically inefficient. The turnover rate between Cyt. c and the electrode was estimated, by chronoamperometric experiments, to be $20\,s^{-1}$. Thus, the cathode is the rate-limiting step for charge transport, and hence, the cathode is, at present, the "bottle-neck" for the extractable current from the biofuel cell. Accordingly, further improvements of the electrical contacting of the biocatalysts associated with the cathode are anticipated to significantly enhance the performance of biofuel cells. (iii) The redox potentials at which the mediated oxidation of glucose, and the reduction of O_2, proceeded are far from the thermodynamic potential values for these reactions. As a result, the ΔV_{ac} is substantially lower than the theoretical value. For example, the PQQ-mediated oxidation of glucose within the GOx monolayer-functionalised electrode proceeded at $-0.15\,V$ $vs.$ SCE, while the thermodynamic redox potential of the process is $ca.$ $-0.55\,V$ $vs.$ SCE.

Also, externally controlled bioelectrocatalytic transformations at electrode surfaces might improve the power output of biofuel cell elements. The application of an external magnetic field, parallel to electrode surfaces, was found to enhance the electron transfer at the electrode-solution interface. Theoretical studies indicated that under a parallelly applied magnetic field, the interfacial electron transfer is controlled by hydrodynamic convection, rather than by diffusion, resulting in the decrease in the diffusion layer at the electrode surface.[45] These theoretical studies were experimentally supported, and enhanced power outputs of biofuel cell elements were achieved.[46] For example, the bioelectrocatalysed oxidation of glucose by an electrode functionalised with GOx reconstituted on the PQQ/FAD monolayer was $ca.$ 2-fold enhanced under an external magnetic field of $0.92\,T$. The effect of an external magnetic field on the power output of a biofuel cell element was demonstrated in a system consisting of an electrically contacted lactate dehydrogenase-functionalised electrode acting as anode, and the Cyt. c/COx-functionalised electrode as cathode. The integrated, electrically contacted lactate dehydrogenase-functionalised electrode was composed of lactate dehydrogenase cross-linked on a PQQ/NAD$^+$ monolayer associated with the electrode, Figure 3.14(A). The Cyt. c/COx integrated electrode served as a biocatalytic cathode for the reduction of

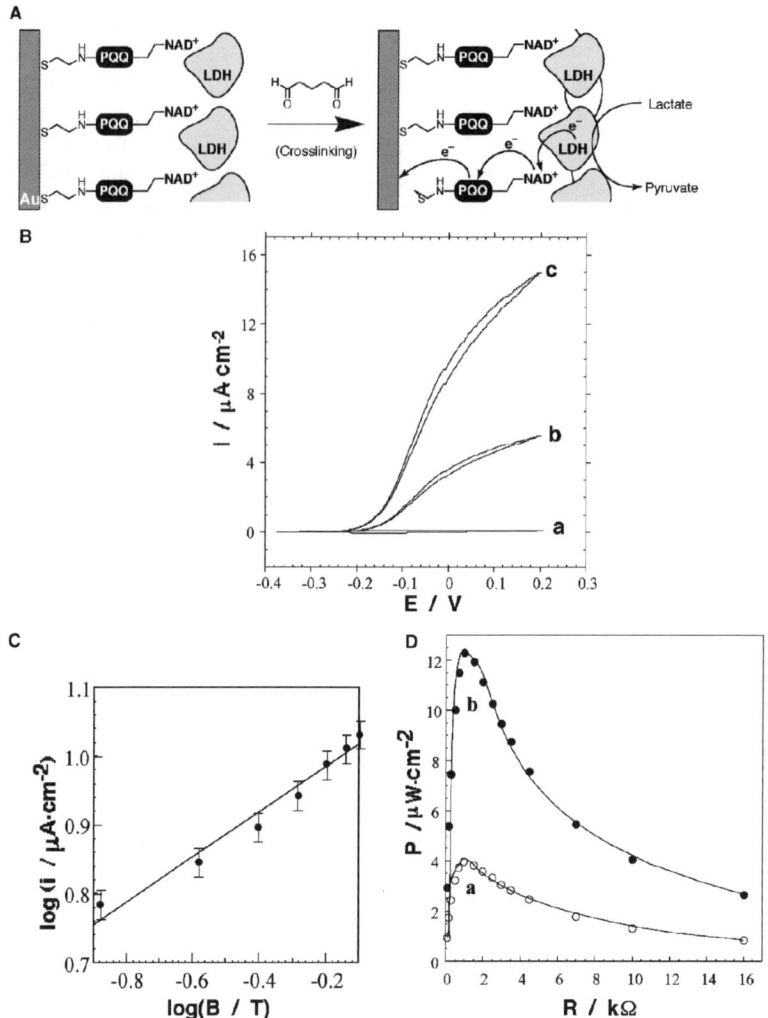

Figure 3.14 (A) The assembly of an integrated LDH monolayer-electrode by the cross-linking of an affinity complex formed between LDH and a PQQ/NAD$^+$ monolayer-functionalised Au-electrode. (B) Cyclic voltammograms of the integrated cross-linked PQQ/NAD$^+$/LDH electrode: (a) in the absence of lactate and in the absence of magnetic field, (b) in the presence of lactate, 20 mM, and in the absence of magnetic field, (c) in the presence of lactate, 20 mM, and in the presence of magnetic field, $B = 0.92$ T. (C) The dependence of the electrocatalytic current density on the magnetic flux density at lactate concentration of 20 mM. (D) The power density output generated by the biofuel cell: (a) in the absence of magnetic field, (b) in the presence of magnetic field, $B = 0.92$ T. The biofuel cell operated upon pumping of a solution (flow rate 1 mL min^{-1}) composed of 0.1 M TRIS-buffer, pH 7.0, containing CaCl$_2$, 10 mM, lactate, 20 mM, and oxygen (the solution was equilibrated with air). (Reprinted with permission from reference 46. Copyright 2005 American Chemical Society.)

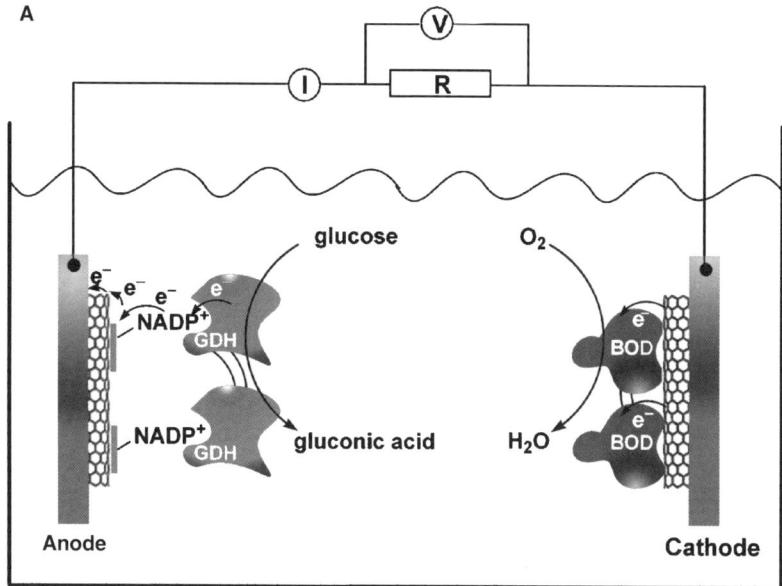

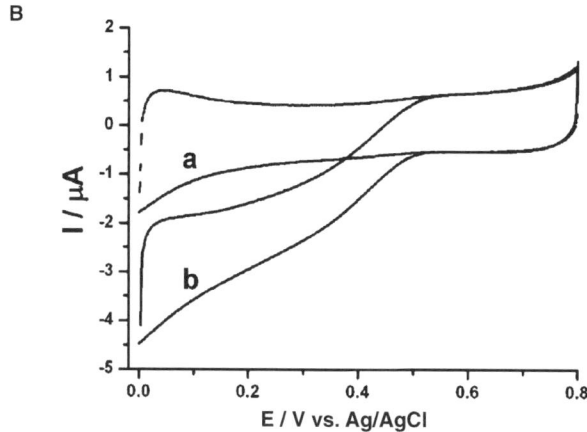

Figure 3.15 (A) Schematic configuration of a glucose/O_2 biofuel cell based on glucose dehydrogenase, GDH, and bilirubin oxidase, BOD, electrically contacted on CNT-modified electrodes. (B) Cyclic voltammograms corresponding to the bioelectrocatalysed reduction of O_2 by the BOD-modified CNTs-associated with a GC electrode: (a) in the absence of O_2, (b) in the presence of O_2. Measurements were performed in phosphate buffer, pH 7.0, at a scan rate of $10\,\text{mV}\,\text{s}^{-1}$. (C) Current–voltage curves (a) and power outputs (b) at different external resistances for the glucose/O_2 biofuel cell that employs GDH and BOD electrically contacted bioelectrocatalytic electrodes. (D) Current–voltage curves (a) and power outputs (b) at different external resistances for the alcohol/O_2 biofuel cell that employs AlcDH and BOD electrically contacted bioelectrocatalytic electrodes. (Reproduced with permission from reference 37. Copyright Wiley-VCH Verlag GmbH & Co. KGaA.)

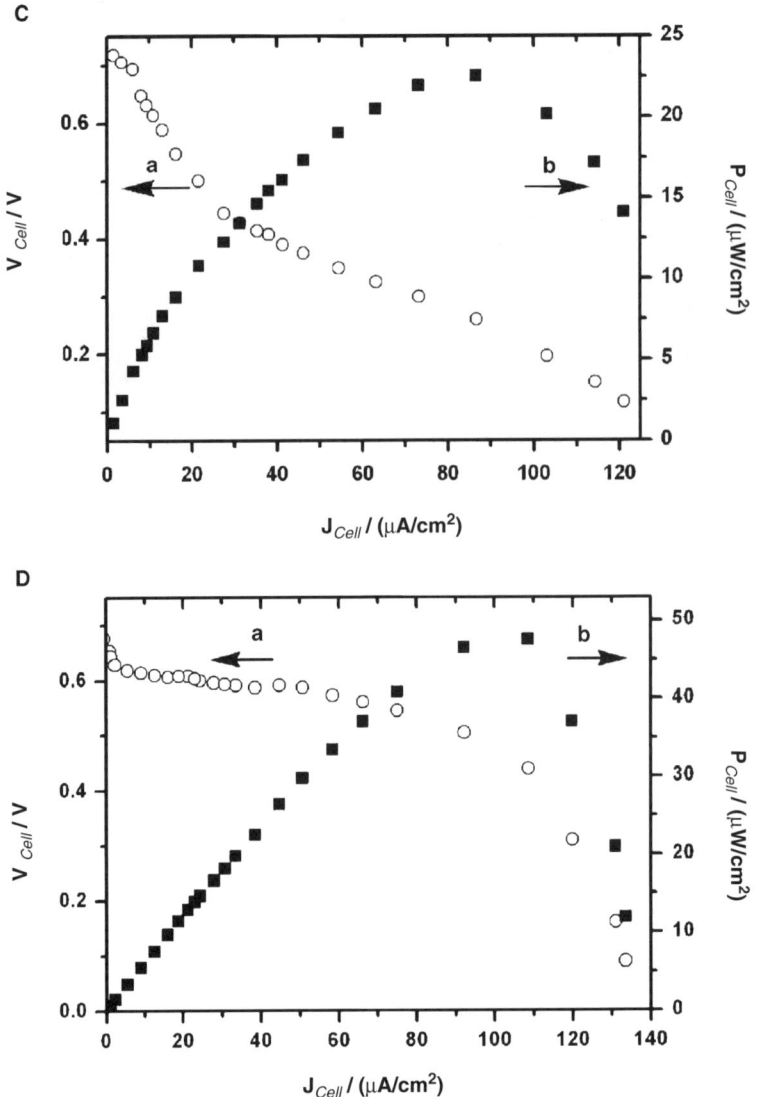

Figure 3.15 Continued.

O_2 (*cf.* Figure 3.12(A)). Figure 3.14(B) shows the cyclic voltammograms corresponding to the bioelectrocatalysed oxidation of lactate in the absence, and the presence, of an external magnetic field. Figure 3.14(C) depicts the bioelectrocatalytic currents generated at different strengths of external magnetic field. The enhancement of the power output of the lactate/O_2 biofuel cell under an applied field of 0.92 T is shown in Figure 3.14(D). A 3-fold higher power output is observed for an external resistance of 1.8 kΩ, and an applied magnetic field of 0.92 T.

The bioelectrocatalytic electrodes consisting of glucose dehydrogenase, GDH, or alcohol dehydrogenase, AlcDH, reconstituted on the $NADP^+$ or NAD^+ cofactors linked to Nile-blue associated with the CNTs, respectively, (cf. Figure 3.11) were used as the anodes of biofuel cells that utilised glucose or ethanol as fuels, respectively.[37] The cathode in these biofuel cells, Figure 3.15(A), consisted of bilirubin oxidase, BOD, adsorbed onto the CNTs. The cyclic voltammograms of BOD on the CNTs indicated that the adsorption of BOD onto the CNTs resulted in direct electrical contact between the electrode and the redox centres of BOD. Indeed, the enzyme stimulated the biocatalysed reduction of O_2 to H_2O, Figure 3.15(B). The current–voltage relationship of the glucose/O_2 biofuel cell, at different external resistances, and the respective power outputs of the cell, are depicted in Figure 3.15(C). The maximum power extracted from the cell corresponded to $24\,\mu W\,cm^{-2}$. Similarly, Figure 3.15(D) shows the current–voltage curve at different external resistances, for the ethanol/O_2 biofuel cell. The maximum power output for the cell corresponded to $48\,\mu W\,cm^{-2}$. Although the power values generated by the two cells are quite low, the results demonstrate that the reconstitution of the enzymes on CNT-functionalised electrodes yields active biocatalytic electrodes that convert the chemical energy, stored in the fuels, into electrical power. The detailed analysis of the bioelectrocatalytic functions of the anodes and cathodes in the two biofuel cell configurations indicated that the bioelectrocatalysed reduction of O_2 at the cathode is the power-limiting process. Hence, improving the cathode material could significantly improve the performance of these, or related, biofuel cells.

3.7 Conclusions and Perspectives

This chapter has reviewed the advances in the electrical contacting of redox enzymes with electrodes by means of the reconstitution principle. The reconstitution paradigm provides an experimental approach to shorten and vectorially control the electron transfer between the redox centres of the proteins and the electrode surface. It represents a major advance in bioelectronics, and provides the most efficient means to electrically communicate redox proteins with electrodes. The advantages of the reconstitution process rest on the fact that all of the enzyme redox centres are aligned in an optimal configuration with respect to the molecular or polymer electron mediator units that transport the electrons between the enzyme and the electrode. Particularly promising are the systems where nano-elements such as NPs or CNTs electrically contact redox enzymes with electrodes. The effective electrical contact between the enzyme and the electrode enabled the development of sensitive, and selective, amperometric biosensors, and the construction of non-compartmentalised biofuel cell elements. The high current densities generated by the reconstituted enzyme electrode will enable the design of miniaturised implantable amperometric sensing electrodes for the continuous monitoring of analytes, such as glucose or lactate. Also, the resulting biofuel cells could yield implantable devices that provide electrical power extracted from body fluids (*e.g.* glucose in blood). This electrical power could activate hearing aids, prosthetic elements and, eventually, pacemakers.

The reconstitution paradigm was, until now, applied only for a limited number of cofactors to fabricate integrated, electrically contacted enzyme electrodes. The extension of this method to other cofactors will yield new functional bioelectrocatalytic electrodes. Also, the improvement of the charge transport properties of the matrices that couple the reconstituted biocatalysts with the electrode may substantially enhance the bioelectrocatalytic functions of the resulting enzyme electrodes. Recent studies reported on the enhanced charge transport properties of conductive polymers/Au nanoparticles[47] or polymer/carbon nanotubes[48] composite systems. The reconstitution of enzymes on these matrices is anticipated to yield enzyme electrodes with improved bioelectrocatalytic properties. Finally, the reconstitution paradigm could be applied to fabricate other bioelectronic devices. Redox enzymes were immobilised on the gate surfaces of field-effect transistor devices, and the systems were used as bioelectronic sensing systems.[49] The reconstitution of proteins on the transistors is anticipated to yield biocatalytic interfaces with improved gating functions, thus yielding bioelectronic devices of enhanced sensitivities.

Acknowledgement

This research is supported by the BioMedNano EC project.

References

1. (a) L. Murphy, *Curr. Opin. Chem. Biol.*, 2006, **10**, 177; (b) I. Willner, *Science*, 2002, **298**, 2407; (c) E. Katz and I. Willner, *Angew. Chem. Int. Ed.*, 2004, **43**, 6042.
2. S. C. Barton, J. Gallaway and P. Atanassov, *Chem. Rev.*, 2004, **104**, 4867.
3. (a) B. Willner, E. Katz and I. Willner, *Curr. Opin. Biotechnol.*, 2006, **17**, 589; (b) A. Heller, *Phys. Chem. Chem. Phys.*, 2004, **6**, 209.
4. R. A. Marcus and N. Sutin, *Biochim. Biophys. Acta*, 1985, **811**, 265.
5. (a) I. Willner and E. Katz, *Angew. Chem. Int. Ed.*, 2000, **39**, 1180; (b) A. Heller, *J. Phys. Chem.*, 1992, **96**, 3579.
6. (a) P. N. Bartlett, P. Tebbutt and R. G. Whitaker, *Prog. React. Kinetics*, 1991, **16**, 55; (b) D. L. Williams, A. P. Doig Jr and A. Korosi, *Anal. Chem.*, 1970, **42**, 118; (c) P. Janada and J. Weber, *J. Electroanal. Chem.*, 1991, **300**, 119; (d) T. Matsue, N. Kasai, M. Narumi, M. Nishizawa, H. Yamada and I. Uchida, *J. Electroanal. Chem.*, 1991, **300**, 111.
7. (a) W. Schuhmann, T. J. Ohara, H.-L. Schmidt and A. Heller, *J. Am. Chem. Soc.*, 1991, **113**, 1394; (b) Y. Degani and A. Heller, *J. Am. Chem. Soc.*, 1988, **110**, 2615; (c) W. Schuhmann, *Biosens. Bioelectron.*, 1995, **10**,

181; (d) I. Willner, N. Lapidot, A. Riklin, R. Kasher, E. Zahavy and E. Katz, *J. Am. Chem. Soc.*, 1994, **116**, 1428; (e) I. Willner, A. Riklin, B. Shoham, D. Rivenson and E. Katz, *Adv. Mater.*, 1993, **5**, 912.
8. (a) R. Maidan and A. Heller, *Anal. Chem.*, 1992, **64**, 2889; (b) B. A. Gregg and A. Heller, *J. Phys. Chem.*, 1991, **95**, 5970.
9. I. Willner and B. Willner, *Trends Biotechnol.*, 2001, **19**, 222.
10. F. Davis and S. P. J. Higson, *Biosens. Bioelectron.*, 2007, **22**, 1224.
11. I. Willner, V. Heleg-Shabtai, R. Blonder, E. Katz, G. Tao, A. F. Bückmann and A. Heller, *J. Am. Chem. Soc.*, 1996, **118**, 10321.
12. H. G. Eisenwiener and G. V. Schultz, *Naturwissenschaften*, 1969, **56**, 563.
13. M. Zayats, E. Katz and I. Willner, *J. Am. Chem. Soc.*, 2002, **124**, 14724.
14. E. Katz, O. Lioubashevsky and I. Willner, *Am. Chem. Soc.*, 2004, **126**, 15520.
15. E. Katz, L. Sheeney-Haj-Ichia and I. Willner, *Angew. Chem. Int. Ed.*, 2004, **43**, 3292.
16. Y. Xiao, F. Patolsky, E. Katz, J. F. Hainfeld and I. Willner, *Science*, 2003, **299**, 1877.
17. (a) F. Remacle, I. Willner and R. D. Levine, *J. Phys. Chem. B*, 2004, **108**, 18129; (b) F. Remacle and R. D. Levine, *Chem. Phys. Lett.*, 2004, **383**, 537.
18. M. Zayats, E. Katz, R. Baron and I. Willner, *J. Am. Chem. Soc.*, 2005, **127**, 12400.
19. O. Lioubashevski, V. I. Chegel, F. Patolsky, E. Katz and I. Willner, *J. Am. Chem. Soc.*, 2004, **126**, 7133.
20. J. E. Fischer, H. Dai, A. Thess, R. Lee, N. M. Hanjani, D. L. Dehaas and R. E. Smalley, *Phys. Rev. B*, 1997, **55**, R4921.
21. (a) E. Katz and I. Willner, *ChemPhysChem*, 2004, **5**, 1184; (b) B. L. Allen, P. D. Kichambare and A. Star, *Adv. Mater.*, 2007, **19**, 1439.
22. (a) J. J. Davis, K. S. Coleman, B. R. Azamian, C. B. Bagshaw and M. L. H. Green, *Chem. Eur. J.*, 2003, **9**, 3732; (b) J. Wang, G. Liu and M. R. Jan, *J. Am. Chem. Soc.*, 2004, **126**, 3010; (c) K. Besteman, J. O. Lee, F. G. M. Wiertz, H. A. Heering and C. Dekker, *Nano Lett.*, 2003, **3**, 727; (d) E. S. Jeng, A. E. Moll, A. C. Roy, J. B. Gastala and M. S. Strano, *Nano Lett.*, 2006, **6**, 371; (e) P. W. Barone and M. S. Strano, *Angew. Chem. Int. Ed.*, 2006, **45**, 8138.
23. P. Avouris, *Acc. Chem. Res.*, 2002, **35**, 1026.
24. (a) A. Star, J.-C. P. Gabriel, K. Bradley and G. Grüner, *Nano Lett.*, 2003, **3**, 459; (b) K. Bradley, M. Briman, A. Star and G. Grüner, *Nano Lett.*, 2004, **4**, 253; (c) K. Keren, R. S. Berman, E. Buchstab, U. Sivan and E. Braun, *Science*, 2003, **302**, 1380; (d) K. Keren, M. Krueger, R. Gilad, G. Ben-Yoseph, U. Sivan and E. Braun, *Science*, 2002, **297**, 72.
25. (a) C. Cai and J. Chen, *Anal. Biochem.*, 2004, **325**, 285; (b) L. Zhang, G.-C. Zhao, X.-W. Wei and Z.-S. Yang, *Chem. Lett.*, 2004, **33**, 86; (c) J. J. Gooding, R. Wibowo, J. Liu, W. Yang, D. Losic, S. Orbons, F. J. Mearns, J. G. Shapter and D. B. Hibbert, *J. Am. Chem. Soc.*, 2003, **125**, 9006.

26. F. Patolsky, Y. Weizmann and I. Willner, *Angew. Chem. Int. Ed.*, 2004, **43**, 2113.
27. A. Heller, *Acc. Chem. Res.*, 1990, **23**, 128.
28. R. A. Bullen, T. C. Arnot, J. B. Lakeman and F. C. Walsh, *Biosens. Bioelectron.*, 2006, **21**, 2015.
29. O. A. Raitman, E. Katz, A. F. Bückmann and I. Willner, *J. Am. Chem. Soc.*, 2002, **124**, 6487.
30. O. A. Raitman, F. Patolsky, E. Katz and I. Willner, *Chem. Commun.*, 2002, 1936.
31. (a) J. Moiroux and P. J. Elving, *J. Am. Chem. Soc.*, 1980, **102**, 6533; (b) H.-L. Schmidt and W. Schuhmann, *Biosens. Bioelectron.*, 1996, **11**, 127.
32. (a) I. Katakis and E. Dominguez, *Mikrochim. Acta*, 1997, **126**, 11; (b) L. Gorton, B. Persson, P. D. Hale, L. I. Boguslavsky, H. I. Karan, H. S. Lee, T. A. Skotheim, H. L. Lan and Y. Okamoto, in *Biosensors and Chemical Sensors*, ed. P. G. Edelman and J. Wang, American Chemical Society, Washington, 1992, p. 56.
33. (a) H. Jaegfeldt, T. Kuwana and G. Johansson, *J. Am. Chem. Soc.*, 1983, **105**, 1805; (b) L. L. Miller and J. R. Valentine, *J. Am. Chem. Soc.*, 1988, **110**, 3982.
34. (a) L. Gorton, *J. Chem. Soc., Faraday Trans. 1*, 1986, **82**, 1245; (b) B. Persson and L. Gorton, *J. Electroanal. Chem.*, 1990, **292**, 115; (c) D. D. Schlereth, E. Katz and H.-L. Schmidt, *Electroanalysis*, 1995, **7**, 46.
35. (a) I. Willner and A. Riklin, *Anal. Chem.*, 1994, **66**, 1535; (b) E. Katz, T. Lötzbeyer, D. D. Schlereth, W. Schuhmann and H.-L. Schmidt, *J. Electroanal. Chem.*, 1994, **373**, 189.
36. (a) Y. M. Yan, W. Zheng, L. Su and L. Q. Mao, *Adv. Mater.*, 2006, **18**, 2639; (b) Y. M. Yan, M. N. Zhang, K. P. Gong, L. Su, Z. X. Guo and L. Q. Mao, *Chem. Mater.*, 2005, **17**, 3457; (c) A. Malinauskas, T. Ruzgas and L. Gorton, *J. Electroanal. Chem.*, 2000, **484**, 55; (d) F. D. Munteanu, N. Mano, A. Kuhn and L. Gorton, *Bioelectrochemistry*, 2002, **56**, 67; (e) F. D. Munteanu, N. Mano, A. Kuhn and L. Gorton, *J. Electroanal. Chem.*, 2004, **564**, 167; (f) A. A. Karyakin, E. E. Karyakina and H. L. Schmidt, *Electroanalysis*, 1999, **11**, 149.
37. Y. M. Yan, O. Yehezkeli and I. Willner, *Chem. Eur. J.*, 2007, **13**, 10168.
38. V. Pardo-Yissar, E. Katz, I. Willner, A. B. Kotlyar, C. Sanders and H. Lill, *Faraday Discussions*, 2000, **116**, 119.
39. I. Willner, V. Heleg-Shabtai, E. Katz, H. K. Rau and W. Haehnel, *J. Am. Chem. Soc.*, 1999, **121**, 6455.
40. R. Rajagopalan, A. Aoki and A. Heller, *J. Phys. Chem.*, 1996, **100**, 3719.
41. C. R. Hess, G. A. Juda, D. M. Dooley, R. N. Amii, M. G. Hill, J. R. Winkler and H. B. Gray, *J. Am. Chem. Soc.*, 2003, **125**, 7156.
42. I. Willner, G. Arad and E. Katz, *Bioelectrochem. Bioenerg.*, 1998, **44**, 209.
43. E. Katz, I. Willner and A. B. Kotlyar, *J. Electroanal. Chem.*, 1999, **479**, 64.

44. N. Mano, F. Mao and A. Heller, *J. Am. Chem. Soc.*, 2002, **124**, 12962.
45. O. Lioubashevsky, E. Katz and I. Willner, *J. Phys. Chem. B*, 2004, **108**, 5778.
46. E. Katz, O. Lioubashevsky and I. Willner, *J. Am. Chem. Soc.*, 2005, **127**, 3979.
47. E. Granot, E. Katz, B. Basnar and I. Willner, *Chem. Mater.*, 2005, **17**, 4600.
48. E. Granot, B. Basnar, Z. Cheglakov, E. Katz and I. Willner, *Electroanalysis*, 2006, **18**, 26.
49. (a) A. B. Kharitonov, M. Zayats, A. Lichtenstein, E. Katz and I. Willner, *Sensor. Actuar. B-Chem.*, 2000, **70**, 222; (b) S. P. Pogorelova, M. Zayats, A. B. Kharitonov, E. Katz and I. Willner, *Sensor. Actuar. B-Chem.*, 2003, **89**, 40; (c) K. Besteman, J. Lee, F. G. M. Wiertz, H. A. Heering and C. Dekker, *Nano Lett.*, 2003, **3**, 727; (d) B. L. Allen, P. D. Kichambare and A. Star, *Adv. Mater.*, 2007, **19**, 1439.

CHAPTER 4
Single-wall Carbon Nanotube Forests in Biosensors

JAMES F. RUSLING,[a,b,c] XIN YU,[a,d] BERNARD S. MUNGE,[e] SANG N. KIM[b] AND FOTIOS PAPADIMITRAKOPOULOS[a,b]

[a] Department of Chemistry, University of Connecticut, Storrs, CT; [b] Institute of Materials Science, University of Connecticut, Storrs, CT; [c] Department of Cell Biology, University of Connecticut Health Center, Farmington, CT; [d] Schering Plough Corporation, Union, New Jersey; [e] Salve Regina University, Newport, Rhode Island

4.1 Unique Properties of Carbon Nanotubes

4.1.1 Introduction

Carbon nanotubes (CNTs) are novel nanowires with metallic or semi-conducting properties[1] and are supplied commercially as mixtures of these two types. CNTs are typically grown from catalytic metal nanoparticles using a carbon source at $>600\,°C$.[2] In their pristine form, they are composed solely of carbon atoms, and can be made in single wall (SWNTs) and nested, multiwall (MWNTs) versions (Figure 4.1). Carbon nanotube structures can be visualised as cylindrical rolled-up graphene sheets containing carbon atoms in honeycombed 6-membered rings.[1] Nanotubes grow outward from these nanoparticles as SWNTs or MWNTs depending on experimental conditions. Diameters range from 0.4 to 3 nm for SWNTs and 2 to 100 nm for MWNTs. Lengths from 10 nm to several micrometers can be controlled by adjustment of growth conditions or by oxidative shortening.[3]

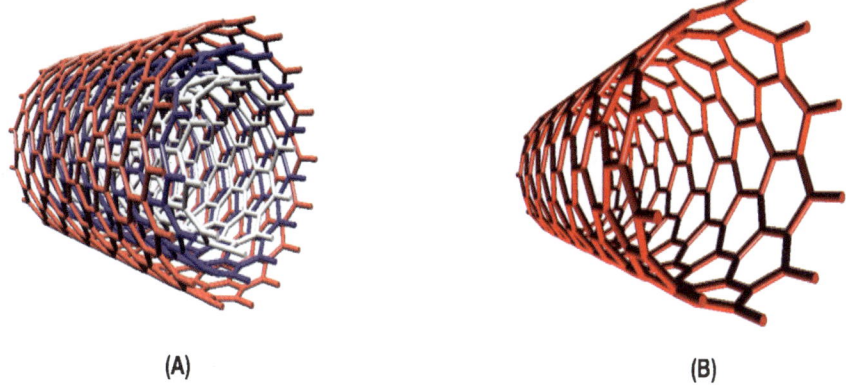

Figure 4.1 Representations of multi-wall (A) and single-wall (B) carbon nanotubes. Points of intersection in the images represent carbon atoms.

The unprecedented electrical properties and mechanical strength of CNTs have contributed to emerging technologies ranging from high-strength composites, miniature electrical devices[4–8] and biosensors.[9–16] In biosensor applications, sensing motifs can be coupled to optical[17] and electronic properties.[12,18] There is considerable interest in the electrocatalytic properties of carbon nanotubes[19] as well as speculation that SWNTs with diameters on the order of 1 nm can pierce polypeptide "shells" of proteins to electrically address individual redox enzyme cofactors.[20] This chapter focuses on *electrochemical* biosensors, *i.e.* those deriving their signals from active oxidation and reduction processes, as opposed to nanotube field effect transistor (FET) sensors.[18]

The unique properties of carbon nanotubes suggest the potential to provide new approaches and critical improvements in electrochemical biosensors. The high surface area of electrically conductive SWNTs, estimated as high as $1\,600\,m^2\,g^{-1}$,[21] can lead to high densities of attached biomolecules, facilitating high sensitivity and device miniaturisation. Ballistic transport of electrons or holes along the length of SWNTs with current densities as high as $10^9\,A\,cm^{-2}$ in the presence of oxygen[22] can enhance biodevice performance. Facile electron transfer between nanotubes ends and protein redox cofactors[12] opens up opportunities for sensing strategies based on direct electron transfer rather than mediation.

CNTs can be made into electrodes by combination with mineral oil into pastes, direct dispersion on conductive surfaces with or without binders, screen-printing, self-assembly into vertically aligned "forests" or vertical growth from conductive substrates.[12,13,15,16,18,19] Furthermore, an assortment of covalent and non-covalent chemical functionalisation strategies have been developed to link virtually any molecule onto CNTs.[23–26] This means that biorecognition molecules such as enzymes or antibodies can be easily attached to carboxylated carbon nanotube ends and/or sidewalls forming the basis for a wide variety of biosensors. These features of CNTs allow them to be readily integrated into biorecognition and enzyme-based sensors with high sensitivity as has already been demonstrated in many cases. The small size and patternability of CNTs also permits them to be employed in micro- and nanosensor arrays that may result in devices capable of multiplexed analysis on biomedical samples such as serum or tissue lysates.

This chapter represents a selective summary of recent application of carbon nanotubes in electrochemical biosensors. Section 4.1.2 summarises current views concerning electrocatalysis by CNTs. Section 4.2 summarises recent biosensor applications of non-oriented CNT electrodes. Section 4.3 discusses fabrication, characterisation and application of biosensors based on SWNT forest electrodes, and Section 4.4 attempts to predict future horizons.

4.1.2 Electrocatalytic Properties

Given the many possible types of CNT electrodes featuring vertically oriented or disordered nanotube macro-structures, it is useful to briefly consider the effects of electrode configuration on electrocatalysis with CNT electrodes. Electrocatalysis is defined as the promotion of electron transfer rates and efficiencies at electrodes by the properties of the electrode surface, and has been widely attributed to CNTs used as electrodes in voltammetry.[12,13,19] Orientation issues also impact ballistic electron transport that proceeds down the length of a SWNT,[22] and it is expected that such longitudinal electron transport would be beneficial in sensor design. The growing literature of molecular electrochemistry on nanotube electrodes reveals that the degree of electrocatalysis depends on nanotube orientation and amount of oxygenation. There is strong evidence that enhancement of electron transfer rates to or from reactants depends on the amount of surface oxides on the nanotubes.[27] The emerging view based on studies with model small molecules and ions is that electrocatalysis of carbon nanotubes is due to the often oxygenated edge plane sites at nanotube tips similar to those found on edge plane pyrolytic graphite.[19] These conclusions suggest that disordered nanotube electrode configurations do not take full advantage of the electrocatalytic properties of the nanotube tips as do the forest organisations. It has been suggested that edge plane pyrolytic graphite electrodes should be used in control experiments for CNT electrodes to establish whether specific electrocatalytic reactions are truly unique to the CNTs.[19]

4.2 Biosensors Using Non-oriented Carbon Nanotube Electrodes

Non-oriented SWNTs or MWNTs were first used as electrodes after casting onto conductive surfaces in a mat configuration (Figure 4.2A),[28-31] or packing into a micropipette. Micropipette CNT electrodes were used by Hill *et al.* in the first report of CNTs for reversible protein cyclic voltammetry (CV) of cytochrome *c* (cyt *c*) and azurin.[32] Other methods developed to prepare disordered

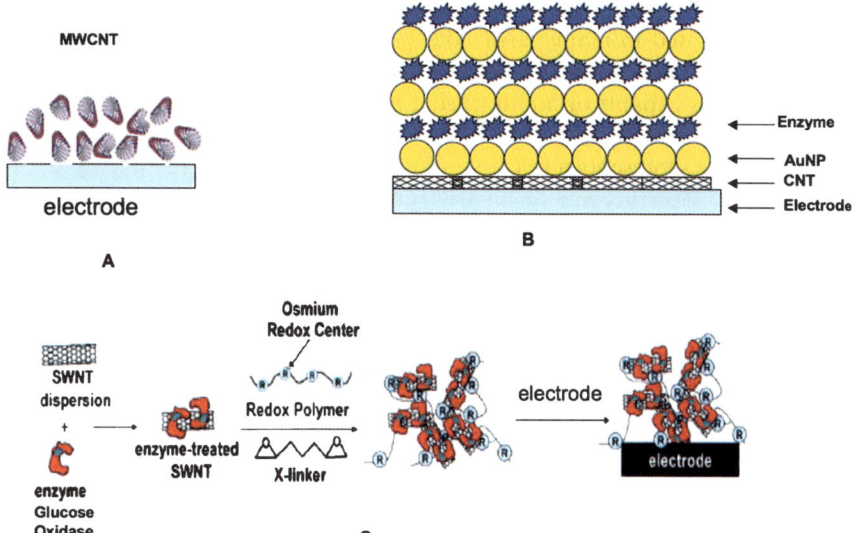

Figure 4.2 Examples of non-oriented nanotube electrodes: (A) diagram of a random MWNT mat electrode formed by casting the DMF dispersion of MWCNT onto the surface of glassy carbon electrode and evaporating the solvent; (B) diagram of composite multilayer films of MWNTs, gold nanoparticles and enzymes on an electrode; (C) scheme showing the fabrication of an amperometric sensor in which SWNTs were first incubated with glucose oxidase solution, then were incorporated into the redox hydrogel. The redox hydrogel solution containing the enzyme-modified SWNTs was then spread on top of a bare glassy carbon electrode. Adapted with permission from reference 61, copyright 2005 American Chemical Society.

CNT electrodes include self-assembly,[33–36] sol-gel procedures[37–40] and compositing with polymers,[41–45] biomacromolecules[46,47] or nanoparticles[48–53] to form CNT nanocomposites. The key to fabrication of non-oriented, random networks of nanotubes on underlying Au, glassy carbon and Pt electrodes is to prepare soluble nanotube dispersions and follow through with proper post-treatment such as coating with Nafion or polymer composite, or acid reflux. The resulting network or "mat" electrodes require only low potentials for the detection of NADH and H_2O_2,[54] which are cofactors or products in numerous enzyme reactions. Irrespective of mechanisms for electrocatalysis, nanotube electrodes typically provide fast responses and good sensitivities and detection limits.[54]

CNT-nanoparticle electrodes provide enhanced multi-functionality and versatility. A good example involves multilayer assemblies of nanotubes with metal nanoparticles, which have been used for immobilisation of enzymes in biosensors (Figure 4.2B).[55,56] CNTs promise the exciting possibility of ultrasensitive, electrochemical biosensors[12,13,18,57] by coupling CNT electrodes to electroanalytical methods. Many biosensors have been based on enzyme electrodes coupling a conducting CNT layer for transduction and a specific enzyme layer for selectivity.

Over the last two decades, a variety of thin film technologies has been developed to facilitate electron transfer between enzymes and electrodes.[58] However, for some important enzymes, redox cofactors are too deeply buried in the enzyme structure and direct electron transfer is not readily achievable or is too slow for sensor applications.[59] A well-studied example of this type of enzyme is glucose oxidase (GOx), which is a key component in commercial blood glucose sensors used by diabetic patients. In these sensors, redox mediators are used to shuttle electrons between the active centre of the enzyme and the electrodes.[60]

GOx has been a popular enzyme for studies with carbon nanotubes. It is stable, and does not readily exchange electrons with most macroscopic electrodes. Guiseppi-Elie *et al.* first reported direct electron exchange between adsorbed GOx and FAD onto SWNT mat electrodes.[20] Annealing at 450 °C was utilised to clean SWNTs and promote adsorption of GOx, achieving an electron transfer rate of $1.7\,s^{-1}$. GOx maintained its catalytic activity, and it was suggested that the tubular nanotubes could be positioned within the electron tunnelling distance of the FAD cofactor. In another example, an amperometric biosensor was constructed by incorporating SWNT modified with glucose oxidase into a redox polymer hydrogel (Figure 4.2C).[61] Compared to bare electrodes without SWNTs, this resulted in a 2- to 10-fold increase in the oxidation and reduction peak currents during cyclic voltammetry while the glucose electro-oxidation current increased 3-fold to $\sim 1\,mA\,cm^{-2}$. In another study, oligonucleotide-modified yeast cyt *c* was adsorbed to SWNTs *via* the oligonucleotide, and gave small reversible voltammetry peaks after background subtraction.[62] Other examples of fast electron transfers of redox proteins with MWNT mat electrodes include ferritin, cytochrome *c*, haemoglobin and azurin.[32,63–65] However, except in a few special cases (*e.g.* glucose oxidase), nanotube mat electrodes do not seem to provide significant advantages in reversibility or signal to noise compared to the best redox protein films on conventional electrodes such as pyrolytic graphite or gold with properly prepared surfaces.[58]

Immunosensors made with CNT electrodes were recently reviewed.[66] Highlights for disordered CNT systems include a label-free, Pt microelectrode with dispersed SWNTs to detect total prostate specific antigen (T-PSA).[67] The label-free electrochemical signals result from intrinsic oxidation of the protein. Here, the increased current observed as the concentration of T-PSA was increased may result from conformational changes of T-PSA during binding to the antibody on the nanotubes.[67] A CNT-based immuno-electrochemiluminescence sensor for α-fetoprotein was also developed,[68] as well as a highly sensitive CNT immunosensor for cholera toxin using conductive polymer-coated CNTs.[69] Strategies using polylysine-coated CNTs were evaluated to improve the temperature stability of antibodies in immunosensor formats.[70]

Carbon nanotubes have also proved valuable when used in secondary amplification strategies to achieve ultrasensitive electrochemical detection of proteins and DNA.[9,71] There have been excellent advances in DNA hybridisation for the detection of genetic disease.[72–76] Wang et al. capitalised on such approaches by demonstrating a novel method[74] using nanotubes to enhance the sensitivity and stability of enzyme-labelled electrochemical bioassays for DNA hybridisation. CNTs were used as carriers for multiple enzyme labels and for accumulating the products of the enzyme reactions, playing a dual amplification role in both recognition and transduction. The amplified signal reflects interfacial accumulation of phenolic products of the alkaline phosphatase enzyme label onto the CNT layer. The attractive performance characteristics of the multi-amplification electrochemical detection of DNA hybridisation were reported in connection to the detection of a breast cancer BRCA1 gene. In addition, a new strategy for dramatically amplifying enzyme-linked electrical detection of proteins and DNA using carbon nanotubes loaded with multiple enzymes was made by layer-by-layer (LbL) fabrication.[75] Such a CNT-derived, double-step amplification pathway (again, of recognition and transduction events) allowed the detection of DNA down to 5 aM and proteins to 67 aM in 25-μL samples, suggesting promise for ultrasensitive detection of disease biomarkers. In another example, a novel electrochemical detection of DNA based on the oxidation of guanine bases on a CNT-modified electrode was reported.[76] In this approach, a CNT-modified electrode was combined with $Ru(bpy)_3^{2+}$-mediated electrochemical guanine oxidation resulting in the detection of a few attomoles of oligonucleotide target. The sensitivity of DNA detection was improved by several orders of magnitude compared to methods where the DNA was immobilised using self-assembled monolayers on conventional electrodes. Gooding et al. demonstrated that better sensitivity and lower detection limits than for normal nanotubes can be achieved using bamboo-like nanotubes.[77] These bamboo-like nanotubes are made by milling and annealing graphite powder or by pyrolysis of iron phthalocyanine. They resemble bamboo in that they feature transverse walls at regular intervals along the nanotube. These transverse walls result in edge plane graphene along the length of the nanotube, and should provide better electrocatalysis than conventional nanotubes.[77]

4.3 Biosensors Utilising Vertically Aligned Carbon Nanotube Forests

4.3.1 CNT Forest Fabrication

Compared to disordered CNT electrodes, vertically aligned *CNT forest* electrodes can take advantage of directional electron transport between electrochemically reported biorecognition events and external circuits, and electrocatalysis at the nanotube ends. These types of CNT electrodes promise lower detection limits, increased signal-to-noise ratio, high density of biomolecules and enhanced electron transfer rates due to direct access to redox centres in enzymes.[57] Vertically aligned CNT arrays have been used to fabricate many novel biosensors over the last several years.[9,10,12,13,18,19,34,78–83]

There are two major approaches to make vertically aligned CNT forests on electrode surfaces. One is to grow aligned nanotubes *directly* from the surface, using chemical vapour deposition, arc discharge synthesis or pyrolysis of hydrocarbons. The other is to chemically prepare the underlying conductive surface, then *self-assemble* the CNTs by electrostatic interactions, acid–base chemistry or covalent linkage of end-functionalised CNTs to functional groups on the surface.[27,83,84] Several possible nanotube "array" or forest structures are depicted in Figure 4.3. Table 4.1 summarises different types of vertically aligned CNT forests utilised in biosensor applications with attendant nanotube type, surface coverage, average height of the nanotubes in the forest and hydrodynamic stability. The last feature refers to how long the forest can be used on a rotating disk electrode or in a flowing system.

Densely-packed SWNT forests self-assembled on ultrathin Nafion-FeO(OH)/FeOCl layers have been developed by our group as platforms for biosensors, and have excellent hydrodynamic stability (Table 4.1). Fabrication and applications are described in more detail below. Carbodiimide-based covalent linking of SWNTs to amine-functionalised conductive underlayers gives a significantly less dense SWNT forest coverage compared to our Fe^{3+}-assisted electrostatic assembly. Both methodologies have provided new opportunities for advanced electrochemical biosensors through ease of fabrication and patterning,[86,89] carboxyl groups at nanotube ends for convenient bioconjugation of key molecules,[9,12,18,27,87] SWNT length-fractionation to tune forest height,[34,90] vectorial electron transport along the nanotube length to attached redox enzymes[34,87,91] and hydrodynamic stability at electrode rotation rates of several thousand *rpm*.[91,92]

Liu *et al.*[93] reported thiolated SWNTs that could be self-assembled normal to Au surfaces. The same group later reported an alternative method in which SWNTs were cut into short tubes by chemical oxidation, producing end –COOH groups. These nanotubes were then self-assembled vertically onto an Ag surface based on electrostatic interaction between negatively charged carboxylic groups and the positively charged Ag surface. These SWNT forests were characterised by AFM and TEM. Liu and coworkers also reported making forests by using Zn^{2+} as bridging ion between a carboxylic acid modified surface and CNTs end-functionalised with a carboxylate group.[94] Gooding and coworkers[95,96] oxidised SWNTs in acid to shorten and functionalise them with carboxyl groups and converted them to carbodiimide esters using dicyclohexylcarbodiimide (DCC). A gold electrode with a self-assembled monolayer (SAM) of cysteamine was then placed in a DMF suspension of these functionalised SWNTs for 4 hours. The amines at the end of the SAM formed amide linkages with the tubes, resulting in SWNT forests used for biomolecule electrochemistry. Willner and coworkers[12,34] adopted a similar approach to achieve long-range electrical communication with glucose oxidase by SWNT connectors of different lengths. Huang *et al.*[97] demonstrated that the vertical alignment of oxidised SWNTs on

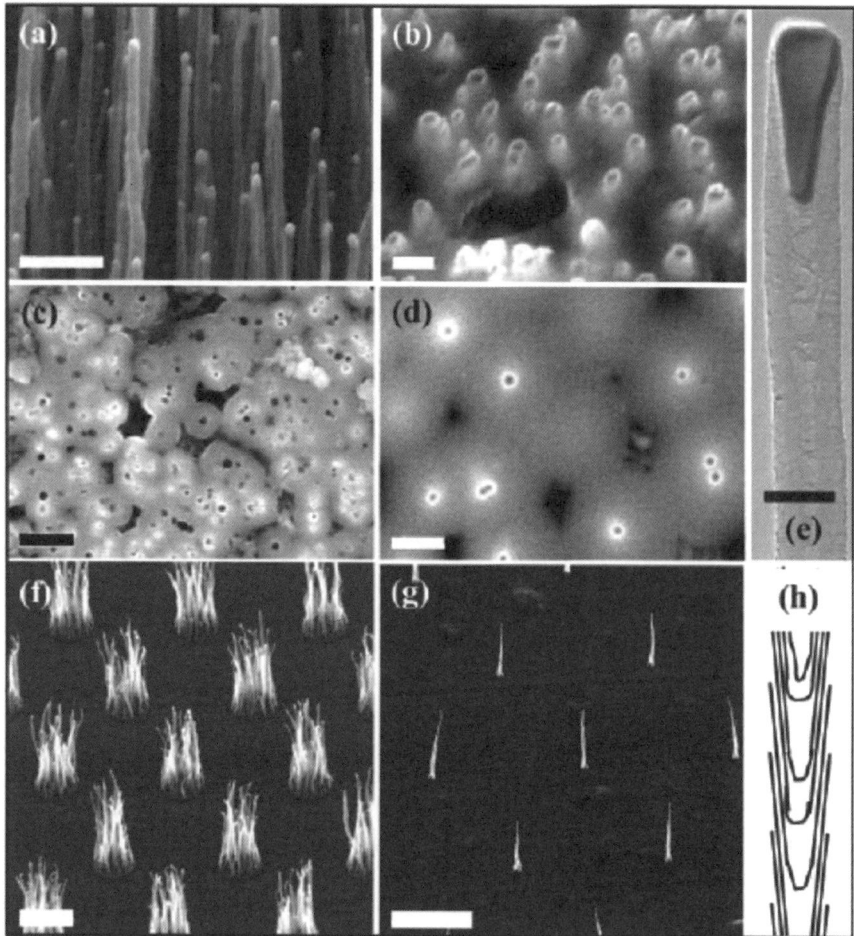

Figure 4.3 Examples of SEM images of CNT arrays: (a) an as-grown MWCNT array, (b) the surface of a polished MWNT array embedded in an SiO_2 matrix, (c) a high-density MWNT nanoelectrode array (hd, $\sim 2 \times 10^9$ electrodes cm^{-2}), (d) a low-density MWNT nanoelectrode array (ld, $\sim 7 \times 10^7$ electrodes cm^{-2}), (f) an as-grown MWNT array on 2 mm Ni spots patterned with UV lithography, (g) an as-grown MWNT array on ~ 100 nm diameter Ni spots patterned with e-beam lithography and (h) a schematic of the defective bamboo-like structure. (a), (b), (f) and (g) are 45° perspective views while (c) and (d) are top views. (e) is the TEM view of a single MWCNT. The scale bars in (a)–(g) are 500 nm, 200 nm, 500 nm, 500 nm, 50 nm, 2 mm and 5 mm, respectively. Reproduced with permission from reference 112, copyright 2004 Royal Society of Chemistry, UK.

cysteamine SAM on Au can be facilitated by dispersing SWNTs in acetone instead of DMF. In addition, MWNT forests were made by allowing shortened, oxidised nanotubes to self-assemble on a polycationic poly(diallyldimethylammonium chloride) (PDDACl) layer adsorbed on a silicon wafer.[98]

Table 4.1 Major methods for fabricating carbon nanotube forests with relevant properties.

Substrate chemistry	CNT type	Surface coverage ($tubes\,cm^{-2}$)	Height	Hydrodynamic Stability	References
Nafion-FeO(OH)/FeOCl	Carboxy-functionalised SWNTs	2×10^{13}	30–250 nm	Weeks	9,18,85–87
Amide-linked alkylthiols on Au	Carboxy-functionalised SWNTs	10^{11}–10^{12}	25–50 nm	Not reported	10,34
CVD grown (Fe/Co)	MWCNTs	10^7–10^9	2–4 µm	Unstable as grown. Needs support by impregnating with SiO_2 or conducting polymers	78,88

Growing carbon nanotubes directly on conductive surfaces using chemical vapour deposition (CVD) displays the lowest surface coverage (10^7–10^9 tubes cm^{-2}), and on drying these forests may collapse.[88] Impregnation strategies to overcome the instability are discussed below. Dai and coworkers[99] demonstrated the growth of vertically aligned MWNT on a photoresist pre-patterned quartz plate. The quartz plate was first heated at high temperature under argon to carbonise the photoresist polymer into a carbon layer, followed by pyrolysis of iron(II) phthalocyanine (FePc) under Ar/H_2 at 800–1100 °C. Iron particles are surrounded by carbon formed on the surface of the quartz plate upon the thermal decomposition of FePc. Segregation of Fe then occurs and leads to an increase in the size of the catalytic centre, and then the surrounding carbon starts to transform into nanotubes once the Fe particle reaches an optimal size for carbon nanotube nucleation. Decomposition of FePc provides carbon to the tubule segment already formed on the Fe particle, allowing continuous vertical growth of the carbon nanotube. Finally, immersing the quartz plate with CNTs into an aqueous hydrofluoric acid solution (10–40% w/w) results in a "floating film" that separates from the plate and can be transferred to the surface of a gold electrode. The same group also demonstrated the feasibility of electronically depositing a concentric layer of the conductive polymer polyaniline onto each of the aligned carbon nanotubes to make novel conductive polymer-CNT coaxial nanowires.[100]

Takedaa et al.[101] used an Si wafer with a thin thermally grown SiO_2 layer as substrate. A polymethylmethacrylate (PMMA) film was formed on this substrate by spin coating, followed by electron beam lithography to generate pairs of holes with diameter 3 µm at a spacing of 3 µm in the PMMA film. A 10-nm-thick Si film, a 10-nm-thick Mo film and a 3-nm-thick Fe film were sequentially deposited on this patterned substrate by electron beam deposition. After the PMMA film was stripped off, ordered catalyst patterns resulted upon which SWNTs were grown vertically by CVD. The patterned substrate was heated to 900 °C under Ar flow, which was replaced by methane and hydrogen after reaching 900 °C to grow the SWNTs forest.

An interesting variation of the CVD method was proposed by Ren and coworkers, where the major goal was to generate controlled, low-density, vertically oriented CNT arrays.[102–104] Tuning the nanotube diameter, length and density in the CNT forest is considered important because these properties could have a critical influence for some applications. The diameter and length of grown CNTs can be controlled by changing the catalyst particle size and the growth time, but the majority of methods used to reduce site density of CNTs arrays involve electron-beam- or UV-lithography. Ren *et al.* introduced a new density control method in which Ni nanoparticles were randomly electro-deposited on $1\,cm^2$ Cr-coated silicon substrate by applying a current pulse to the substrate in $NiSO_4$ solution. By controlling the amplitude and duration of the current pulse, the size and density of the Ni nanoparticles were controlled. Vertical CNT growth was then done by PECVD at $<660\,°C$ for 5–8 min with NH_3 and C_2H_2 gases as precursor. This group demonstrated that by changing the density of Ni nanoparticles, MWCNT site density ranging from a single standing CNT to $3 \times 10^8\,cm^{-2}$ could be achieved.[102]

Iron-assisted electrostatic assembly of dense SWNT forests is now described in more detail. This process was invented and optimised by the Papadimi-trakopoulos group and transferred to the Rusling group for biosensor development, and the reproducibility of the methodology has been tested by over a dozen researchers in our labs. The first step is to carboxyl-functionalise and shorten pristine SWNTs by oxidation in $3:1\,HNO_3/H_2SO_4$ under sonication for 4 hours at $70\,°C$ (Figure 4.4).[85,86] The nanotube dispersions are filtered, washed extensively with water, dried in vacuum overnight and dispersed in slightly basic DMF by sonication. In our hands, the nanotube dispersions give best results for forest fabrication when aged by standing at room temperature for one week, then used prior to a "sell-by" date of about 4 months after preparation.

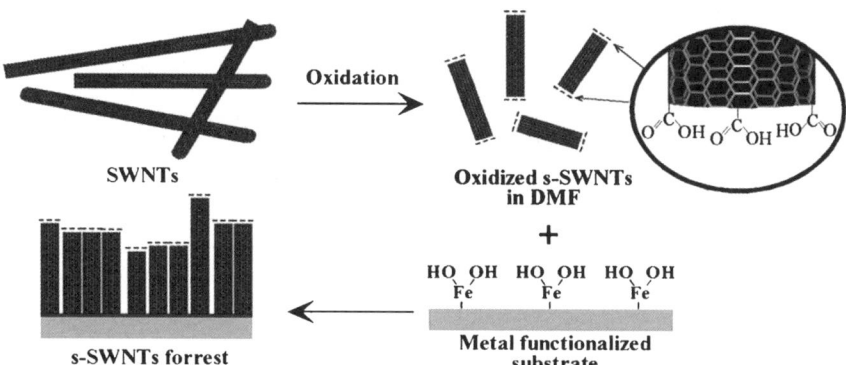

Figure 4.4 Schematic representation of steps in the self-assembly of SWNT forests onto Nafion/FeO(OH)–FeOCl functionalised surfaces from shortened, oxidised nanotube dispersion in DMF (see references 85–87 for details).

Nanotubes forests can be assembled onto any properly prepared surface from these dispersions. Pyrolytic graphite or gold have been used as substrates for biosensors and flat Si wafers or mica surfaces were used for AFM and Raman analyses. Surfaces are prepared for forest assembly by adsorption of a thin layer of Nafion. The substrates are then sequentially dipped into aqueous acidic 30 mM $FeCl_3$ solution (15 min) and the SWNT-DMF dispersion (30 min) with an intermediate DMF wash.[85,86] Control of the acidity of the $FeCl_3$ solution is critical for precipitation of flat FeO(OH)–FeOCl crystallites conducive to reproducible SWNT forest assembly, and solution pH should be controlled at 1.7–1.9, although the early papers on the method suggested pH 2.2.

It is recommended that each batch of SWNT forest electrodes made be subjected to analysis by atomic force microscopy (AFM) and resonance Raman spectroscopy for quality control. Typical AFM images show that SWNT forests made from aged dispersions provide nearly complete coverage of the underlying surface (Figure 4.5(a)). The average feature width is $\sim$20 nm. Although most likely influenced by AFM tip broadening, this width is much larger than the $\sim$1 nm diameter of a SWNT. The 20 nm width probably reflects nanotube bundling during forest assembly, which probably begins in the DMF dispersions. Figure 4.4(b) shows an AFM image after the antibody anti-human serum albumin was covalently linked onto the SWNT forest by amidisation of lysines of the protein with carboxylated nanotubes.[87] The needle-like SWNT forest features are replaced by a "rolling hill" AFM topography similar to that of any thin polyion or protein coating.[105]

Resonance Raman spectra show differences for assemblies made from fresh and aged SWNT dispersions (Figure 4.6).[87] Figure 4.6(a) shows the graphite- (G-)

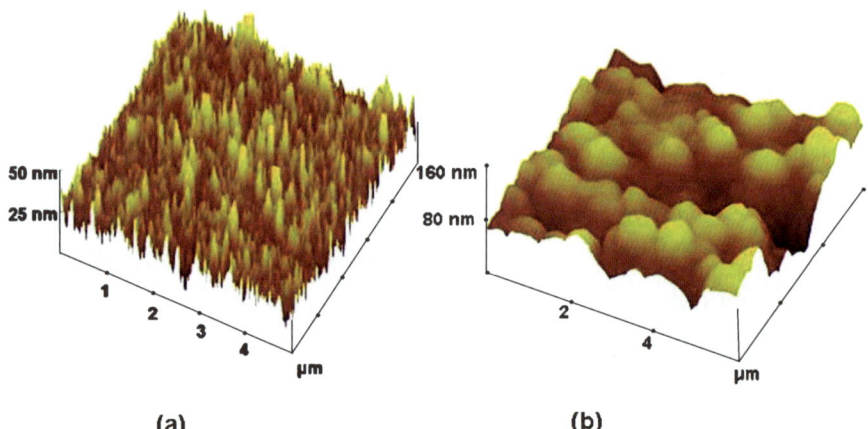

Figure 4.5 Tapping mode atomic force microscopy (AFM) images of: (a) SWNT forest on smooth silicon and (b) anti-biotin antibody functionalised SWNT on smooth silicon. Reproduced with permission from reference 88, copyright 2005 Royal Society of Chemistry, UK.

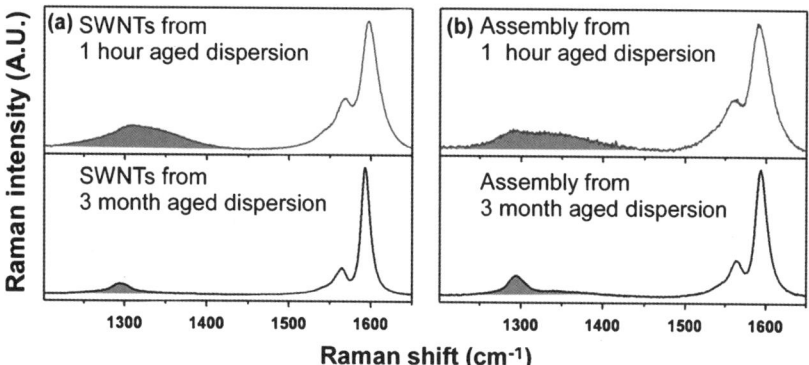

Figure 4.6 Resonance Raman spectra using 785-nm laser for acid treated SWNTs after dispersion in DMF and (a) aging for 1 hour and 3 months, respectively and (b) after self-assembly of SWNT forests on Si substrates from the above DMF/SWNTs dispersions. Shaded regions indicate the locations of defect-induced band (D-band). Reproduced with permission from reference 88, copyright 2005 Royal Society of Chemistry, UK.

and defect- (D-) modes in the spectra that are characteristic of SWNTs.[106] The D-band between 1250 and 1450 cm^{-1} results from scattering by in-plane heteroatoms, grain boundaries and other defects.[106] The aged SWNT/DMF dispersion shows large decreases in D-band width compared to fresh SWNT/DMF dispersion, suggesting decreased defects on SWNTs upon aging of the SWNT/DMF dispersion by allowing it to stand in a closed vessel at room temperature. Correspondingly, forests assembled from aged dispersions show much smaller D-band widths than the assemblies made from nanotubes aged for 1 hour (Figure 4.6(b)). Since fewer defects mean better conductivity, this "defect healing" effect may be related to the better performance found when using SWNT-DMF dispersion aged for 1 week to 4 months.

Another excellent quality control experiment for SWNT forest fabrication involves linkage of peroxidase enzymes onto the forests by amidisation, and observation of sensitivity and detection limit for the amperometric determination of hydrogen peroxide. Peroxide converts the iron heme peroxidase enzymes to a ferryloxy form that can be reduced at relatively low applied voltage. Figure 4.7(a) shows direct, reversible cyclic voltammograms of the iron heme enzymes myoglobin and horseradish peroxidase (HRP). These electrodes were stable for more than one month.[87] Figure 4.7(b) shows the sensitive response of an HRP/SWNT electrode to hydrogen peroxide in the nM range, providing a detection limit of ~40 nM as three times the average noise (Table 4.2). Addition of the electron-transfer mediator hydroquinone to the solution gave no change in amperometric current, suggesting that all of the HRP is in direct electronic communication with the forest.[87] Table 4.2 provides sensitivities and detection limits that can be used for quality control assessment of SWNT forests.

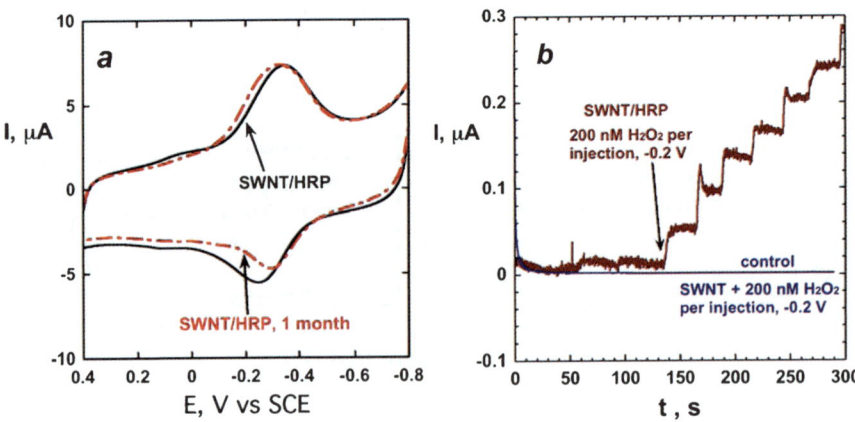

Figure 4.7 Electrochemical characterisation of SWNT forests on pyrolytic graphite disk surfaces after covalent attachment of enzyme horseradish peroxidase (HRP): (a) Cyclic voltammograms at 300 mV s^{-1} of SWNT/HRP in pH 6.5 buffer before and after 1 month storage; (b) Catalytic amperometry for reduction of hydrogen peroxide on SWNT/HRP electrodes at -0.2 V vs. SCE and 2000 rpm showing stepwise increases in reduction current after 200 nM peroxide injections into the buffer. Rotating disk amperometric results also showed that the SWNT forest electrodes are stable under hydrodynamic conditions. Reproduced with permission from reference 88, copyright 2005 Royal Society of Chemistry, UK.

Table 4.2 Electrochemical characteristics of peroxidases on SWNT forests[a].

Film structure	Conc. of electroactive enzyme (nmol cm^{-2})	Sensitivity for H_2O_2 ($\mu A \mu M^{-1}$)	Approx. detection limit (nM H_2O_2)
SWNT/Mb	0.195 ± 0.014	0.108 ± 0.006	80
SWNT/HRP	0.080 ± 0.008	0.179 ± 0.009	40

[a]Data from reference 87. Concentration of electroactive protein estimated by integration of CV reduction peak at scan rates less than 50 mV s^{-1}; sensitivity determined from slope of calibration curve from rotating disk amperometry at -0.2 V and 2000 rpm; detection limits as three times the average noise.

Future applications of biosensors are expected to involve electronic arrays with individually addressable electrodes, e.g. for multiplexed detection of suites of protein biomarkers. If SWNT forests are to be used in such arrays, selective patterning on microelectronic chips will be necessary. Papadimitrakopoulos and coworkers have developed schemes to selectively pattern SWNT forests on gold elements of arrays utilising deposition of iron oxide layers with and without Nafion.[86,89]

4.3.2 Biosensor Applications of SWNT Forests

4.3.2.1 SWNT Immunosensors

In addition to the peroxide sensors on SWNT forest electrodes (see Figure 4.7 and Table 4.2), research has been directed toward developing SWNT immunosensors for proteins.[18,66] These approaches feature enzyme-linked immunosorbent assays (ELISA) built onto SWNT forests. Primary antibodies (Ab_1) that capture the antigen (*i.e.* the protein analyte) are attached to the nanotube ends, and antigen (Ag) binding, washing and enzyme label detection are all done on the sensor surface.

Protein detection is based on sandwich immunoassays with good inherent selectivity and sensitivity.[107] Here, Ab_1 attached to the sensor selectively captures the antigen from the sample (Figure 4.8). Then, a secondary antibody (Ab_2) labelled with a redox enzyme (*i.e.* Ab_2-HRP) binds to the antigen. Specific binding of Ab_2-HRP to the antigen on the sensor surface provides an electroactive HPR label whose concentration depends on the amount of antigen. An alternative high-sensitivity strategy involves attaching many HRP labels and a few Ab_2 to a CNT to make an Ab_2-HRP-CNT bioconjugate to replace the single label Ab_2-HRP (Figure 4.8(b)). The multiple labels provide a large amplification of the signal. In both protocols, injection of H_2O_2 into an electrochemical cell containing the sensor provides a signal proportional to the amount of bound antigen (Figure 4.8).

For proof of concept, an SWNT immunosensor to detect biotin using capture antibodies adsorbed onto SWNT forest assemblies was first designed.[92] This work pointed out the importance of decreasing non-specific binding (NSB) of labelled species and of using mediators for high sensitivity. Scheme 4.1 illustrates the hydroquinone (H_2Q) mediated efficient electron transfer from the electrode to the oxidised HRP label. Detection of the catalytic HRP label is achieved by addition of H_2O_2 to convert the $HRPFe^{III}$ form to the easily reducible $\cdot HRPFe^{IV}=O$, referred to as compound I. In the scheme, the H_2O_2 added to the analysis solution oxidises $HRPFe^{III}$ to compound I (eqn (2)), which is reduced to ferric HRP by H_2Q (eqn (3)). Reduction of quinone (Q) at the electrode (eqn (1)) provides the immunosensor signal.

SWNT forest immunosensors were then designed to detect human serum albumin (HSA) in a sandwich immunoassay (Figure 4.9).[87] In these electrochemical immunosensors, the carboxylated ends of the nanotube forests provide a high surface area for attachment of capture antibodies. Using hydroquinone mediation, a detection limit (three times the noise) for HSA of $1\,pmol\,mL^{-1}$ (1 nM) was achieved. Controls showed that SWNT forests increased the mediated amperometric signal 10- to 16-fold compared to that of planar electrodes.

SWNT forest sensors were then developed for prostate specific antigen (PSA), a clinical biomarker for diagnosing and monitoring prostate cancer.[108] The most sensitive detection strategy employed the multi-label Ab_2-HRP-CNT bioconjugate (Figures 4.10 and 4.11). This amplification strategy improved detection limit 100-fold and sensitivity 800-fold compared to using the single-label Ab_2-HRP.

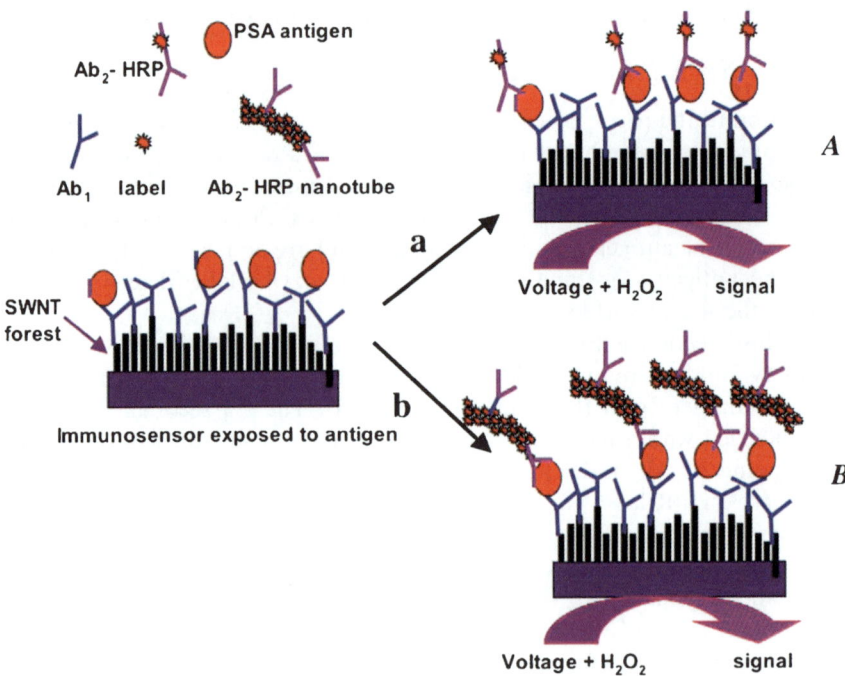

Figure 4.8 Detection principles of SWNT forest immunosensors. On the left is an SWNT immunosensor that has been equilibrated with an antigen sample, along with biomaterials used for fabrication (HRP is the enzyme label). Picture (**A**) on the right shows the immunosensor after treating with a conventional HRP-Ab_2 providing one label per binding event. Picture (**B**) on the right shows the immunosensor after treating with HRP-CNT-Ab_2 to obtain amplification by providing numerous enzyme labels per binding event. The final detection step involves immersing the immunosensor after secondary antibody attachment into a buffer containing mediator in an electrochemical cell, applying voltage and injecting a small amount of hydrogen peroxide. Adapted with permission from reference 9, copyright 2006 American Chemical Society.

$$Q + 2e^- + 2H^+ \rightleftarrows H_2Q \qquad (1)$$

$$H_2O_2 + HRPFe^{III} \rightleftarrows \bullet HRPFe^{IV}=O + H_2O \qquad (2)$$

$$\bullet HRPFe^{IV}=O + H_2Q \rightleftarrows HRPFe^{III} + Q + 2H^+ \qquad (3)$$

Scheme 4.1 Pathway for mediated electrochemical detection of HRP labels.

The detection limit was $4 \, pg \, mL^{-1}$ PSA in $10 \, \mu L$ serum, or a mass detection of $40 \, fg$. The Ab_2-HRP-CNT bioconjugates were assayed for enzyme activity and antibodies (imaged after labelling with quantum dots) to estimate ~ 82 HRPs and 30 ± 15 secondary antibodies per $100 \, nm$ of antibody-HRP-CNT bioconjugate.[109]

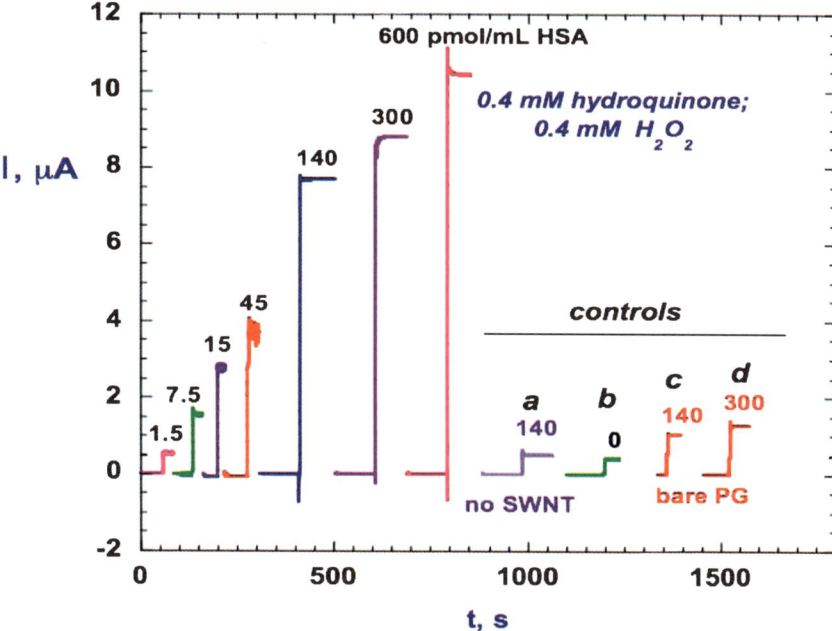

Figure 4.9 Mediated amperometric sandwich assays at −0.2 V and 2000 rpm for human serum albumin (HSA) in which SWNT/anti-HSA immunosensors were incubated with HSA (concentration in pmol mL^{-1} labelled on curves) in 20 µL 2% Casein + 0.05% Tween-20 in PBS for 1 hr followed by 10 µL 0.6 nmol mL^{-1} anti-HSA-HRP for 1 hr. Currents shown after placing electrodes in buffer containing 0.4 mM hydroquinone (H$_2$Q), then injecting H$_2$O$_2$ to 0.4 mM. Controls are shown on right with HSA concentrations: (a) anti-HSA treated Nafion-iron oxide-coated PG electrode, (b) SWNT-anti-HSA immunosensor omitting addition of HSA, (c) and (d) anti-HSA treated bare PG electrode for two different HSA levels. Reproduced with permission from reference 88, copyright 2005 Royal Society of Chemistry, UK.

Using this amplification strategy, it was possible to quantitatively measure PSA in lysates from 1000 prostate cancer cells from frozen human tissue (Figure 4.11). In addition, excellent correlation of PSA detection in human patient serum using the SWNT forest immunosensors was found compared to the established enzyme-linked immunosorbent assay (ELISA).[9] These results suggest a promising future for SWNT forest electrochemical immunosensors for reliable point-of-care cancer diagnostics. Multiplexed biomarker detection for prostate and other cancers to provide statistically reliable early cancer detection are currently being pursued in our laboratory.

4.3.2.2 Other Biosensors Using SWNT Forests

Many other sensor applications have been reported for SWNT forest electrodes made by various methods (Table 4.1). A sensitive, selective glucose sensor was

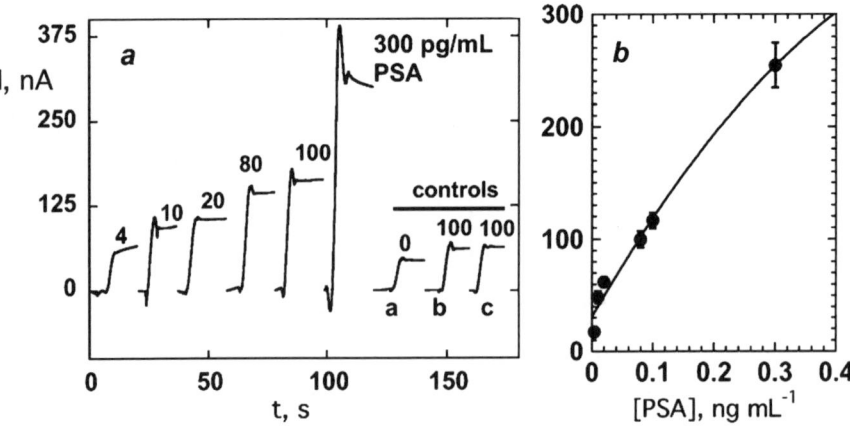

Figure 4.10 Amperometric response for SWNT immunosensors incubated with PSA (concentration in pg mL^{-1} labelled on curves) in 10 μL undiluted newborn calf serum for 1.25 hr: (a) current at −0.3 V and 3000 rpm using the Ab$_2$-CNT-HRP bioconjugate (11 pmol mL^{-1} in HRP). Controls shown on right with PSA concentrations: (a) full SWNT immunosensor omitting addition of PSA, (b) immunosensor built on bare PG surface for 100 pg mL^{-1} PSA, (c) immunosensor built on Nafion-iron oxide-coated PG electrode for 100 pg mL^{-1} PSA. (b) Influence of PSA concentration on steady-state current for immunosensor using Ab$_2$-CNT-HRP bioconjugate. Error bars in (b) represent device-to-device standard deviations ($n = 3$). Adapted with permission from reference 9, copyright 2006 American Chemical Society.

developed using polypyrrole to form the conductive polymer-CNT coaxial nanowire.[78] He and Dai fabricated a CNT-DNA electrochemical sensor[79] whereby the alignment of the gold-supported CNTs after grafting with ssDNA chains was retained. The ssDNA probes were grafted onto both tips and walls of plasma-activated aligned CNTs by amide linkages. An approach to protect the fragile structure of CVD fabricated forests was proposed by Li et al.,[88] where gaps between the nanotubes were filled with spin-on glass (SOG) prior to the oxidative reaction and subsequent immobilisation of DNA chains. In this work, an MWNT forest was grown by plasma-enhanced chemical vapour deposition (PECVD) from 10- to 20-nm-thick Ni catalyst films embedded in 200-nm-thick Cr patterned by lithography. Oligonucleotide probes were then selectively functionalised at the open ends of nanotubes, and the hybridisation of subattomole DNA targets was detected by Ru(bpy)$_3^{2+}$-mediated guanine oxidation, with a detection limit of ~6 attomole. A similar approach was used by this group to design and characterise CNT nanoelectrode arrays.[110,111] CVD-generated CNT forests have also been used for the sensitive detection of biomolecules including blood cholesterol,[112] glucose,[113] prostate specific antigen[114] and anti-hemagglutin.[101]

Electron transfer kinetics have been shown to strongly depend on orientation of nanotubes.[115,116] With this in mind, significant effort has been made to enhance

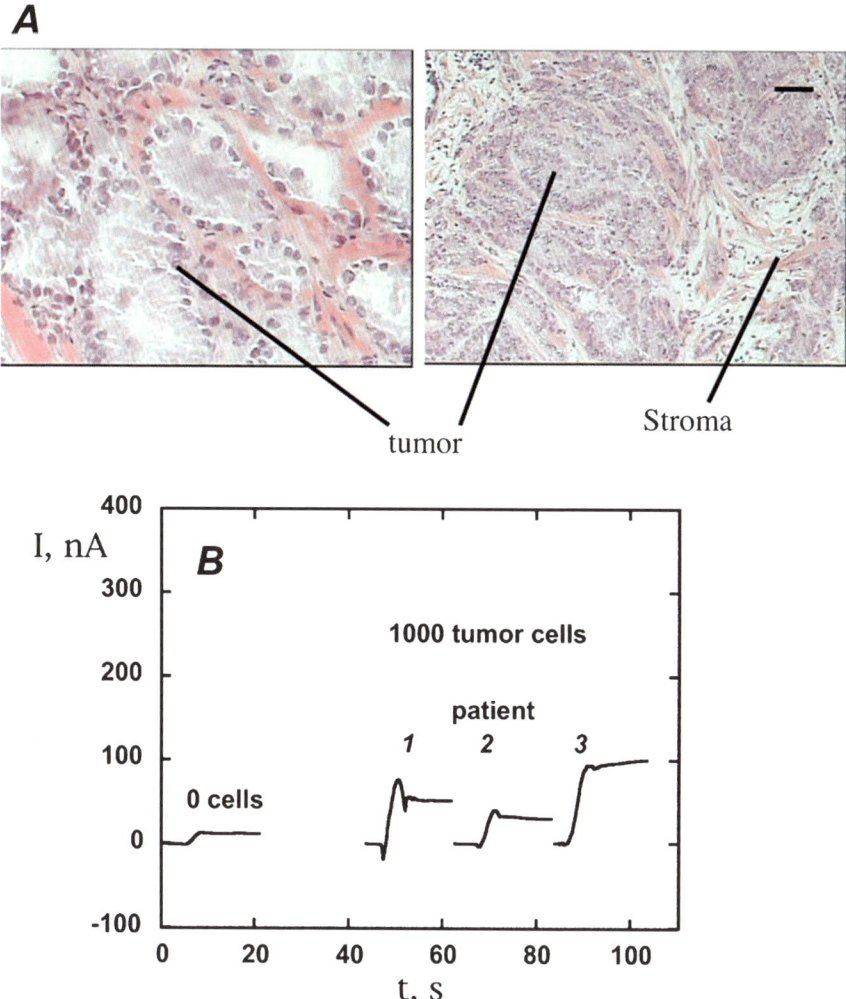

Figure 4.11 Measurement of PSA in prostate cancer cells: (A) Micrograph of representative prostate cancer biopsy stained with haemotoxylin and eosin to reveal areas of tumour and stromal cells (scale bar at upper right is 100 microns). Cells were subsequently procured by laser microdissection and processed for PSA detection by SWNT immunosensors. Non-malignant normal cells (not shown) were processed similarly. (B) Amperometric current -0.3 V and 3000 rpm in which SWNT immunosensors were incubated with prostate tissue lysates ($\sim$1000 cells) in 10 µL buffer for 1.25 hours followed by 10 µL anti-PSA-CNT-HRP bioconjugate (11 pmol mL^{-1} in HRP) in 0.05% Tween-20 for 1.25 hrs. Adapted with permission from reference 9, copyright 2006 American Chemical Society.

the electron turnover rate of GOx on nanotube forest electrodes in order to increase sensitivity to glucose. Covalent linkage of intact GOx onto the tips of vertically aligned MWNT forests achieved an electron turnover rate of $1500\,s^{-1}$.[88] Willner *et al.* further increased the electron turnover rate to $4100\,s^{-1}$ by covalently linking the FAD cofactors to the tips of an SWNT forest, then capturing apo-glucose oxidase on the nanotubes by assimilation of the cofactor.[34] This approach ensures close proximity of the enzyme redox site with the nanotubes to facilitate fast electron transfer. A linear correlation of electron transfer rate *vs.* the inverse of SWNT length in this study suggested that length-dependent back scattering and nanotube resistance are directly coupled to biocatalysis.[34]

4.4 Outlook for the Future

Clearly, CNT-based electrochemical biosensors have already revealed excellent promise as future tools for biomedical research and diagnostics. Ordered nanotube arrays exemplified by SWNT forests will be necessary to obtain the best biosensor sensitivities and detection limits in order to take advantage of the electronic coupling of biomolecules with the nanotube *molecular wires*. However, for less-demanding applications disordered nanotube electrodes may provide reliable tools with electrocatalytic properties similar to edge plane pyrolytic graphite, but with the possibility of a higher density of catalytic sites. The full power of these nanotube biodevices will be realised by their fabrication into individually addressable arrays for multiplexed biomolecule determinations. Such devices should be readily adaptable to modern biomedical analyses in cancer diagnostics, genomics, proteomics and systems biology.

An important practical milestone for single SWNT immunosensors was reached by the accurate detection of PSA in cancer patient serum and tissue.[9] Here the nanotube forests serve as high surface area substrates for the attachment of a high density of capture antibodies, and in this way help to provide high sensitivity. SWNT immunosensors can be adapted easily for detection of other relevant biomarkers, as is now being pursued in our laboratories. Application to detection of bacteria and biopathogens should be possible. The ability to selectively deposit SWNT forests onto patterned metal arrays[86] indicates the potential for easy fabrication of multiplexed arrays that should facilitate simultaneous measurement of multiple biomarkers for reliable point-of-care diagnostics of cancer and other diseases.

Further improvement in sensitivity and detection limits of SWNT biosensors may be on the horizon. Electrochemical detection from nanotube forest arrays might be improved by utilising *fully conductive* nanotubes, and by incorporation of conductive polymers at resistive junctions in SWNT forest devices. Creative approaches to incorporate multi-label and pre-concentration-based amplification strategies could significantly advance biodetection devices. Gaining a practical insight into control of non-specific binding for multiple biomarkers is also essential. SWNT-based biosensor devices are expected to play a significant role in the future of bioanalytical science.

References

1. (a) R. Saito, G. Dresselhaus and M. S. Dresselhaus, *Physical Properties of Carbon Nanotubes,* Imperial Coll. Press, London, 1998; (b) M. S. Dresselhaus, G. Dresselhaus and P. C. Eklund, *Science of Fullerenes and Carbon Nanotubes,* Academic Press, San Diego, 1996; (c) H. Dai, *Acc. Chem. Res.*, 2002, **35**, 1035.
2. J.-C. Charlier and S. Iijima, *Top. Appl. Phys.*, 2001, **80**, 55.
3. (a) J. Liu, A. G. Rinzler, H. Dai, J. H. Hafner, R. K. Bradley, P. J. Boul, A. Lu, T. Iverson, K. Shelimov, C. B. Huffman, F. Rodriguez-Macias, Y.-S. Shon, T. R. Lee, D. T. Colbert and R. E. Smalley, *Science*, 1998, **280**, 1253; (b) J. J. Gooding, R. Wibowo, J. Liu, W. Yang, D. Losic, S. Orbons, F. J. Mearns, J. G. Shapter and D. B. Hibbert, *J. Am. Chem. Soc.*, 2003, **125**, 9006.
4. P. M. Ajayan, *Chem. Rev.*, 1999, **99**, 1787.
5. P. M. Ajayan, J. C. Charlier and A. G. Rinzler, *Proc. Natl. Acad. Sci. USA*, 1999, **96**, 14199.
6. C. N. R Rao, B. C. Satishkumar, A. Govindararaj and M. Nath, *ChemPhysChem*, 2001, **2**, 78.
7. (a) Z. Yao, C. L. Kane and C. Dekker, *Phys. Rev. Lett.*, 2000, **84**, 2941; (b) Z. Wu, Z. Chen, X. Du, J. M. Logan, J. Sippel, M. Nikolou, K. Kamaras, J. R. Reynolds, D. B. Tanner, A. F. Hebard and A. G. Rinzler, *Science*, 2004, **305**, 1273; (c) K. H. An, W. S. Kim, Y. S. Park, Y. C. Choi, S. M. Lee, D. C. Chung, D. J. Bae, S. C. Lim and Y. H. Lee, *Adv. Mater.*, 2001, **13**, 497; (d) S. Fan, M. G. Chapline, N. R. Franklin, T. W. Tombler, A. M. Cassell and H. Dai, *Science*, 1999, **283**, 512; (e) R. H. Baughman, C. Cui, A. A. Zakhidov, Z. Iqbal, J. N. Barisci, G. M. Spinks, G. G. Wallace, A. Mazzoldi, D. De Rossi, A. G. Rinzler, O. Jaschinski, S. Roth and M. Kertesz, *Science*, 1999, **284**, 1340.
8. D. P. Burt, N. R. Wilson, J. M. R. Weaver, P. S. Dobson and J. V. Macpherson, *Nano Lett.*, 2005, **5**, 639.
9. X. Yu, B. Munge, V. Patel, G. Jensen, A. Bhirde, J. D. Gong, S. N. Kim, J. S. Gillespie, J. S. Gutkind, F. Papadimitrakopolous and J. F. Rusling, *J. Am. Chem. Soc.*, 2006, **128**, 11199.
10. J. J. Gooding, R. Wibowo, J. Q. Liu, W. R. Yang, D. Losic, S. Orbons, F. J. Mearns, J. G. Shapter and D. B. Hibbert, *J. Am. Chem. Soc.*, 2003, **125**, 9006.
11. J. Wang, R. P. Deo, P. Poulin and M. Mangey, *J. Am. Chem. Soc.*, 2003, **125**, 14706.
12. E. Katz and I. Willner, *ChemPhysChem*, 2004, **5**, 1085.
13. J. Wang, *Analyst*, 2005, **130**, 421.
14. D. M. Guldi, G. M. A. Rahman, N. Jux, N. Tagmatarchis and M. Prato, *Angew. Chemie., Int. Ed.*, 2004, **43**, 5526.
15. K. Balasubramanian and M. Burghard, *Anal. Bioanal. Chem.*, 2006, **385**, 452.

16. M. Pumera, S. Sanchez, I. Ichinose and J. Tang, *Sens. & Actuators B.*, 2007, **123**, 1195.
17. (a) P. Cherukuri, S. M. Bachilo, S. H. Litovsky and R. B. Weisman, *J. Am. Chem. Soc.*, 2004, **126**, 15638; (b) P. W. Barone, S. Baik, D. A. Heller and M. S. Strano, *Nat. Mater.*, 2005, **4**, 86.
18. S. N. Kim, J. F. Rusling and F. Papadimitrakopolous, *Adv. Mater.*, 2007, **19**, 3214.
19. G. G. Wildgoose, C. E. Banks, H. C. Leventis and R. G. Compton, *Microchim. Acta*, 2006, **152**, 187.
20. A. Guseppi-Elie, C. Lei and R. H. Baughman, *Nanotechnology*, 2002, **13**, 559.
21. M. Cinke, J. Li, B. Chen, A. Cassell, L. Delzeit, J. Han and M. Meyyappan, *Chem. Phys. Lett.*, 2002, **365**, 69.
22. (a) S. Heinze, J. Tersoff, R. Martel, V. Derycke, J. Appenzeller and P. Avouris, *Phys. Rev. Lett.*, 2002, **89**, 106801/1; (b) B. Q. Wei, R. Vajtai and P. M. Ajayan, *Appl. Phys. Lett.*, 2001, **79**, 1172.
23. R. J. Chen, S. Bangsaruntip, K. A. Drouvalakis, N. W. S. Kam, M. Shim, Y. Li, W. Kim, P. J. Utz and H. Dai, *Proc. Nat. Acad. Sci. USA*, 2003, **100**, 4984.
24. D. A. Britzab and A. N. Khlobystov, *Chem. Soc. Rev.*, 2006, **35**, 637.
25. A. Bianco, K. Kostarelos, D. Partidos and M. Prato, *Chem. Commun.*, 2005, 571.
26. A. Bianco, K. Kostarelos and M. Prato, *Curr. Opin. Chem. Biol.*, 2005, **9**, 674.
27. J. J. Gooding, *Electrochim. Acta*, 2005, 50, 3049 and references therein.
28. H. X. Luo, Z. J. Shi, N. Q. Li, Z. N. Gu and Q. K. Zhuang, *Anal. Chem.*, 2001, **73**, 915.
29. J. X. Wang, M. X. Li, Z. J. Shi, N. Q. Li and Z. N. Gu, *Anal. Chem.*, 2002, **74**, 1993.
30. K. P. Gong, X. Z. Zhu, R. Zhao, S. X. Xiong, L. Q. Mao and C. F. Chen, *Anal. Chem.*, 2005, **77**, 8158.
31. J. Chen, D. Du, F. Yan, H. M. Ju and H. Z. Lian, *Chem.-Eur. J.*, 2005, **11**, 1467.
32. J. J. Davis, R. J. Coles and H. A. O. Hill, *J. Electroanal. Chem.*, 1997, **440**, 279.
33. M. Burghard, G. Duesberg, G. Philipp, J. Muster and S. Roth, *Advanced Mater.*, 1998, **10**, 584.
34. Y. Patolsky, Y. Weizmann and I. Willner, *Angew. Chem. Int. Ed.*, 2004, **43**, 2113.
35. P. Diao and Z. F. Liu, *J. Phys. Chem. B*, 2005, **109**, 20906.
36. C. Song, Y. Y. Xia, M. W. Zhao, X. D. Liu, B. D. Huang, F. Li and Y Ji, *J. Phys. Rev. B*, 2005, **72**, 165430.
37. K. P. Gong, M. N. Zhang, Y. M. Yan, L. Su, L. Q. Mao, Z. X. Xiong and Y. Chen, *Anal Chem.*, 2004, **76**, 6500.
38. V. G. Gavalas, S. A. Law, J. C. Ball, R. Andrews and L. G. Bachas, *Anal. Biochem.*, 2004, **329**, 247.

39. Q. C. Shi, T. Z. Peng, Y. N. Zhu and C. F. Yang, *Electroanalysis*, 2005, **17**, 857.
40. M. H. Yang, Y. H. Yang, Y. L. Liu, G. L. Shen and R. Q. Yu, *Biosens. Bioelectron.*, 2006, **21**, 1125.
41. X. Luo, A. Killard, A. Morrin and M. R. Smyth, *Anal. Chim. Acta*, 2006, **575**, 39.
42. F. Qu, M. Yang, J. Jiang, G. Shen and R. Yu, *Anal. Biochem.*, 2005, **344**, 108.
43. X. F. Zhang, T. Liu, T. V. Sreekumar, S. Kumar, V. C. Moore, R. H. Hauge and R. E. Smalley, *Nano Lett.*, 2003, **3**, 1285.
44. J. Wang and M. Musameh, *Anal. Chem.*, 2003, **75**, 2075.
45. P. P. Joshi, S. A. Merchant, Y. D. Wang and D. W. Schmidtke, *Anal. Chem.*, 2005, **77**, 3183.
46. A. Ortiz-Acevedo, H. Xie, V. Zorbas, W. M. Sampson, A. B. Dalton, R. H. Baughman, R. K. Draper, I. H. Musselman and G. R. Dieckmann, *J. Am. Chem. Soc.*, 2005, **127**, 9512.
47. E. K. Hobbie, B. J. Bauer, J. Stephenes, M. L. Becker, P. McGuiggan, S. D. Hudson and H. Wang, *Langmuir*, 2005, **21**, 10284.
48. S. Hrapovic, Y. L. Liu, K. B. Male and J. H. T. Luong, *Anal. Chem.*, 2004, **76**, 1083.
49. W. Wiyaratn, S. Hrapovic, Y. L. Liu, W. Surareungchai and J. H. T. Luong, *Anal. Chem.*, 2005, **77**, 5742.
50. C. Ou, Y. Ruo, C. Yaqin, T. Mingyu, C. Rong and H. Xiulan, *Anal. Chim. Acta*, 2007, **603**, 205.
51. C. Hui-Fang, Y. Jian-Shan, Z. Wei-De, L. Chan-Ming and J. H. T. Luong, *Anal. Chim. Acta*, 2007, **594**, 175.
52. B. Wu, S. Hou, F. Yin, Z. Zhao, Y. Wang, X. Wang and Q. Chen, *Biosens. Bioelectron.*, 2007, **22**, 2854.
53. P. Young, Y. Lu, R. Terrill and L. Li, *J. Nanosci. Nanotech.*, 2005, **5**, 1509.
54. J. Wang, T. Tangkuaram, S. Loyprasert, T. Vazquez-Alvarez, W. Veerasai, P. Kanatharana and P. Thavarungkul, *Anal. Chim. Acta*, 2007, **581**, 1.
55. X. B. Yan, J. Chen, B. K. Tay and K. A. Khoret, *Electrochem. Commun.*, 2007, **9**, 1269.
56. M. Yang, Y. Yang, H. Yang, G. Shen and R. Yu, *Biomaterials*, 2006, **27**, 246.
57. Y. Yun, Z. Dong, V. Shanov, W. R. Heineman, B. Halsall, A. Bhattacharya, L. Conforti, R. K. Narayan, W. S. Ball and M. J. Shulz, *Nanotoday*, 2007, **2**, 30.
58. (a) J. F. Rusling and Z. Zhang, in *Biomolecular Films*, ed. J. F. Rusling, Marcel Dekker, New York, 2003, pp. 1–64; (b) J. F. Rusling (ed.), *Biomolecular Films: Design, Function and Applications*, Marcel Dekker, New York, 2003, Chapters 2, 7 and 10.
59. A. Heller, *Acc. Chem. Res.*, 1990, **23**, 128.
60. G. Ramsay, *Commercial Biosensors*, Wiley, New York, 1998.

61. P. P. Joshi, S. A. Merchant, Y. Wang and D. W. Schmidke, *Anal. Chem.*, 2005, **77**, 3183.
62. H. A. Heering, K. A. Williams, S. de Vries and C. Dekker, *ChemPhysChem*, 2006, **7**, 1705.
63. Y. Yin, P. Wu, Y. Lü, P. Du, Y. Shi and C. J. Cai, *Solid State Electrochem.*, 2007, **11**, 390.
64. H. Zhang, L. Fan and S. Yan, *Chem.–Eur. J.*, 2006, **12**, 7161.
65. R. P. Deo, J. Wang, I. Block, A. Mulchandani, K. A. Joshi, M. Trojanowicz, F. Scholz, W. Chen and Y. Lin, *Anal. Chim. Acta*, 2005, **530**, 185.
66. J. V. Veetil and K. Ye, *Biotechnol. Prog.*, 2007, **23**, 517.
67. J. Okuno, K. Maehashi, K. Kerman, Y. Takamura, K. Matsumoto and E. Tamiya, *Biosens. Bioelectron.*, 2006, **22**, 2377.
68. J. N. Wohlstadter, J. L. Wilbur, G. B. Sigal, H. A. Biebuyck, M. A. Billadeau, L. Dong, A. B. Fischer, S. R. Gudibande, S. H. Jameison, J. H. Kenten, J. Leginus, J. K. Leland, R. J. Massey and S. J. Wohlstadter, *Adv. Mater.*, 2003, **15**, 1184.
69. S. Viswanathan, L. Wu, M.-R. Huang and J. A. Ho, *Anal. Chem.*, 2006, **78**, 1115.
70. V. Cataldo, A. Vaze and J. F. Rusling, *Electroanalysis*, 2008, **20**, 115.
71. J. Wang, G. Liu and R. M. Jan, *J. Am. Chem. Soc.*, 2004, **126**, 3010.
72. H. H. Thorp, *Trends Biotechnol.*, 1998, **16**, 117.
73. E. Palecek and M. Fojta, *Anal. Chem.*, 2001, **73**, 74A.
74. J. Wang, A. N. Kawde and M. R. Jan, *Biosensor Bioelectron.*, 2004, **20**, 995.
75. B. Munge, G. Liu, G. Collins and J. Wang, *Anal. Chem.*, 2005, **77**, 4662.
76. J. Wang, A. N. Kawde and M. Musameh, *Analyst*, 2003, **128**, 912.
77. L. Y. Heng, A. Chou, J. Yu, Y. Chen and J. J. Gooding, *Electrochem. Commun.*, 2005, **7**, 1457.
78. M. Gao, L. Dai and G. G. Wallace, *Electroanalysis*, 2003, **15**, 1089.
79. P. He and L. Dai, *Chem. Comm.*, 2004, **1**, 348.
80. S. Takedaa, A. Sbagyoa, Y. Sakodab, A. Ishiic, M. Sawamurad, K. Sueokaa, H. Kidab, K. Mukasaa and K. Matsumotoe, *Biosens. Bioelectronics*, 2005, **21**, 201.
81. G. D. Withey, A. D. Lazareck, M. B. Tzolov, A. Yina, P. Aich, J. I. Yeh and J. M. Xu, *Biosens. Bioelectronics*, 2006, **21**, 1560.
82. J. Okunoa, K. Maehashi, K. Kerman, Y. Takamura, K. Matsumoto and E. Tamiya, *Biosens. Bioelectronics*, 2007, **22**, 2377.
83. G. A. Rivas, M. D. Rubianes, M. C. Rodriguez, N. F. Ferreyra, G. L. Luque, M. L. Pedano, S. A. Miscoria and C. Parrado, *Talanta*, 2007, **74**, 291.
84. L. Dai, A. Patil, X. Gong, Z. Guo, L. Liu, Y. Liu and D. Zhu, *ChemPhysChem*, 2003, **4**, 1150.
85. D. Chattopadhyay, I. Galeska and F. Papadimitrakopolos, *J. Am. Chem. Soc.*, 2001, **123**, 9451.

86. H. Wei, S. Kim, S. N. Kim, B. D. Huey, F. Papadimitrakopolos and H. L. Marcus, *J. Mater. Chem.*, 2007, **17**, 4577.
87. X. Yu, S. N. Kim, F. Papadimitrakopoulos and J. F. Rusling, *Mol. BioSyst.*, 2005, **1**, 70.
88. J. Li, H. T. Ng, A. Cassell, W. Fan, H. Chen, Q. Ye, J. Koehne, J. Han and M. Meyyappan, *Nano Lett.*, 2003, **3**, 597–602.
89. H. Wei, S. N. Kim, H. L. Marcus and F. Papadimitrakopoulos, *Chem. Mater.*, 2006, **18**, 1100.
90. D. Chattopadhyay, S. Lastella, S. Kim and F. Papadimitrakopoulos, *J. Am. Chem. Soc.*, 2002, **124**, 728.
91. X. Yu, D. Chattopadhyay, I. Galeska, F. Papadimitrakopoulos and J. F. Rusling, *Electrochem. Comm.*, 2003, **5**, 408.
92. M. O'Connor, S. N. Kim, A. J. Killard, R. J. Forster, M. R. Smyth, F. Papadimitrakopoulos and J. F. Rusling, *Analyst*, 2004, **129**, 1176.
93. Z. Liu, Z. Shen, T. Zhu, S. Hou, L. Ying, Z. Shi and Z. Gu, *Langmuir*, 2000, **16**, 3569.
94. X. F. Yu, T. Mu, H. Z. Huang, Z. F. Liu and N. Z. Wu, *Surf. Sci.*, 2000, **461**, 199.
95. J. J. Gooding, R. Wibowo, J. Liu, W. Yang, D. Losic, S. Orbons, F. G. Mearns, J. G. Shapter and D. B. Hibbert, *J. Am. Chem. Soc.*, 2003, **125**, 9006.
96. J. Liu, A. Chou, W. Rahmat, M. N. Paddon-Row and J. J. Gooding, *Electroanalysis*, 2005, **17**, 38.
97. X. J. Huang, H. S. Im, O. Yarimaga, J. H. Kim, D. Y. Jang, D. H. Lee, H. S. Kim and Y. K. Choi, *J. Electroanal. Chem.*, 2006, **594**, 27.
98. B. Kim and W. M. Sigmund, *Langmuir*, 2003, **19**, 4848.
99. Y. Yang, S. Huang, H. He, A. W. H. Mau and L. Dai, *J. Am. Chem. Soc.*, 1999, **121**, 10832.
100. M. Gao, S. Huang, L. Dai, G. Wallace, R. Gao and Z. Wang, *Angew. Chem. Int. Ed.*, 2000, **39**, 3664.
101. S. Takedaa, A. Sbagyoa, Y. Sakodab, A. Ishiic, M. Sawamurad, K. Sueokaa, H. Kidab, K. Mukasaa and K. Matsumoto, *Biosens. Bioelectron.*, 2005, **21**, 201.
102. Z. P. Huang, D. Z. Wang, J. G. Wen and Z. F. Ren, *App. Phy. Lett.*, 2002, **80**, 4018.
103. Y. Yu, Y. H. Lin and Z. F. Ren, *Nano Lett.*, 2003, **3**, 107.
104. Y. H. Lin, F. Lu, Y. Tu and Z. F. Ren, *Nano Lett.*, 2004, **4**, 191.
105. J. B. Schenkman, I. Jansson, Y. Lvov, J. F. Rusling, S. Boussaad and N. J. Tao, *Archives Biochem. Biophys.*, 2001, **385**, 78.
106. M. S. Dresselhaus and P. C. Eklund, *Adv. Phys.*, 2000, **49**, 705.
107. D. S. Wilson and S. Nock, *Angew. Chem., Int. Ed.*, 2003, **42**, 494.
108. (a) T. M. Chu, *J. Clin. Lab. Anal.*, 1994, **8**, 323; (b) B. G. Blijenberg, R. Kranse, I. Eman and F. H. Schroder, *Eur. J. Clin. Chem. Clin. Biochem.*, 1996, **34**, 817.

109. G. C. Jensen, X. Yu, B. Munge, A. Bhirde, J. D. Gong, S. N. Kim, F. Papadimitrakopoulos and J. F. Rusling, *J. Nanosci. Nanotechnol.*, 2008, in press.
110. J. Koehne, J. Li, A. M. Cassell, H. Chen, Q. Ye, H. T. Ng, J. Han and M. Meyyappan, *J. Mater. Chem.*, 2004, **14**, 676.
111. C. V. Nguyen, L. Delzeit, A. M. Cassell, J. Li, J. Han and M. Meyyappan, *Nano Lett.*, 2002, **2**, 1079.
112. S. Roy, H. Vedala and W. Choi, *Nanotechnology*, 2006, **17**, S14.
113. J.-S. Ye, Y. Wen, W. D. Zhang, L. M. Gan, G. Q. Xu and F.-S. Sheu, *Electrochem. Comm.*, 2004, **6**, 66.
114. J. Okunoa, K. Maehashi, K. Kerman, Y. Takamura, K. Matsumoto and E. Tamiya, *Biosens. Bioelectron.*, 2007, **22**, 2377.
115. J. J. Gooding, A. Chou, J. Liu, D. Losic, G. J. Shapter and B. D. Hibbert, *Electrochem. Commun.*, 2007, **9**, 1677.
116. W. Yang, P. Thordarson, J. J. Gooding, S. P. Ringer and F. Braet, *Nanotechnol.*, 2007, **18**, 412001.

CHAPTER 5
Activating Redox Enzymes through Immobilisation and Wiring

H.A. HEERING AND G.W. CANTERS

Leiden University, PO Box 9502, 2300 RA Leiden, The Netherlands

5.1 Introduction

This contribution deals with investigations of the activity of electron transfer (*ET*) proteins and redox enzymes that have been either co-crystallised with or linked covalently to a partner protein or that have been immobilised onto an electrode surface. There are at least two reasons why this subject has gained so rapidly such increasing popularity over the past few years:

- it provides insight not easily obtainable otherwise into the way enzymes operate and into the physical-chemical characteristics of proteins;
- it shows how proteins may be implemented in the design of electronic devices such as chips, sensors and information storage and processing hardware.

Electron transfer in nature is realised in the overwhelming majority of cases by the formation of transient complexes between proteins. The transient nature of these complexes hampers the detailed study of their structural and mechanistic properties. At an early stage, therefore, researchers have tried to limit or abolish the mobility of the proteins in these complexes. One way to achieve this is to immobilise the reaction partners either by co-crystallising them or by covalently connecting them. Another way is to immobilise them on an electrode and study them by electrochemical methods. As we shall see, minimising protein mobility in these complexes abolishes the very feature that is most important for their functioning.

In Section 5.2 the research on co-crystallised and covalently linked protein redox partners will be reviewed briefly. The remainder of the chapter will then be devoted to investigations of proteins immobilised on electrodes.

Section 5.3 contains a short overview of the theory of electron transfer between an electrode and an immobilised redox species and the techniques to measure ET rates.

When brought into contact with a bare metal, a metal oxide or a solid surface in general, proteins tend to deform or even to denature. Electrode preparation is therefore an important issue, which will be dealt with in Section 5.4.

Various ways to immobilise proteins onto electrodes have been developed. They are reviewed in Section 5.5. As some of these techniques make use of organic linkers it is important to understand the conducting properties of these organic molecules, too. They will be reviewed in the same section.

The chapter finishes with a few conclusions presented in Section 5.6.

5.2 Protein Complexes

5.2.1 Co-crystallisation

Crystallographic studies have been important because they showed what the distinguishing features are of a complex of two or more proteins that participate in biological ET. They provided important indications on how enzymes and ET proteins might be used in the construction of man-made bioelectronic interfaces. This will be summarised at the end of Section 5.2.

Pelletier and Kraut in 1992 were the first to report on a co-crystal of two redox partners, cytochrome c (Cc) and cytochrome c peroxidase (CcP).[1] They were followed by Mathews and coworkers who reported on the co-crystallisation of methylamine dehydrogenase (MADH) and its physiological partner, the blue copper protein amicyanin (ami), and on the ternary complex of MADH, ami and cytochrome-c_{551} (Cc551).[2] The overview by Crowley and Carrondo from 2004 lists these results, together with later successful attempts at co-crystallisation of redox partners.[3] The general hope was that the configuration of the protein partners in the crystals would (i) show the arrangement of the cofactors in the solution complex, (ii) possibly reveal the electron transfer paths and (iii), in general, would provide structural details of the encounter complexes in solution.

> i. After almost three decades of theoretical and experimental research it is clear now that the mutual orientation of the cofactors in most cases is not an important parameter in determining the rate of interprotein ET. The distance between the cofactors often is so large that direct overlap

between them is negligible. Instead coupling with the nearest neighbour in the *ET* path is of much greater importance.

ii. *ET* paths have been frequently delineated on the basis of the crystal structure of protein complexes. The calculated *ET* rates on the basis of the pathway model,[4] however, often give equally good results as the continuum model of Dutton et al.[5] that does not make use of pathways.

iii. The structural information about the interfaces did lead to new insights, in line with the results of more recent structural research on transient complexes in solution. In fact, despite their transient nature, *ET* protein complexes appear amenable to structural research by advanced NMR methods as convincingly shown by Ubbink and coworkers[6-8] and Hoffman and coworkers,[9-11] among others. Crystallographic research on co-crystallised redox partners and NMR studies of transient complexes show that the contact area between the partners is relatively small (700–2300 Å^2), and that the contact surface is often made up of a central hydrophobic patch surrounded, first, by a ring of polar residues and then, at the rim, by a ring of charged residues[6,7] (Figure 5.1). From the crystallographic studies it appears, moreover, that the hydrophobic patches are relatively flat and that they pack poorly in the complex leaving room for a considerable number of non-structured water molecules at the interface (35–100).[3] It has been made plausible that these features are crucial in shortening the lifetime of the protein complex so that the rate of protein–protein dissociation after *ET* remains compatible with physiological needs.

Another experimental challenge in the study of these co-crystals is the observation of the actual process of *ET* and of enzymatic activity. Rossi and

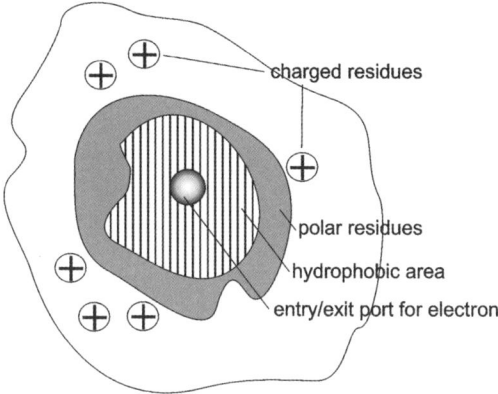

Figure 5.1 Schematic representation of the interface of a transient complex of two redox proteins. The entry/exit port for electrons is surrounded by a set of hydrophobic residues, which are lined with polar residues. Charged residues are found on the edge of the binding site.[7]

coworkers have met this problem by soaking crystals of MADH/ami or of MADH/ami/Cc551 in solutions containing the substrate methylamine.[2,12,13] ET from MADH to ami and Cc551 was slow enough to be followed by optical and EPR spectroscopy.[13] Gray[14,15] and Crane and coworkers[16–18] replaced the iron in the heme groups of some of the Cc or the CcP proteins in their crystals by Zn^{2+}. The Zn-porphyrin (ZnP) can be selectively excited, optically, resulting in a metastable triplet state which is a strong reductant. ET to a partner protein in the crystal as well as the back ET can be followed optically, which provides information on the ET kinetics. One of the main conclusions of this work as well as of the work by Rossi is that in solution small motions and movements within the (partners of the) complex probably have a strong bearing on the rate of ET. This confirms the results from direct studies of transient complexes in solution. These have shown that the protein partners in a transient complex often maintain a high degree of mobility with respect to each other.[6–8] This allows the sampling of a large ensemble of configurations of which only a few allow for efficient ET.

5.2.2 Covalent Complexes

Cross-linking can be used to identify the partner of a redox protein. For instance plastocyanin forms an association complex with photosystem I (PSI) in cyanobacteria to which it can be covalently attached by the application of carbodiimide chemistry.[19,20] Cross-linking proved impossible to PSI particles that were lacking the PsaF subunit demonstrating that the docking site for plastocyanin on the PSI particle is the PsaF subunit.

In the majority of studies, however, redox partners are cross-linked to study their ET reaction. The technique has the advantage that it allows some motion of the partners within the complex, the amount of motion depending on the nature of the cross-link. Secondly the proteins can be studied in solution which is a more natural environment than a crystal and, finally, ET and enzymatic reactions can be easily studied by standard rapid kinetics techniques. The disadvantage is of course that structural details of the cross-linked complex are missing.

The interest in protein complexes grew in the 1980s. Kostic and coworkers were one of the first groups to study complexes of plastocyanins and cytochromes that were covalently linked by carbodiimide chemistry.[21] The drawback of the use of EDC chemistry (EDC = 1-ethyl-3-(3-dimethylaminopropyl)carbodiimide HCl) is that it is reactive towards carboxyl and amino groups and that it is not very specific, therefore. The use of cysteines, either naturally occurring or engineered, allows for better defined cross-linked complexes. Poulos created S–S linked complexes of CcP and Cc in this way[22] and also Brunori employed cysteines to create cross-linked complexes of cytochrome-c oxidase and cytochrome-c.[23] From the studies of all these complexes it gradually transpired, among others, that cross-linking does not always result in ET-competent complexes, contrary to expectations based on modelling of these complexes, and that movement of the redox partners is needed for them to find a conformation that allows fast *ET*.

A case in point is provided by studies of dimers of azurin from *Pseudomonas aeruginosa*. Azurin is a blue copper *ET* protein that exhibits a high rate of electron self exchange (*ese*) in solution ($1 \times 10^6 \text{M}^{-1}\text{s}^{-1}$ at room temperature).[24] It has a single hydrophobic patch located around copper ligand His117, which protrudes through the hydrophobic patch and provides a conducting link with the Cu inside the protein. In azurin crystals the protein molecules crystallise as dimers associating along their hydrophobic patches (Figure 5.2).[25] It has been assumed that in solution the azurin molecules form transient complexes in a similar manner, which would explain the relatively high rate of the ese reaction.

Introduction of a charged residue in the hydrophobic patch by site-directed mutagenesis (M44K or M64E) lowers the ese rate by 2–3 orders of magnitude; neutralisation of the charge by a pH change increases the ese rate back (M64E) or almost back (M44K) to the level of the *wt* protein.[26,27] Modelling made it plausible that introduction of a cysteine at position 42 in the amino acid chain (N42C mutation) would allow the construction of an S–S coupled dimer with a configuration of the monomers as found in the crystal and exhibiting, therefore,

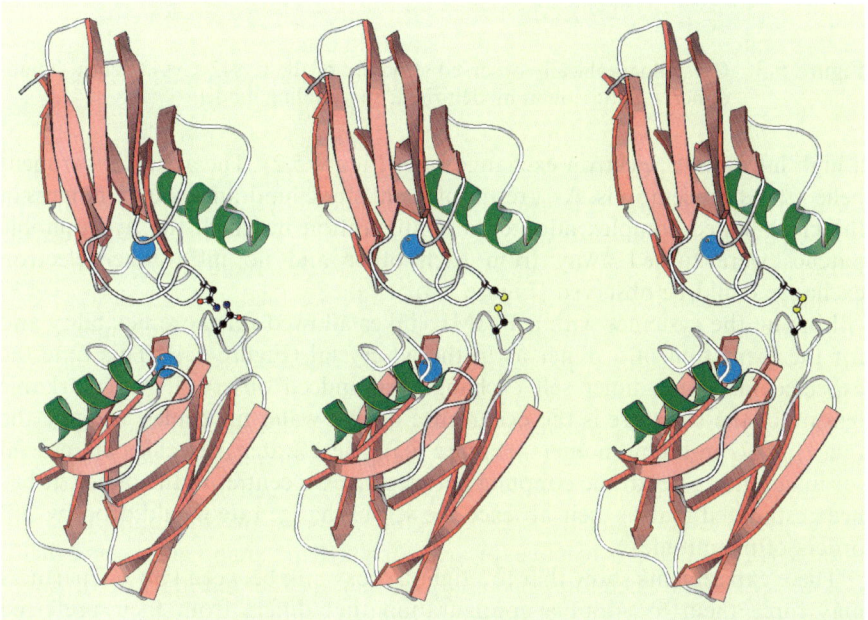

Figure 5.2 Left: ribbon structure of azurin from *P. aeruginosa* as observed in the crystal structure.[25] The Cu in the active site is coloured blue. The side chains of Asn42 in both molecules are represented in ball-and-stick format. Middle: model in which Asn42 has been replaced by a cysteine. Right: model in which the cysteines have been covalently connected.

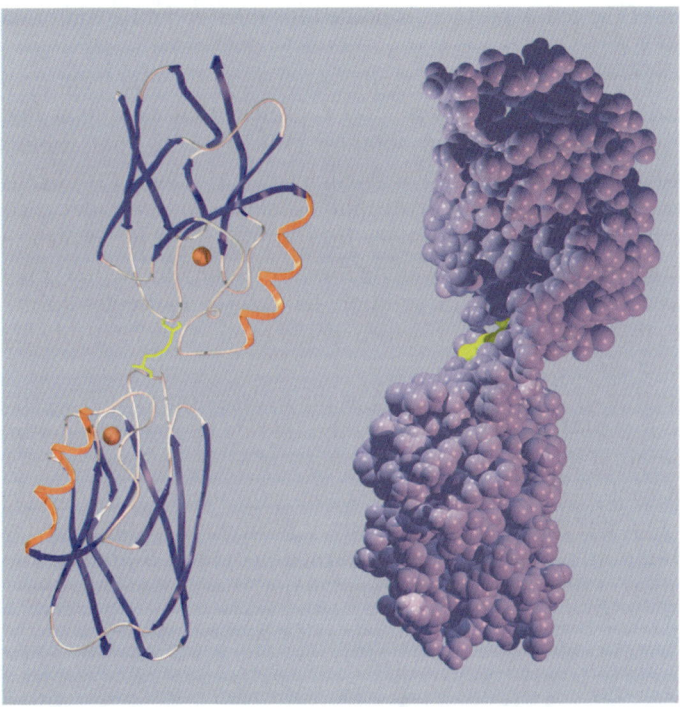

Figure 5.3 Crystallographically observed structure of the Cys42–Cys42 azurin homodimer.[29] Left: ribbon model; right: space-filling model.

a high intra-dimer electron exchange rate (Figure 5.2). The actual experiment belied these expectations. As a result of slight steric hindrance the monomers in the cross-linked complex adopted a configuration in which the hydrophobic patches were turned away from each other and no intra-dimer electron exchange could be observed (Figure 5.3).[28,29]

Linking the cysteines with a BMME linker allowed for more flexibility and for the formation of a dimer as in the *wt* crystal (Figure 5.4). This time the expected fast intra-dimer self exchange was indeed observed. A remarkable feature of the structure is the occurrence of two water molecules bridging the two His117s in the monomers. (Figure 5.5) These water molecules are crucial for maintaining electronic coupling between the Cu centres in the dimer. It has been estimated that in their absence the self exchange rate would drop by 2–3 orders of magnitude.[30]

These experiments show that too tight a cross-link between two *ET* partners may force them to adopt a configuration that differs from their preferred configuration in a transient complex. A tight cross-link may even lead to structural deformations as illustrated by the S118C dimer of azurin.[31] Serine 118 is located on the so-called Cu ligand loop in azurin next to Cu ligand His117. It exhibits limited solvent accessibility and when it is replaced by a cysteine this limited accessibility is reflected in the poor yield of dimers when

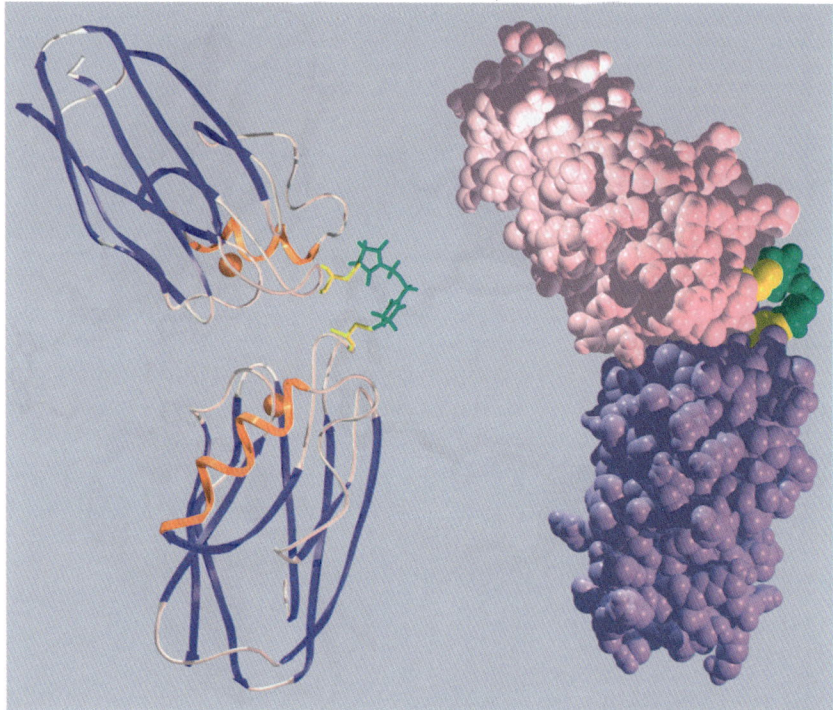

Figure 5.4 Crystallographically observed structure of the azurin homodimer in which the monomers have been connected by a *bis*-maleimidomethylether (BMME) linker attached to the Cys42 in each monomer.[29] Left: ribbon model; right: space-filling model.

one tries to oxidatively link the cysteines in two monomers. The stress that is exerted on the ligand loops by the Cys–Cys bridge causes a shift in the optical absorption maximum of the Cu(II) spectrum and a shift in the midpoint potential of the semi-reduced dimer of about 33 mV as compared to *wt* azurin.

The role of the hydrophobic patch in the *ET* mechanism was further investigated by studying N42C Azu dimers in which a surface charge had been applied in the hydrophobic patch (M64E). Again the directly S–S coupled dimer showed a configuration in which the hydrophobic patches of the monomers had rotated away from each other. The BMME coupled dimer, on the other hand, despite the presence of a charge in the hydrophobic patch exhibited the same conformation in the crystal as in the absence of the M64E mutation. Surprisingly the intra-dimer electron exchange was almost absent. It appeared tempting to correlate this with the lack of structured water in the interface as observed in the crystals.

In conclusion, the work on co-crystals and covalent complexes of redox partner proteins, in combination with what is known from more recent studies of transient complexes of electron transfer proteins (not reviewed here), confirms the architecture of the surface patches, the features of which are crucial

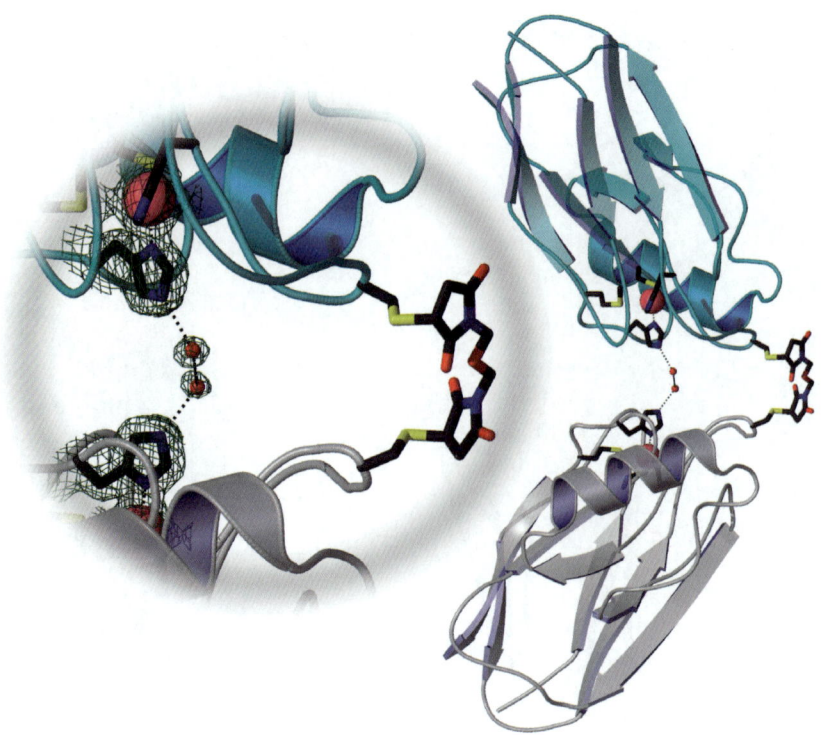

Figure 5.5 Close-up of the hydrophobic patch in the structure depicted in Figure 5.4. The Cu atoms have been coloured red and the side chains of the His117 ligands of the two coppers have been drawn as stick models. Notice the two water molecules (small red spheres) that are connected to each other and the Nε atoms of the histidines-117 by hydrogen bridges.[29]

for the formation of *ET* competent complexes: they consist of a central hydrophobic patch surrounded by a ring of polar residues with charged residues farther towards the rim. Packing of redox partners along their hydrophobic patches is relatively poor, leaving room for an appreciable number of mostly non-structured water molecules, some of which may play a crucial role in *ET*. When covalently linking *ET* partners it may be important to provide for some flexibility in the linker in order to allow the proteins in the complex to sample configurational space and find positions and mutual orientations that are favourable for fast *ET*. A tight link may even deform the partners and affect the mechanistic properties of the proteins in the complex.

5.3 Electron Transfer at Electrodes

For the application of proteins in electronic devices or for the electrochemical study of immobilised proteins the crucial parameter is the rate of electron exchange between electrode and protein. Two experimental approaches to determine this so-called heterogeneous rate are common. The first method, which has been the focus of considerable attention over the past few years, consists of immobilising the molecule of interest between two metal contacts either in solution or *in vacuo* and studying the *conductance* as a function of applied voltage. This method has been extensively explored for organic molecules while the application to proteins is slowly gaining ground. This topic will not be considered further here.

The second experimental approach makes use of *electrochemical* techniques. An electrochemical cell commonly consists of a working electrode, a counter electrode and a reference electrode (Figure 5.6). In 1977, both Eddowes and Hill[32] and Yeh and Kuwana[33] developed methods to prevent denaturation of proteins at the electrode surface (see Section 5.4). This led the way to the direct observation of electron exchange between an electrode and proteins, and of enzymatic substrate turnover as a function of applied potential.[34–37] However, diffusion of the enzyme to the electrode surface, and subsequent transient interactions with the surface prior to electron exchange, obscured the kinetic information.

A second breakthrough came with the discovery that many *ET* proteins and redox enzymes can be immobilised on an electrode surface in such a way that their redox-active centres undergo fast interfacial electron exchange, yet still

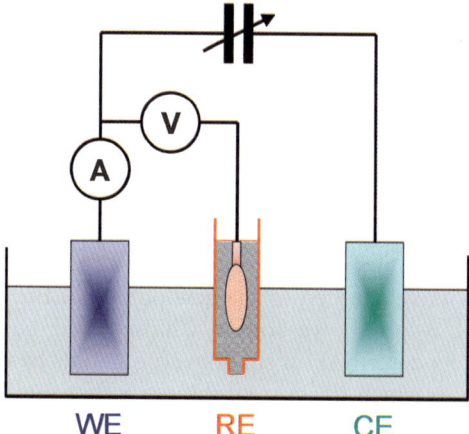

Figure 5.6 *Three-electrode electrochemical cell.* The redox protein is reduced or oxidized depending on the potential, V, between the working electrode (WE) and the reference electrode (RE). The potentials between the WE and the RE, and between the WE and the counter electrode (CE), which carries the current, are controlled by a potentiostat. The current that flows between CE and WE is measured while a feedback circuit maintains the desired (constant or time-dependent) potential between WE and RE. Only with (sub)micron-sized WEs, is the current small enough to be directly loaded onto the reference electrode without damage, and can the CE be left out.

exhibit their native biological activity. Immobilisation removes limitations due to protein diffusion, thus bringing into focus the intrinsic electron-transfer characteristics and coupled reactions.[38–45] In essence, the enzyme molecules are "wired" to the electronic circuit. Moreover, enzyme molecules stay in direct contact with the solution, ensuring optimal accessibility of the active sites to substrate. The electrode surface can be tailored to suit a specific protein by the formation of a densely packed but thin monolayer of a bifunctional linker or promoter, as described in Section 5.4.

5.3.1 Voltammetry

Recording the current as a function of the potential in a cyclic manner (*cyclic voltammetry* or "CV") results in a voltammogram that exhibits peaks in the forward and backward direction around the midpoint potential of the redox active group (Figure 5.7). With increasing sweep rate the peaks start moving apart. Plotting the distance between them as a function of sweep rate results in a "trumpet plot", the shape of which provides the electron transfer rate between electrode and redox active group according to Laviron's theory (Figure 5.8).[46,47] With an ac version of this technique somewhat higher ET rates can be measured.[48]

Measurement of the current as a function of both time and applied voltage yields a detailed kinetic and thermodynamic landscape of the enzyme that cannot easily be obtained by spectroscopic methods alone. Moreover, the immobilised enzyme can be rapidly transferred between solutions, resulting in "instant dialysis" of critical components such as ligands, substrates, inhibitors, buffer pH or ionic strength.

5.3.2 Chronoamperometry

In a *chronoamperometric* experiment the protein is first poised at a fixed potential, say its midpoint potential, E^0, which is subsequently changed stepwise to a new value V. Current will flow until the concentrations of oxidised and reduced species have adapted to the new potential. For the description of the electrochemical response we follow Chidsey and coworkers.[49,50] This response is adequately described for small overpotentials by the Butler–Volmer equation according to which the current decays with a rate given by

$$k = k_f + k_b = k_0 \exp[\alpha e(V - E^0)/k_B T] + k_0 \exp[-(1 - \alpha)e(V - E^0)/k_B T] \quad (1)$$

where k is the sum of the forward and backward reaction, k_f and k_b, and α is the transfer coefficient.[51] The method and the theoretical analysis apply both to immobilised proteins as well as to simple molecules containing a redox active group. With the customary value of $\alpha = 0.5$[51] the expression for k becomes symmetrical around $V = E^0$ and on a semi-logarithmic scale predicts a straight relationship between k and overpotential, in agreement with observations made

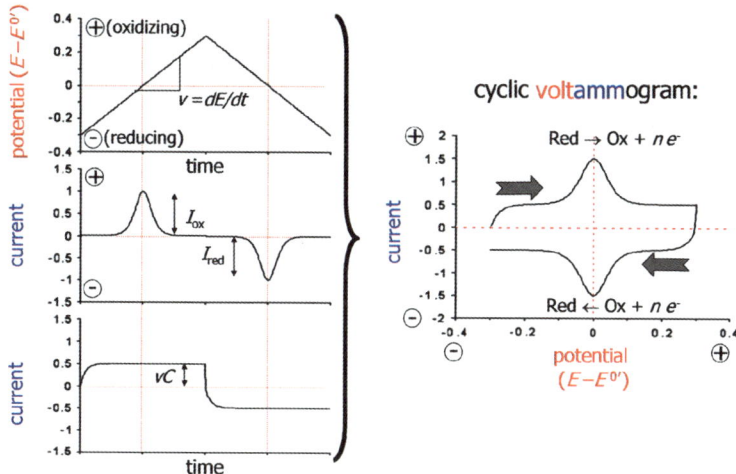

Figure 5.7 Measurement of a cyclic voltammogram, recorded with a surface-confined monolayer of a redox species in an apparatus as depicted in Figure 5.8. The potential between the working and reference electrodes (E) is ramped back and forth with a rate v around the midpoint potential ($E^{0\prime}$) of the redox species (top left). This results in two charging processes: 1) a Faradaic process, *i.e.* the oxidation of the redox molecules upon sweeping from low to high potential and reduction back on the return. This results in a positive and a negative current peak at $E = E^{0\prime}$ (middle left); 2) the charging of the electrode surface and the ensuing accumulation of counter ions at the surface. This results in a diffuse double layer with a capacitance C. Charging by a linear voltage sweep results in a constant current $I = vC$. Due to the ionic resistance of the solution (in series with the double layer capacitance), a characteristic rise-time (RC-time) is observed at the turning points (bottom left). The cyclic voltammogram thus is the sum of these two currents, plotted as a function of the applied potential at the right. The width of the Faradaic peaks is $3.5 kT/n$, where n is the number of electrons per molecule involved in the redox reaction. The area under the peak is proportional to the amount of adsorbed redox species. The oxidation and reduction peaks are both centred at $E = E^{0\prime}$ at low sweep rates, but start to separate at high sweep rates, and the interfacial electron transfer rate can be deduced from a plot of the peak positions as function of log(v) ("trumpet plot", see Figure 5.8).

for small overpotentials (Figure 5.9). At high overpotentials, however, pronounced curvature in the ln k vs. ($V - E^0$) plots becomes noticeable, experimentally, at variance with the predictions of eqn (1).

An improved expression covering the range of zero to large overpotentials is obtained by considering that when ET between protein and electrode is governed by weak electronic coupling, the ET reaction takes place in the non-adiabatic limit and can be described by Marcus theory.[50] Because the electronic coupling occurs between the redox centre and a continuum of metal states the total probability for ET is obtained by integrating over the collection of metallic states. The final expressions for forward and back ET

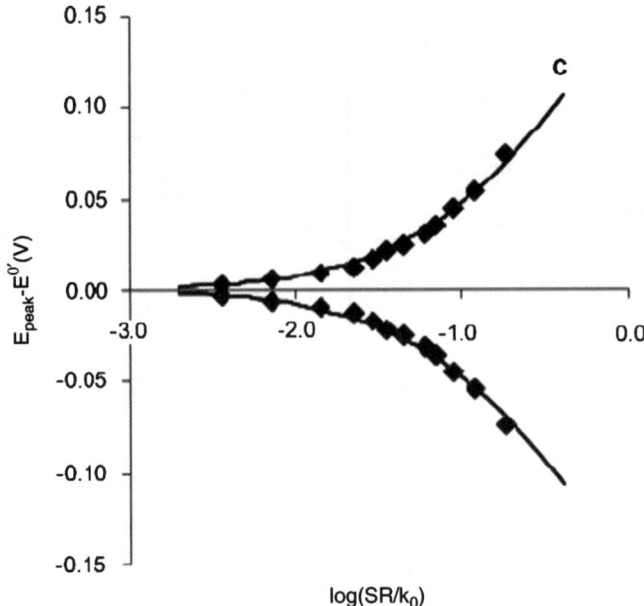

Figure 5.8 Example of a trumpet plot measured on cytochrome-*c* immobilised through an undecane-pyridyl linker on a decanethiol SAM on gold. The peak separation in the cyclic voltammogram is plotted *vs.* the log of the scan rate (SR) divided by the standard electron transfer rate k_0.[95]

rates are given by[48,49]

$$k_{f,b} = (H_{DA}^2 \rho/\hbar)(\pi/\lambda k_B T)^{1/2} \\ \times \int \exp - [\{(\varepsilon - \varepsilon_F) - (\lambda \pm \Delta G)^2\}/(4\lambda k_B T)] f_F(\varepsilon) d\varepsilon \quad (2)$$

in which H_{DA} is the electronic coupling between electrode and redox centre, ρ is the density of states in the metal, λ is the reorganisation energy and ΔG is the driving force

$$\Delta G = e(V - E^0)$$

The integral over ε derives from the integration over all metal states. Fermi–Dirac statistics is taken into account by the presence of the Fermi function, $f_F(\varepsilon)$, in the integral

$$f_F(\varepsilon) = [1 + \exp(\varepsilon - \varepsilon_F)/k_B T]^{-1}$$

with ε_F being the Fermi energy. ρ and V are taken independent of the electronic states of the metal. The fit with experiment now is satisfactory (Figure 5.9).

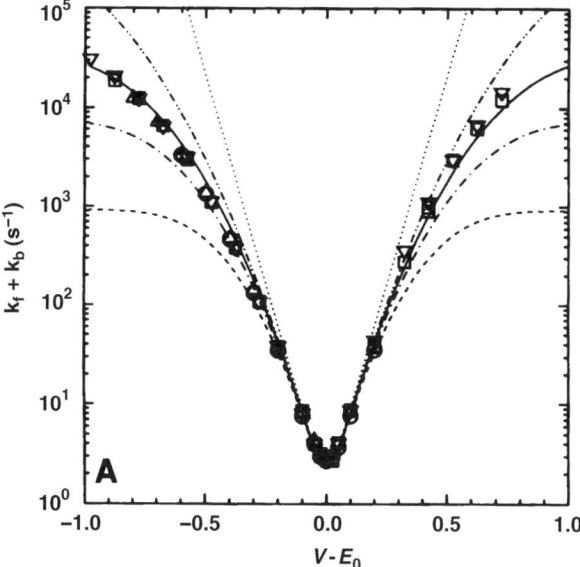

Figure 5.9 Sum of the forward and backward rates of *ET* between a Au electrode and a ferrocene group tethered by a linker to the Au surface as a function of overpotential $(V - E^0)$ (Tafel plot). Straight dotted line: rate according to eqn (1), with $k_0 = 1.25\,\text{s}^{-1}$ and $\alpha = 0.5$. Other lines calculated according to eqn (3) for various values of the matrix coupling element, H_{DA}, and the reorganisation energy, λ. Symbols: experimental points observed at 25 °C.[49]

Contrary to the case of *ET* between two redox centres where the *ET* rate drops at high overpotential ('Marcus inverted region'), here the rate reaches a plateau at high overpotential. This reflects the presence of a continuum of empty states in the metal to which the electron may transfer. The standard electron transfer rate, k_0, obtains when the applied potential equals the midpoint potential, $V = E^0$, and is given by

$$k_0 = (H_{DA}^2 \rho/\hbar)(\pi k_B T/\lambda)^{1/2} \int \exp{-[(x - \lambda^*)^2/4\lambda^*]}/[1 + \exp(x)]\,dx \quad (3)$$

with $\lambda^* = \lambda/k_B T$ and in which the substitution $(\varepsilon - \varepsilon_F)/k_B T \to x$ has been applied.

When H_{DA} is large enough so that the adiabatic limit applies, the transfer rate becomes independent of the strength of the coupling and instead depends on the rate of the nuclear reorganisation (often solvent reorganisation) that accompanies electron transfer. The rate equation then becomes[48]

$$k_0 = \kappa_{el} \nu_n \int \exp{-[(x - \lambda^*)^2/4\lambda^*]}/[1 + \exp(x)]\,dx \quad (4)$$

with κ_{el} an electronic transmission factor which represents the chance for barrier crossing once the electron has reached the barrier and ν_n represents the (nuclear) frequency by which the top of the barrier is reached. The value of κ_{el} is close to 1 and for practical purposes is set equal to this value.

An indication about whether the adiabatic or non-adiabatic regime applies can be obtained according to Sutin[52,53] by comparing the electronic frequency ν_{el}, given by

$$\nu_{el} = (H_{DA}^2/\hbar)(\pi/\lambda k_B T)^{1/2}$$

with ν_n. With $\nu_{el} \gg \nu_n$, electron transfer is adiabatic, with $\nu_{el} \ll \nu_n$ non-adiabatic. For instance, with $\lambda = 0.5$ eV, $\nu_n = 10^{12}$ Hz and $T = 300$ K the critical value of H_{DA} that distinguishes the two regimes is about 50 cm^{-1}. Slightly different criteria to distinguish the adiabatic from the non-adiabatic limit have been suggested,[54,55] but in practice the numerical outcome does not differ much from what is calculated here.

5.4 Surface Preparation

5.4.1 Carbon

The most common electrode materials for protein electrochemistry are carbon and gold and, to a lesser extent, semi-conductors like indium tin oxide (ITO) and silicon. The surface of basal plane graphite is hydrophobic while edge plane graphite exhibits oxygen functionalities like carboxyl and carbonyl groups, hydroxides, phenols, quinols, *etc.* (see Figure 5.10). The basal plane of graphite is suitable for direct contact with hydrophobic protein surfaces, the edge plane for contact with polar and charged (patches of) protein surfaces. Other carbon materials used for electrodes are glassy carbon (GC), highly oriented pyrolytic graphite (HOPG), fullerenes (*e.g.* C_{60}, carbon nanotubes), carbon paste and even boron-doped diamond. Sometimes 'promoter' molecules (also known as 'facilitators') are used to ease the immobilisation of the protein onto the carbon surface. Promoters are usually small molecules, like oligosaccharides, multivalent metal ions or organic multivalent ions such as aminoglycosides (*e.g.* neomycin) and peptides (*e.g.* polymyxin, polylysin) (see Figure 5.11) that are not redox active but modify the surface characteristics of the carbon so that a stable protein film is formed on the electrode.[35,36,56]

Ideally, a promoter stabilises the protein on the electrode, prevents its denaturation and promotes *ET* by optimising the orientation of the protein. A mono-functional modifier interacts only with the electrode to render it inert or to compensate the surface charge, a bifunctional modifier interacts with the electrode as well as the protein.

Due to the many carboxylate groups, the estimated pK of oxygen-rich edge-plane graphite is 5.6.[34] The resulting overall negative charge at physiological pH of these electrodes is optimal for direct, unmediated electrochemistry of positively charged proteins.[34,57] However, by simply adding multivalent positive ions, negatively charged electrodes can also be made suitable for direct electrochemistry of negatively charged proteins. The cations bridge between the

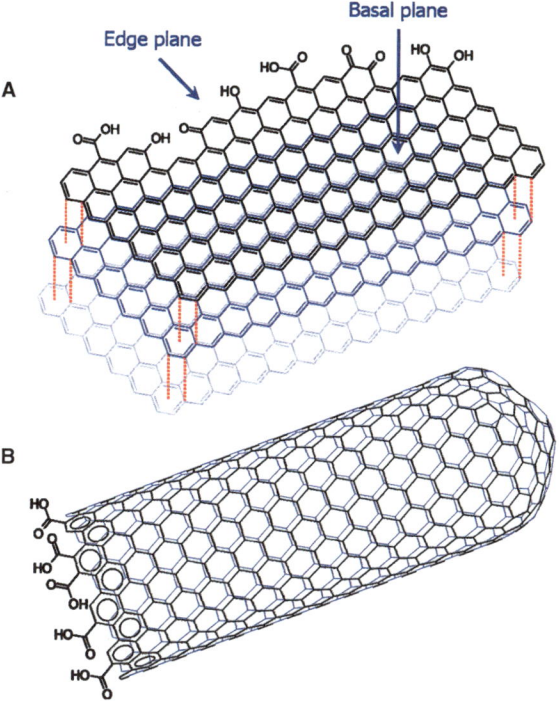

Figure 5.10 Some carbon materials used for electrodes. A. *Graphite*: the edge of the upper sheet contains examples of oxygen functionalities that are created when the material is cut perpendicular to the sheets. B. *Carbon nanotube*: example of a cut single-walled nanotube (metallic "armchair" type) with carboxylic acid functionalities. The sp^2 character is only indicated for the edge but extends throughout the nanotube.

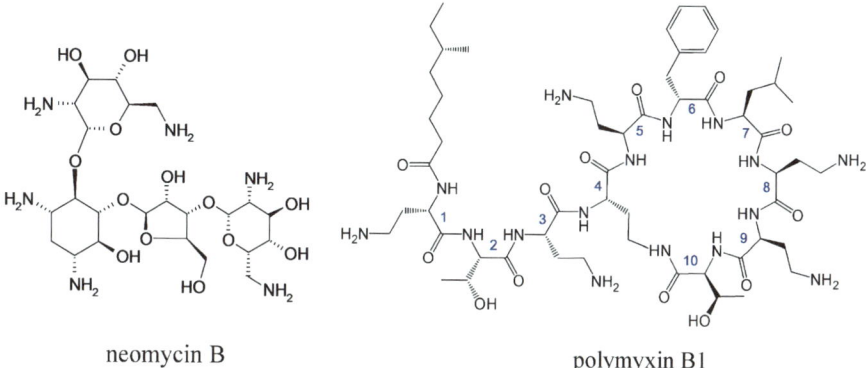

neomycin B   polymyxin B1

Figure 5.11 Examples of promoters that have been used successfully to modify electrode surfaces and make them suitable for immobilisation of proteins.

negatively charged surfaces of both electrode and protein.[34] Although specific interactions (H-bonding, salt-bridging) may play a role, these are not crucial, because multivalent counter ions ($|Z| \geq 3$) tend to form a spatially correlated double layer that can overcompensate the original charge, resulting in charge-inversion even at sub-millimolar concentrations.[58,59]

5.4.2 Gold

Bare gold may cause denaturation of proteins and efficient electronic contact between metal and protein is difficult to establish although there have been reports of electrochemistry, for instance of cytochrome-c and azurin directly on gold or indium tin oxide (ITO).[60–62] The Au surface needs to be modified, therefore. The classical example is the report in the late 1970s of the first reversible voltammograms of cytochrome-c on 4,4-bipyridyl-modified gold.[32,33] For the modification of Au electrodes nowadays often self-assembled monolayers (SAMs) are employed. An extensive review on the subject by Whitesides and coworkers is available[63]. n-Alkanes with a thiol group at one end and a methyl, carboxyl or hydroxyl functionality at the other end have been used frequently to create a hydrophobic, negatively charged or hydrophilic surface, respectively. The carboxyl charge can be reversed by supplying electrolyte containing doubly charged metal ions, like Mg^{2+}.[64,65] Phosphonate or imidazole have been used as end groups as well.[66] An interesting variation consists of the activation of a SAM with chelating groups like 1-acetato-4-benzyl-triaza-cyclononane.[67] In the presence of Ni it forms a stable bond with a His-tag and in this way can assist in the immobilisation of His-tagged proteins on SAMs (see Figure 5.12). SAMs on Au have been constructed also from linear molecules derived from adamantane[68], conjugated aryl thiols[69,70] and oligophenylene-vinylene (OPV).[71,72]

5.4.3 Other Methods

For surface preparation of ITO and silicon surfaces often silane-based chemistry is used (Figure 5.13)[73–76] although direct immobilisation of proteins[77,78] and functionalisation of ITO by aziridine have also been successful.[79]

Other techniques to prepare electrode surfaces relate to the use of 'molecular landers'[80] and the application of a combination of DNA and streptavidin/biotin technology.[81] Short amphiphylic synthetic peptides have been applied to modify the weakly negatively charged surface of SiO_2.[82] These peptides exhibit a tendency to form either α-helical or β-sheet structures on the SiO_2 surface depending on concentration (see Figure 5.14). The strength of the binding to the SiO_2 surface can be regulated by the pH. A different class of promising amphipathic peptides are formed by the 'hydrophobins', which can be applied to change surface properties from hydrophobic to hydrophilic and vice versa.[83–85] The electrode surface can also be modified by the absorption of

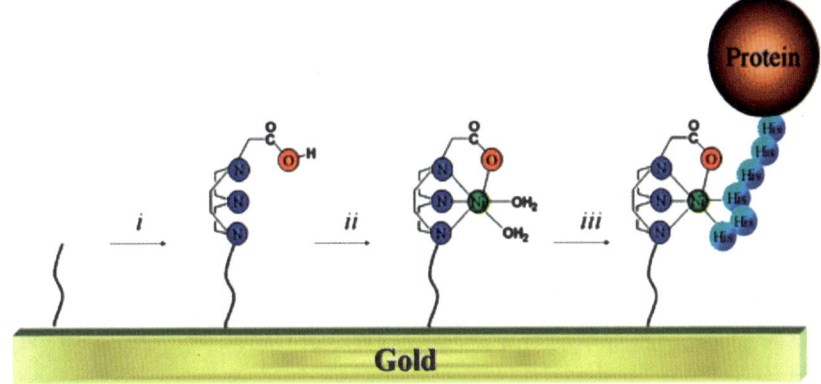

Figure 5.12 The chelator 1-acetato-4-benzyl-triazacyclononane is immobilised on a Au surface (step *i*), incubated with Ni^{2+} (step *ii*) after which a His-tagged protein binds to the Ni (step *iii*).[67]

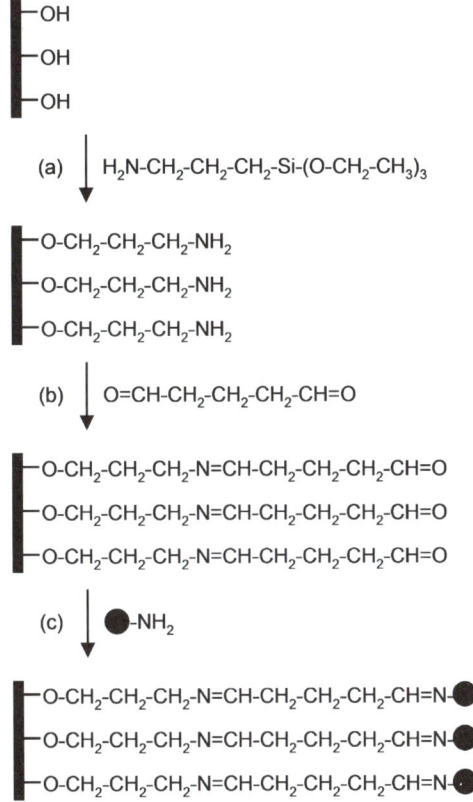

Figure 5.13 Example of a scheme to immobilise a protein (black dot) carrying an NH_2-functionality on an SiO_2 surface by means of silane chemistry. After Pompa et al.[75]

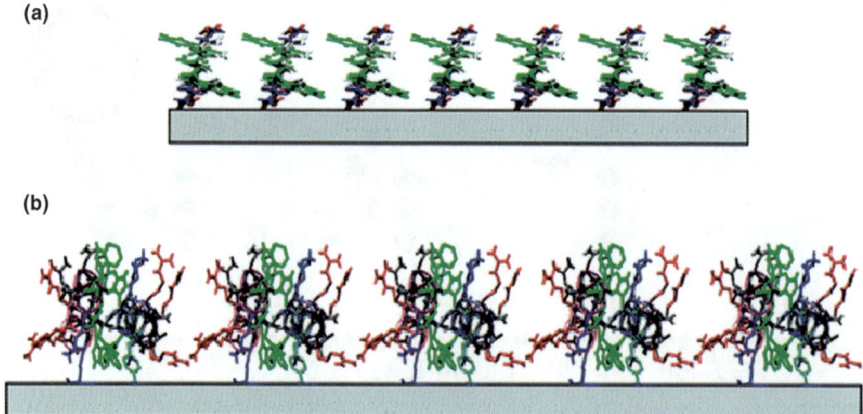

Figure 5.14 Peptides with the amino acid sequence W-V-N-A-K-Q-Y-W-R-I-L-K-R-R-W adsorbed on an SiO_2 surface by the interaction of the negatively charged groups of the substrate and a positively charged side chain (Arg (R)) and a Tyr (Y) side chain of the peptides at low (0.01 wt %) and high (0.1 wt %) concentration of peptide. Colour code: green: W; red: R; blue: K; cyan: Y. a). β-sheet conformation; b). α-helix conformation; peptides cluster in the form of dimers in the latter case.[82]

carefully engineered peptides. A few cysteines suffice to bind the peptide to a Au surface (see Figure 5.15) while the character of the other side chains in the peptide can be adapted to the special application for which the peptide is intended.[86] Surfactants and lipid membranes have been used successfully also to adapt solid surfaces to protein immobilisation.[87,88]

Finally, an electrode surface can be made 'friendly' to a particular enzyme by immobilising its physiological partner on the electrode (see Figure 5.16). In this case the enzyme does not necessarily have to be immobilised but the immobilised partner protein should be in conducting contact with the electrode.[89] This can be achieved either by connecting the partner protein to the surface by a linker allowing for sufficient flexibility so that the partner protein may act as a shuttle between electrode and enzyme, or by immobilising the partner protein on the electrode so that it allows for efficient electron exchange with the electrode without blocking the recognition site for the enzyme.[90]

5.5 Immobilisation

Immobilisation of a protein can be effected by bringing the protein into direct contact with a bare or prepared electrode. This is discussed in Section 5.5.1. Section 5.5.2 discusses the conducting properties of 'wires'. Proteins can also be 'wired' onto electrodes; Section 5.5.3 deals with this topic.

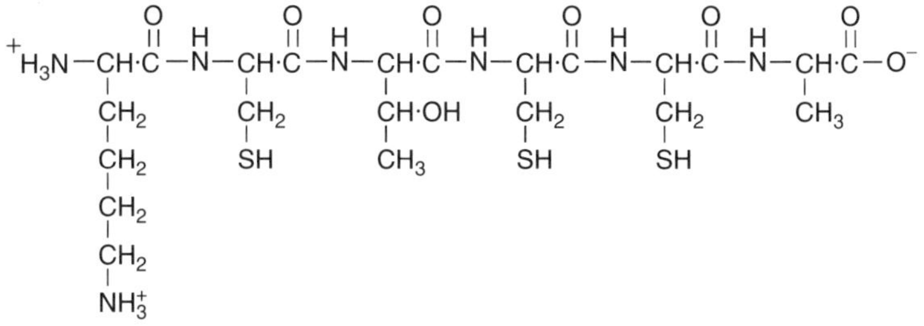

1. Lys-Cys-Thr-Cys-Cys-Ala

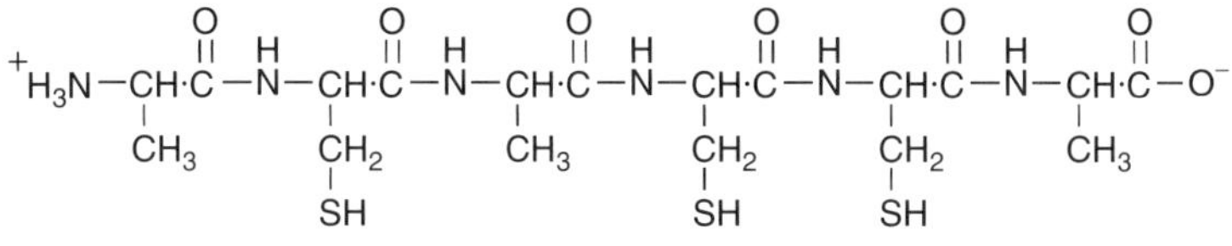

2. Ala-Cys-Ala-Cys-Cys-Ala

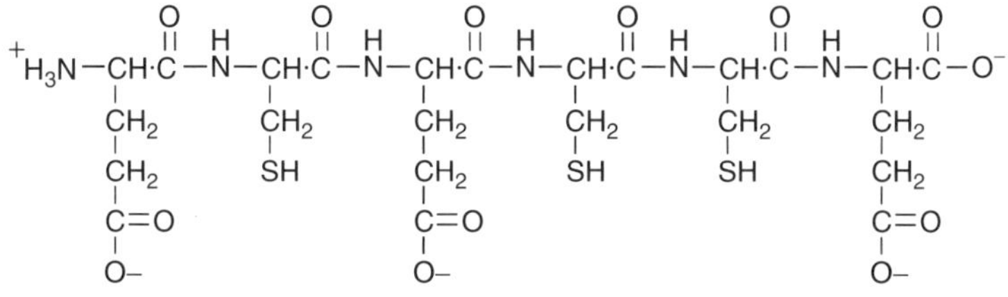

3. Glu-Cys-Glu-Cys-Cys-Glu

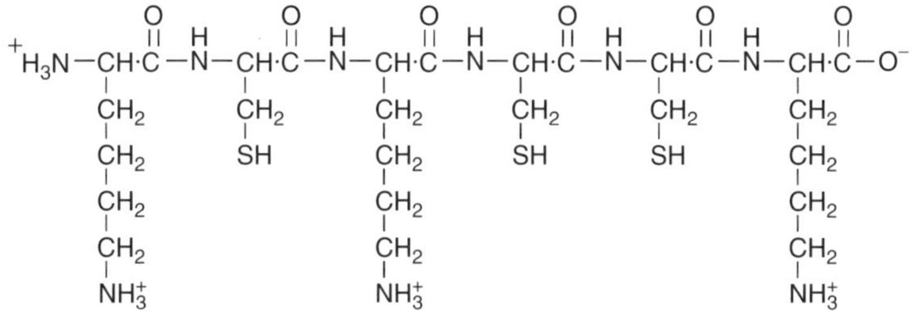

4. Lys-Cys-Lys-Cys-Cys-Lys

Figure 5.15 Examples of peptides used to modify Au electrode surfaces. Cysteines connect the peptides to the Au surface. The surface offered by the peptides after immobilisation is positive (peptide 4), negative (peptide 3), neutral (peptide 2) or of mixed character (peptide 1). After Y. Astier *et al.*[86]

5.5.1 Direct Immobilisation

In Section 5.4 it was explained how electrode surfaces can be modified to make them suitable for protein immobilisation. In particular, electrodes prepared

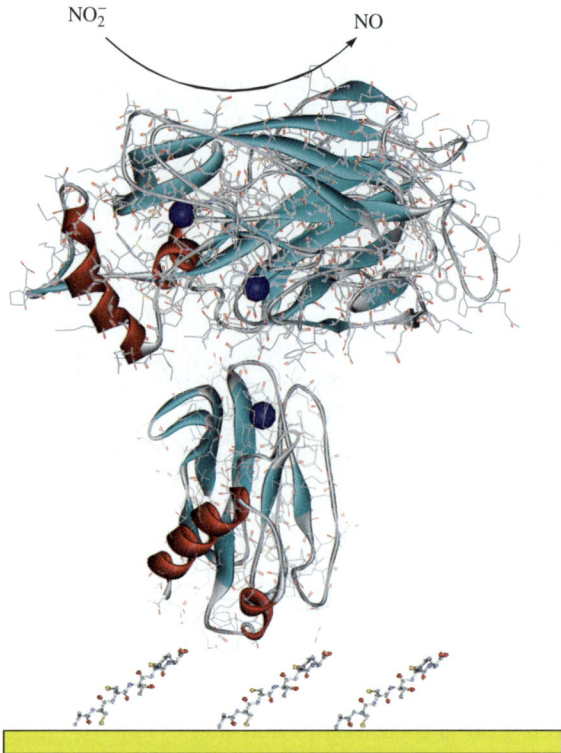

Figure 5.16 Cartoon of nitrite reductase (NiR, top molecule) in contact with its redox partner, pseudo-azurin, on peptide modified Au. NiR is a homotrimer of which only the monomer has been depicted.[89]

with SAMs have been very popular for this purpose. When a protein is immobilised on an electrode covered with a SAM the question is how efficient *ET* between protein and electrode still can be. After all, the protein has to exchange electrons with the electrode across an isolating SAM layer. It has been found that the electron transfer rate between protein and electrode increases exponentially with decreasing thickness of the SAM when using *n*-alkane thiols, down to $n \approx 5-6$[61,91–95]. The exponential decay coefficient amounts to $\beta_n = 1.0 - 1.09$[92–94] in agreement with data obtained for the single molecule conductivity of the *n*-alkanes.[48,50,96–99] For shorter spacers the *ET* rate levels off (Figure 5.17). This has been interpreted as being due to conformational gating or frictional control of the *ET* process.[45,95]

Instead of or in addition to modifying the electrode surface it is also possible to make use of the characteristics of the protein surface or to modify them. For instance, when the negatively charged edge plane of a graphite electrode is used to immobilise an *ET* protein, positive charges – either naturally occurring or engineered – in the surface area that is used by the protein for electron transfer

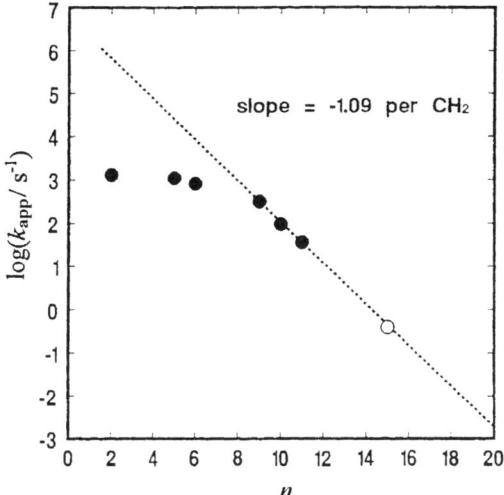

Figure 5.17 Standard heterogeneous ET rates between horse heart cytochrome-c and a Au electrode across SAMs consisting of n-alkane thiols. The length of the alkanes is expressed in the number of CH_2 units, n.[93]

(like the 'hydrophobic patch' in many cases) may promote ET at the electrode.[60] Likewise introduction of a negative charge will slow down the ET reaction.[26] When gold is used as an electrode, cysteines, introduced into the surface by site-directed mutagenesis, may directly couple to the Au surface.[61] In a similar way engineered Cys–Cys bridges have been used for immobilisation of proteins on gold.[100,101]

5.5.2 Wires

When trying to establish conducting contact between a protein and an electrode by means of an organic molecule an obvious question is how much of an advantage conjugated linkers offer over saturated linkers. As an example we compare n-alkanes with oligo-polyphenyl-ethynylenes (OPEs) and oligo-polyphenyl-vinylenes (OPVs) (Figure 5.18). The properties of these molecules have been investigated in a number of ways:

- by attaching a chemically reactive group at one end (like a sulfhydryl, diazonium[102] or carboxyl group[73,103]) and a redox active group like ferrocene[49,50,104–107] or a Ru-compound[107,108] at the other end. The linker is embedded in an n-alkane SAM and the rate of ET between active group and a Au or glassy carbon electrode is established by electrochemical techniques like CV, ac voltammetry,[48] chronoamperometry or the ILIT technique;[50,105,109]
- by immobilising the linkers in a membrane with nanopores and studying their conductivity;[110,111]

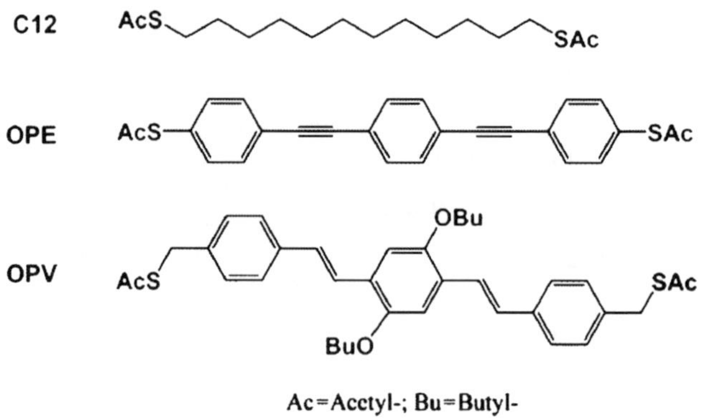

Figure 5.18 Examples of OPE and OPV molecules with $n=3$, compared with a molecule of n-decane. The thiol head groups are protected by acetyl groups and can be activated under alkaline conditions.[111]

- by immobilising the linkers covalently between gold contacts and measuring their i-V characteristics;[112,113]
- by scanning tunnelling spectroscopy (STS) in combination with the two-layer junction model (2LJM) technique whereby the linker is embedded in a SAM of a slightly shorter alkanethiol and the height difference between alkane and linker is measured with an STM tip in constant current mode.[108,114,115]

Reports in which the conducting properties of n-alkanes, OPEs and OPVs are measured and compared in the same study almost exclusively refer to investigations by electrochemical techniques. The data seem to exhibit less spread than the results from studies by the other techniques mentioned above.[116] We therefore limit the following discussion to data derived from electrochemical studies only and consider molecules that were provided at one end with a thiol group and at the other end with a ferrocene (Fc) group. The thiol was bound to a gold surface and the reduction/oxidation of the Fc was studied as a function of the potential of the Au electrode with respect to a reference electrode.

The k_0 values, i.e. the rates at zero overpotential, for ET between the gold electrode and the Fc group are summarised in Figure 5.19 where k_0 is presented as a function of the distance, r, between the gold surface and Fc. In agreement with theory k_0 exhibits an exponential dependence on distance for the n-alkanes and the OPEs

$$k_0 = A\exp(-\beta_r r) = A\exp(-\beta_n n) = A\exp(-\beta_h h)$$

in which r is the length of the molecule measured parallel to the molecular axis (which in most cases equals the distance between the two ends of the molecule),

Activating Redox Enzymes through Immobilisation and Wiring 141

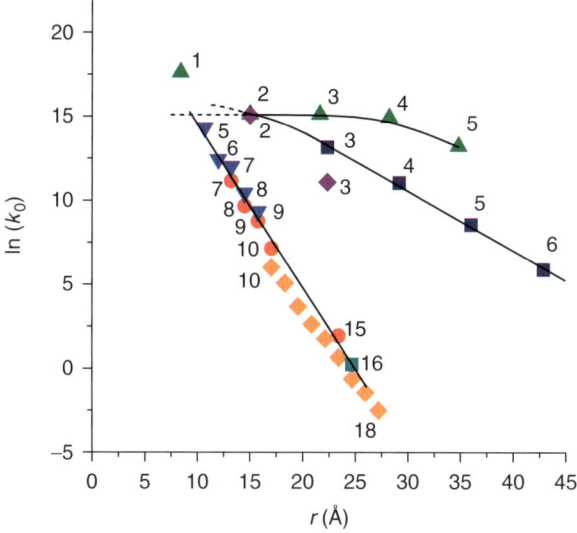

Figure 5.19 Natural logarithm of standard rate of *ET*, k_0, between a Au electrode and a ferrocene head group connected to the Au surface through *n*-alkane-, OPV- and OPE-thiol linkers *vs.* linker length, *r*, in Å. Green triangles: OPV with $n = 1–5$;[105] dark blue squares: OPE with $n = 3–6$;[48] purple diamonds: OPE $n = 2,3$;[97] blue inverted triangles: *n*-alkanes with $n = 5–9$;[50] red circles: *n*-alkanes with $n = 7–10, 15$;[99] blue-green square: *n*-alkane with $n = 16$;[49] orange diamonds: *n*-alkanes with $n = 10–18$.[96] In the latter case the k_0 data were taken from Figure 8 in the cited reference, in all other cases k_0 values were taken directly from the tables or text in the cited references. Distances were calculated with $r = n^* l_{RU} + b_{CS} + b_{Fc}$, with l_{RU} the repeat unit length (6.84, 6.60 and 1.27 Å for OPE, OPV and *n*-alkane, respectively),[48,105] b_{CS} the C–S bond length (1.79 Å)[48] and b_{Fc} an additional contribution for the connection between the linker and the Fc (0, 0 and 2.54 Å for OPE, OPV and *n*-alkane, respectively).[48] In the case of the *n*-alkanes the Fc unit has been connected to the *n*-alkane by an amide[99] or a carboxylester bond.[49,50] Numbers indicate the number of repeat units (*n*). The straight line for the *n*-alkane points is a least squares fit to all the points except the orange diamonds. The straight line through the $n = 3–6$ OPE points (dark blue squares) is a least squares fit. Its extension to the $n = 2$ point has been drawn by hand to guide the eye. The line through the $n = 2–5$ OPV points is cubic spline fit. Notice that the OPV and OPE points for $n = 2$ almost coincide.

n is the number of repeating units along the molecule (when applicable), *h* is the height measured from the top of the molecule to the surface in a direction perpendicular to the surface and β carries the corresponding subscript. In the case of the *n*-alkanes it follows from simple geometric considerations (C–C distance of 1.54 Å, C–C–C angle 109.5°[117]) that $\beta_n/\beta_r = 1.21$. When the molecular axis makes an angle α with the normal to the substrate plane $\beta_r/\beta_h = \cos \alpha$.

From Figure 5.19 one finds $\beta_n = 1.21$ or $\beta_r = 0.9 \text{ Å}^{-1}$ for the *n*-alkanes and $\beta_r = 0.36 \text{ Å}^{-1}$ for the OPEs. These values are in agreement with literature reports where values of $\beta_n = 1.0$–1.2 or $\beta_r = 0.83$–0.99 Å$^{-1,48,50,99,118,119}$ and of $\beta_r = 0.28$–0.57 Å$^{-1,97,120}$ have been quoted for the alkanes and OPEs, respectively. Various data sets have been presented in Figure 5.19 for the *n*-alkanes. The two main data sets exhibit regression lines with slightly different slopes but on the whole exhibit reasonable agreement. The OPE data sets by Sachs *et al.*[97] and Creager *et al.*[48] differ substantially for $n = 3$ for which there seems no obvious explanation.

For OPVs the distance dependence is much shallower and almost non-existent up to 5 OPV units (34 Å).[105] The reason for this distance insensitivity is that the electron transfer no longer takes place in the non-adiabatic but does so in the adiabatic limit and thereby loses its sensitivity to the donor/acceptor matrix coupling element, H_{DA}. This is the parameter that is responsible for the exponential distance dependence of the rate.

That the adiabatic limit applies can be seen by considering, for instance, the observed *ET* rate of $k_0 = 3 \times 10^6$ s^{-1} for the OPV with 4 PV-units.[48,105] Assuming the *ET* would be non-adiabatic, then with eqn (3) and with $\rho = 0.3$ eV^{-1} atom^{-1} and $\lambda = 850$ meV the magnitude of the matrix coupling element would amount to at least 50 cm^{-1}.[99,105] This is too large to be compatible with non-adiabatic *ET*. It is therefore of interest to analyse the data in the *adiabatic* limit. The rate observed for OPV with $n = 1$ amounts to about 5×10^7 s^{-1} (Figure 5.19).[105] With $\rho = 0.3$ eV^{-1} atom^{-1}, $\lambda = 850$ meV and using eqn (4) with $\kappa_{el} = 1$ one finds for v_n a value of 10^{11} s^{-1}, which has been tentatively related to the reorientation of the *Fc* group at the SAM/water interface[105] (see also references[72,121]). It is of interest to note that the rate observed for this OPV falls out of line with the other OPV points (Figure 5.19). Since the OPV is embedded in a SAM of relatively short alkanes the high k_0 value may signal a different activation energy and/or a higher value for $\kappa_{el} v_n$ in eqn (4) than for the other OPV molecules as a result of increased motional flexibility of the SAM and the OPV molecule. In more recent work it has been suggested that in the case of the *n*-alkane linkers structural disorder might limit the rate of *ET* between the *Fc* and the Au.[106,107] An alternative explanation is that transitions between conducting and non-conducting conformations might cause a 'gating effect' that becomes rate limiting at short alkane lengths.[106] Variations in the distribution of backbone torsion angles might be an important factor that limits the conductivity of OPE linkers.[107]

The conducting properties of *n*-alkanes, OPEs and OPVs have also been measured by the nanopore method.[111] For $n = 4$ OPV was found to conduct 10 times better than OPE and 1000 times better than *n*-dodecane. According to Figure 5.19 the numbers are 55 and 670 times, respectively. The nanopore data refer to ensembles of 1000–4000 molecules, the number of which had to be estimated and this may have affected the precision of the data and may explain the difference with the data in Figure 5.19. The *ET* rate has also been measured by means of the *2LJM* method for OPE (2 PE units) with a Ru coordination compound instead of *Fc* at the end of the linker.[108] Two values have been quoted for β (presumably β_h), of 0.88 and 0.59 Å^{-1}, depending on the atom of the Ru compound that was contacted in the measurements. Since there is uncertainty about the angle between the OPE and the surface it is not certain how these values compare with the data presented above (Figure 5.19).

5.5.3 Wiring Proteins

5.5.3.1 "Hot Wiring"

The natural application of conducting molecules is to use them as single wires to connect an enzyme with the electrode. This can be combined with a SAM in which the wire is embedded. The obvious position to anchor the wire onto is somewhere close to or inside the active site of the enzyme or redox protein.[122]

One of the earliest examples demonstrating the feasibility of this approach is provided by a mutated variant of azurin from *P. aeruginosa*.[123,124] By mutagenesis one of the two histidine ligands of the Cu in the active site was replaced by a glycine. In the native protein this histidine provides electronic access to the Cu.[26,27,125] The replacement by a glycine creates a cavity by which the Cu now is directly accessible from the outside. Insertion of an imidazole (*Imz*) or a pyridine (*Py*) restores the spectral and, to some extent, mechanistic characteristics of the blue copper site.[123,124] By attaching a chain-like molecule to the *Imz* or *Py* the protein can be wired. For instance, by using (α,ω)bis-*Imz* n-alkanes it appeared possible to link two molecules of this azurin variant.[126] The X-ray structure of the linked dimer was recently reported (see Figure 5.20).[127] A similar type of modification has been reported for the copper-containing nitrite reductase (NiR) from *Alcaligenes faecalis*-S6.[128]

However, the accessibility of the active site for substrates or partner proteins once such a "hot wire" has been installed remains a concern. A more practicable example is provided by the work of Willner, who used a derivatised flavin to immobilise apo-glucose dehydrogenase (GDH) onto an electrode.[129] The flavin is the natural cofactor of the GDH and reconstitution with the derivatised flavin produced an active enzyme. The construct has been used in the design of glucose sensors.[130] This strategy can be extended by attaching Au nanoparticles to the flavin derivative to extend into or around the curvature of the immobilised enzyme, in effect reducing the distance between the surface and the active site of the enzyme.[131,132]

A third variant on this scheme has been explored by Gray and coworkers,[133] who showed that a diethylaniline inhibitor of amine oxidase can be provided with a conducting OPE wire and attached to an electrode without the enzyme losing its ability to communicate with the electrode (Figure 5.21). A very interesting variation has been elaborated by Waldeck[134,135] and Waldeck and Murgida[136] in a series of studies on cytochrome that was immobilised by pyridine n-alkane thiols on gold (Figure 5.22). The pyridine head group is able to penetrate the protein matrix and replace the native imidazole ligand of the heme group. In this way a direct conducting contact with the electro-active centre of the protein was established.

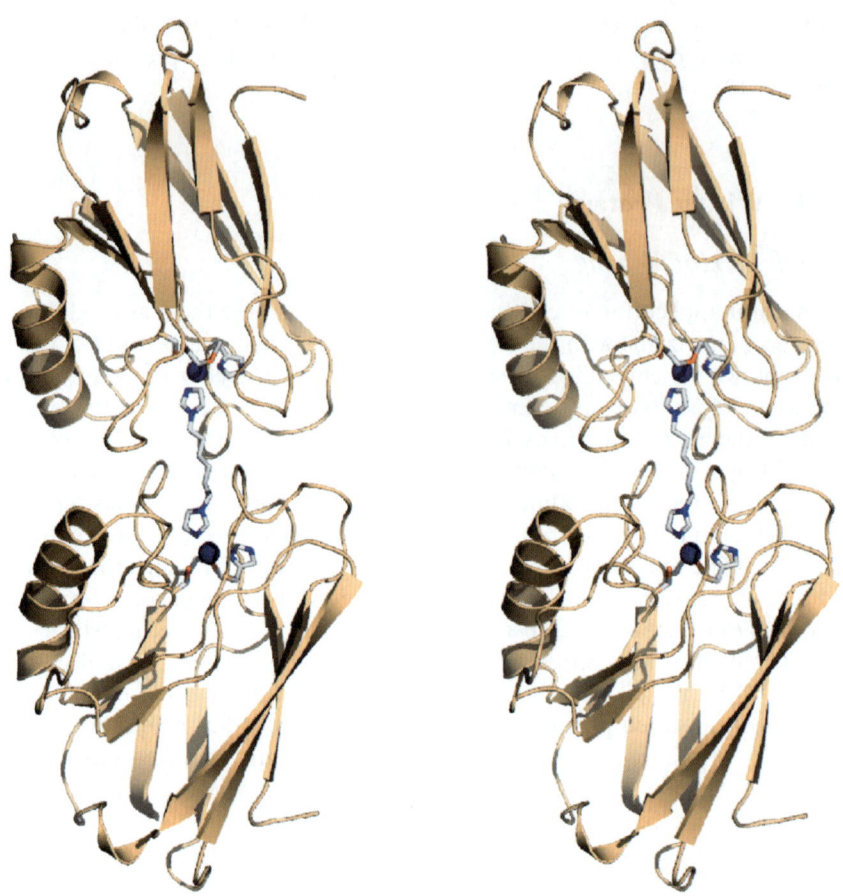

Figure 5.20 Stereo representation of the dimer of the Zn-substituted His117Gly variant of azurin from *P. aeruginosa*. The monomers are connected by an (α,ω)-bis-imidazole-*n*-hexane linker.[127]

5.5.3.2 Other Wiring Schemes

The hot-wiring method of attaching a protein to an electrode is attractive but also has some obvious disadvantages as mentioned above. One of the alternatives is to attach the protein/enzyme covalently to the electrode. A recent report on myoglobin (Mb) and horseradish peroxidase (HRP) seems promising in that respect. A GC electrode was covered with a 17–18-Å-thick SAM of polyethyleneglycol in which 20-Å-long OPE wires were inserted with a carboxyl functionality at their end. By means of conventional *N*-hydroxysuccinimide and carbodiimide chemistry the carboxyl groups were linked to an NH_2 group on the protein. When immobilised in this way on the electrode both Mb and HRP exhibited reversible electrochemistry (2 and $13.4\,s^{-1}$, respectively) while HRP clearly showed catalytic activity towards H_2O_2.[102]

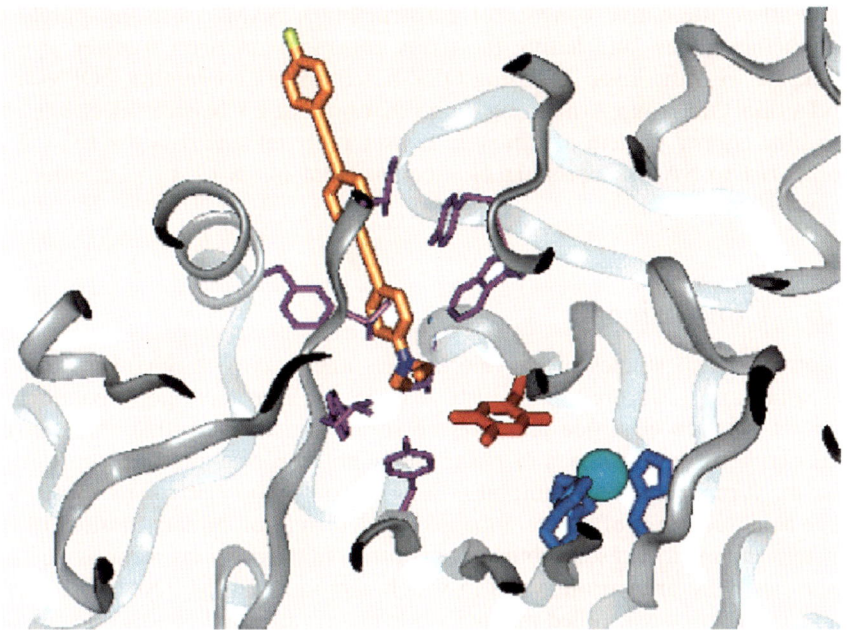

Figure 5.21 Model of an OPE linker (bronze coloured) with a diethylaniline end group inserted into the substrate cavity of amine oxidase from *Arthrobacter globiformis*. The enzyme side chains lining the cavity are in purple, the TPQ co-factor is in red, the Cu atom and its ligands are in blue.[133]

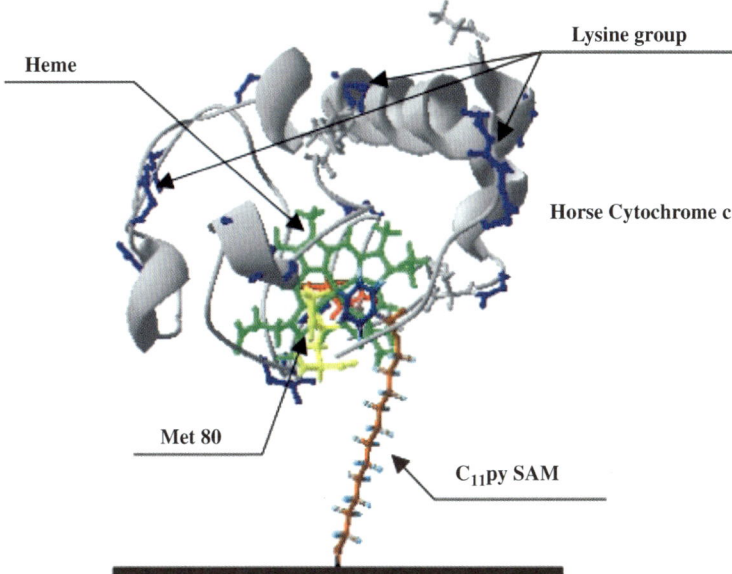

Figure 5.22 Model of horse heart cytochrome-*c* immobilised on Au by a pyridine-terminated undecanethiol linker.[133,135]

Another alternative consists of constructing a SAM consisting purely of conducting linkers and letting the redox enzyme or electron transfer protein associate with the layer. By using OPV-based linkers containing 2 OPV-units and a thiol end group Armstrong found that rates of ET between electrode and the blue copper protein azurin were considerably enhanced ($1.6 \times 10^3 \, \text{s}^{-1}$) as compared to SAMs consisting of n-octane-thiol ($4.8 \times 10^2 \, \text{s}^{-1}$) or n-decane-thiol ($6.0 \times 10^1 \, \text{s}^{-1}$).[71]

5.6 Conclusion

Electrochemical studies of enzyme kinetics and enzyme mechanisms can be revealing and rewarding. Obviously it is desirable to have good conducting contact between electrode and enzyme so as to make sure that the electron transfer between enzyme and electrode is not the rate-limiting step during enzyme turnover. Direct immobilisation of proteins on electrode surfaces has been achieved but is often accompanied by deformation or denaturation of the protein and in most cases is difficult to control. In the previous sections we have seen that there are various ways in which a protein can be immobilised on a solid surface in a controlled manner while preserving good conducting contact with the electrode.

Providing the solid surface with a SAM of n-alkanes is one option, although thick layers will lower the heterogeneous exchange rates to a value that is less than the enzyme turnover rate and thus heterogeneous exchange becomes rate limiting, a situation that is to be avoided. A SAM of conducting organic molecules can efficiently and easily bridge distances of 20–40 Å, whereby OPV-based linkers display the best conducting characteristics so far. Hot wiring directly into the active centre of the enzyme seems ideal provided the substrate binding and conversion process is not disturbed. The covalent attachment of a conducting linker to the protein envelope seems a promising alternative.

References

1. H. Pelletier and J. Kraut, *Science*, 1992, **258**, 1748.
2. A. Merli, D. E. Brodersen, B. Morini, Z. W. Chen, R. C. E. Durley, F. S. Mathews, V. L. Davidson and G. L. Rossi, *J. Biol. Chem.*, 1996, **271**, 9177.
3. P. B. Crowley and M. A. Carrondo, *Proteins: Struct. Funct. Bioinform.*, 2004, **55**, 603.
4. D. N. Beratan, J. N. Bettsand and J. N. Onuchic, *Science*, 1991, **252**, 1285.
5. C. C. Moser, J. M. Keske, K. Warncke, R. S. Farid and P. L Dutton, *Nature*, 1992, **355**, 796.
6. P. B. Crowley and M. Ubbink, *Accounts Chem. Res.*, 2003, **36**, 723.
7. M. Prudencio and M. Ubbink, *J. Mol. Recogn.*, 2004, **17**, 524.
8. M. Ubbink, *Photosynth. Res.*, 2004, **81**, 277.

9. B. M. Hoffman, L. M. Celis, D. A. Cull, A. D. Patel, J. L. Seifert, K. E. Wheeler, J. Y. Wang, J. Yao, I. V. Kurnikov and J. M. Nocek, *Proc. Natl. Acad. Sci. USA*, 2005, **102**, 3564.
10. J. L. Seifert, T. D. Pfister, J. M. Nocek, Y. Lu and B. M. Hoffman, *J. Am. Chem. Soc.*, 2005, **127**, 5750.
11. K. E. Wheeler, J. M. Nocek, D. A. Cull, L. A. Yatsunyk, A. C. Rosenzweig and B. M. Hoffman, *J. Am. Chem. Soc.*, 2007, **129**, 3906.
12. D. Ferrari, A. Merli, A. Peracchi, M. Di Valentin, D. Carbonera and G. L. Rossi, *Biochim. Biophys. Acta Protein Proteomics*, 2003, **1647**, 337.
13. D. Ferrari, M. Di Valentin, D. Carbonera, A. Merli, Z. W. Chen, F. S. Mathews, V. L. Davidson and G. L. Rossi, *J. Biol. Inorg. Chem.*, 2004, **9**, 231.
14. B. R. Crane, A. J. Di Bilio, J. R. Winkler and H. B. Gray, *J. Am. Chem. Soc.*, 2001, **123**, 11623.
15. F. A. Tezcan, B. R. Crane, J. R. Winkler and H. B. Gray, *Proc. Natl. Acad. Sci. USA*, 2001, **98**, 5002.
16. S. A. Kang, P. J. Marjavaara and B. R. Crane, *J. Am. Chem. Soc.*, 2004, **126**, 10836.
17. S. A. Kang and B. R. Crane, *Proc. Natl. Acad. Sci USA*, 2005, **102**, 15465.
18. S. A. Kang, K. R. Hoke and B. R. Crane, *J. Am. Chem. Soc.*, 2006, **128**, 2346.
19. M. Hippler, F. Drepper, J. Farah and J. D. Rochaix, *Biochemistry*, 1997, **36**, 6343.
20. M. Hippler, F. Drepper, W. Haehnel and J. D. Rochaix, *Proc. Natl. Acad. Sci. USA*, 1998, **95**, 7339.
21. L. M. Peerey and N. M. Kostic, *Biochemistry*, 1989, **28**, 1861.
22. H. S. Pappa, S. Tajbaksh, A. J. Saunders, G. J. Pielak and T. L. Poulos, *Biochemistry*, 1996, **35**, 4837.
23. F. Malatesta, F. Antonini, F. Nicoletti, A. Giuffre, E. Ditri, P. Sarti and M. Brunori, *Biochem. J.*, 1996, **315**, 909.
24. C. M. Groeneveld and G. W. Canters, *Eur. J. Biochem.*, 1985, **153**, 559.
25. H. Nar, A. Messerschmidt, R. Huber, M. Vandekamp and G. W. Canters, *J. Mol. Biol.*, 1991, **221**, 765.
26. G. VanPouderoyen, S. Mazumdar, N. I. Hunt, H. A. O. Hill and G. W. Canters, *Eur. J. Biochem.*, 1994, **222**, 583.
27. G. VanPouderoyen, G. Cigna, G. Rolli, F. Cutruzzola, F. Malatesta, M. C. Silvestrini, M. Brunori and G. W. Canters, *Eur. J. Biochem.*, 1997, **247**, 322.
28. I. M. C. van Amsterdam, M. Ubbink, L. J. C. Jeuken, M. P. Verbeet, O. Einsle, A. Messerschmidt and G. W. Canters, *Chem. A Eur. J.*, 2001, **7**, 2398.
29. I. M. C. van Amsterdam, M. Ubbink, O. Einsle, A. Messerschmidt, A. Merli, D. Cavazzini, G. L. Rossi and G. W. Canters, *Nat. Struct. Biol.*, 2002, **9**, 48.
30. K. V. Mikkelsen, L. K. Skov, H. Nar and O. Farver, *Proc. Natl. Acad. Sci. USA*, 1993, **90**, 5443.

31. I. M. C. van Amsterdam, M. Ubbink and G. W. Canters, *Inorg. Chim. Acta*, 2002, **331**, 296.
32. M. J. Eddowes and H. A. O. Hill, *J. Chem. Soc., Chem. Comm.*, 1977, 771.
33. P. Yeh and T. Kuwana, *Chem. Lett.*, 1977, 1145.
34. F. A. Armstrong, P. A. Cox, H. A. O. Hill, V. J. Lowe and B. N. Oliver, *J. Electroanal. Chem.*, 1987, **217**, 331.
35. F. A. Armstrong, H. A. O. Hill and N. J. Walton, *Accounts Chem. Res.*, 1988, **21**, 407.
36. J. J. Davis, H. A. O. Hill and A. M. Bond, *Coord. Chem. Rev.*, 2000, **200**, 411.
37. H. A. O. Hill, *Coord. Chem. Rev.*, 1996, **151**, 115.
38. F. A. Armstrong, H. A. Heering and J. Hirst, *Chem. Soc. Rev.*, 1997, **26**, 169.
39. F. A. Armstrong, R. Camba, H. A. Heering, J. Hirst, L. J. C. Jeuken, A. K. Jones, C. Leger and J. P. Mcevoy, *Faraday Discuss.*, 2000, 191.
40. F. A. Armstrong, *Cur. Opin. Chem. Biol.*, 2005, **9**, 110.
41. P. V. Bernhardt, *Aust. J. Chem.*, 2006, **59**, 233.
42. M. Fedurco, *Coord. Chem. Rev.*, 2000, **209**, 263.
43. G. Gilardi, A. Fantuzzi and S. J. Sadeghi, *Curr. Opin. Struct. Biol.*, 2001, **11**, 491.
44. J. Hirst, *Biochim. Biophys. Acta Bioenerg.*, 2006, **1757**, 225.
45. L. J. C. Jeuken, *Biochim. Biophys. Acta Bioenerg.*, 2003, **1604**, 67.
46. C. Amatore and E. Maisonhaute, *Anal. Chem.*, 2005, **77**, 303A.
47. E. Laviron, *J. Electroanal. Chem.*, 1979, **101**, 19.
48. S. Creager, C. J. Yu, C. Bamdad, S. O'Connor, T. MacLean, E. Lam, Y. Chong, G. T. Olsen, J. Y. Luo, M. Gozin and J. F. Kayyem, *J. Am. Chem. Soc.*, 1999, **121**, 1059.
49. C. E. D. Chidsey, *Science*, 1991, **251**, 919.
50. J. F. Smalley, S. W. Feldberg, C. E. D. Chidsey, M. R. Linford, M. D. Newton and Y. P. Liu, *J. Phys. Chem.*, 1995, **99**, 13141.
51. A. J. Bard and L. R. Faulkner, *Electrochemical Methods, Fundamentals and Applications,* John Wiley & Sons, New York, 1980, p. 103.
52. N. Sutin, *Accounts of Chemical Research*, 1982, **15**, 275.
53. N. Sutin, *Prog. Inorg. Chem.*, 1983, **30**, 441.
54. J. N. Onuchic, D. N. Beratan and J. J. Hopfield, *J. Phys. Chem.*, 1986, **90**, 3707.
55. D. N. Beratan and J. N. Onuchic, *J. Chem. Phys.*, 1988, **89**, 6195.
56. H. A. Heering, Y. B. M. Bulsink, W. R. Hagen and T. E. Meyer, *Biochemistry*, 1995, **34**, 14675.
57. W. R. Hagen, *Eur. J. Biochem.*, 1989, **182**, 523.
58. K. Besteman, M. A. G. Zevenbergen, H. A. Heering and S. G. Lemay, *Phys. Rev. Lett.*, 2004, **93**, 170802.
59. B. I. Shklovskii, *Phys. Rev. E*, 1999, **60**, 5802.
60. M. Collinson and E. F. Bowden, *Langmuir*, 1992, **8**, 2552.
61. J. J. Davis, D. Bruce, G. W. Canters, J. Crozier and H. A. O. Hill, *Chem. Commun.*, 2003, **1**, 576.

62. H. A. Heering, F. G. M. Wiertz, C. Dekker and S. de Vries, *J. Am. Chem. Soc.*, 2004, **126**, 11103.
63. J. C. Love, L. A. Estroff, J. K. Kriebel, R. G. Nuzzo and G. M. Whitesides, *Chem. Rev.*, 2005, **105**, 1103.
64. J. R. Lu, S. Perumal, I. Hopkinson, J. R. P. Webster, J. Penfold, W. Hwang and S. G. Zhang, *J. Am. Chem. Soc.*, 2004, **126**, 8940.
65. M. Rivera, M. A. Wells and F. A. Walker, *Biochemistry*, 1994, **33**, 2161.
66. T. R. Lee, R. I. Carey, H. A. Biebuyck and G. M. Whitesides, *Langmuir*, 1994, **10**, 741.
67. D. L. Johnson and L. L. Martin, *J. Am. Chem. Soc.*, 2005, **127**, 2018.
68. S. Fujii, U. Akiba and M. Fujihira, *J. Am. Chem. Soc.*, 2002, **124**, 13629.
69. B. de Boer, H. Meng, D. F. Perepichka, J. Zheng, M. M. Frank, Y. J. Chabal and Z. N. Bao, *Langmuir*, 2003, **19**, 4272.
70. A. A. Dhirani, R. W. Zehner, R. P. Hsung, P. GuyotSionnest and L. R. Sita, *J. Am. Chem. Soc.*, 1996, **118**, 3319.
71. F. A. Armstrong, N. L. Barlow, P. L. Burn, K. R. Hoke, L. J. C. Jeuken, C. Shenton and G. R. Webster, *Chem. Commun.*, 2004, 316.
72. T. T. Liang, H. Azehara, T. Ishida, W. Mizutani and H. Tokumoto, *Synthetic Met.*, 2004, **140**, 139.
73. J. A. Howarter and J. P. Youngblood, *Langmuir*, 2006, **22**, 11142.
74. G. MacBeath and S. L. Schreiber, *Science*, 2000, **289**, 1760.
75. P. P. Pompa, L. Blasi, L. Longo, R. Cingolani, G. Ciccarella, G. Vasapollo, R. Rinaldi, A. Rizzello, C. Storelli and M. Maffia, *Phys. Rev. E2003*, **67**, 041902.
76. R. Wilson and D. J. Schiffrin, *Analyst*, 1995, **120**, 175.
77. A. P. Fang, H. Ng, X. D. Su and S. F. Y. Li, *Langmuir*, 2000, **16**, 5221.
78. H. T. Ng, A. P. Fang, L. Q. Huang and S. F. Y. Li, *Langmuir*, 2002, **18**, 6324.
79. C. O. Kim, S. Y. Hong, M. Kirn, S. M. Park and J. W. Park, *J. Colloid Interface Sci.*, 2004, **277**, 499.
80. F. Moresco and A. Gourdon, *Proc. Natl. Acad. Sci. USA*, 2005, **102**, 8809.
81. J. Ladd, C. Boozer, Q. M. Yu, S. F. Chen, J. Homola and S. Jiang, *Langmuir*, 2004, **20**, 8090.
82. J. R. Lu, S. Perumal, I. Hopkinson, J. R. P. Webster, J. Penfold, W. Hwang and S. G. Zhang, *J. Am. Chem. Soc.*, 2004, **126**, 8940.
83. S. Askolin, M. Linder, K. Scholtmeijer, M. Tenkanen, M. Penttila, M. L. de Vocht and H. A. B. Wosten, *Biomacromolecules*, 2006, **7**, 1295.
84. H. J. Hektor and K. Scholtmeijer, *Curr. Opin. Biotechnol.*, 2005, **16**, 434.
85. K. Scholtmeijer, M. I. Janssen, M. B. M. van Leeuwen, T. G. van Kooten, H. Hektor and H. A. B. Wosten, *Bio-medical Materials and Engineering*, 2004, **14**, 447.

86. Y. Astier, A. M. Bond, H. J. Wijma, G. W. Canters, H. A. O. Hill and J. J. Davis, *Electroanalysis*, 2004, **16**, 1155.
87. S. Boussaad and N. J. Tao, *J. Am. Chem. Soc.*, 1999, **121**, 4510.
88. A. E. F. Nassar, J. M. Bobbitt, J. D. Stuart and J. F. Rusling, *J. Am. Chem. Soc.*, 1995, **117**, 10986.
89. Y. Astier, G. W. Canters, J. J. Davis, H. A. O. Hill, M. P. Verbeet and H. J. Wijma, *ChemPhysChem*, 2005, **6**, 1114.
90. A. S. Haas, D. L. Pilloud, K. S. Reddy, G. T. Babcock, C. C. Moser, J. K. Blasie and P. L. Dutton, *J. Phys. Chem. B*, 2001, **105**, 11351.
91. A. Avila, B. W. Gregory, K. Niki and T. M. Cotton, *J. Phys. Chem. B*, 2000, **104**, 2759.
92. Q. J. Chi, O. Farver and J. Ulstrup, *Proc. Natl. Acad. Sci. USA*, 2005, **102**, 16203.
93. Z. Q. Feng, S. Imabayashi, T. Kakiuchi and K. Niki, *J. Chem. Soc. Faraday Trans.*, 1997, **93**, 1367.
94. K. Fujita, N. Nakamura, H. Ohno, B. S. Leigh, K. Niki, H. B. Gray and J. H. Richards, *J. Am. Chem. Soc.*, 2004, **126**, 13954.
95. H. J. Yue, D. Khoshtariya, D. H. Waldeck, J. Grochol, P. Hildebrandt and D. H. Murgida, *J. Phys. Chem. B*, 2006, **110**, 19906.
96. L. H. Dubois and R. G. Nuzzo, *Ann. Rev. Phys. Chem.*, 1992, **43**, 437.
97. S. B. Sachs, S. P. Dudek, R. P. Hsung, L. R. Sita, J. F. Smalley, M. D. Newton, S. W. Feldberg and C. E. D. Chidsey, *J. Am. Chem. Soc.*, 1997, **119**, 10563.
98. A. Salomon, D. Cahen, S. Lindsay, J. Tomfohr, V. B. Engelkes and C. D. Frisbie, *Adv. Mater.*, 2003, **15**, 1881.
99. K. Weber, L. Hockett and S. Creager, *J. Phys. Chem. B*, 1997, **101**, 8286.
100. L. Andolfi, B. Bonanni, G. W. Canters, M. P. Verbeet and S. Cannistraro, *Surf. Sci.*, 2003, **530**, 181.
101. L. Andolfi, G. W. Canters, M. P. Verbeet and S. Cannistraro, *Biophys. Chem.*, 2004, **107**, 107.
102. G. Z. Liu and J. J. Gooding, *Langmuir*, 2006, **22**, 7421.
103. M. W. Holman, R. C. Liu and D. M. Adams, *J. Am. Chem. Soc.*, 2003, **125**, 12649.
104. S. P. Dudek, H. D. Sikes and C. E. D. Chidsey, *J. Am. Chem. Soc.*, 2001, **123**, 8033.
105. H. D. Sikes, J. F. Smalley, S. P. Dudek, A. R. Cook, M. D. Newton, C. E. D. Chidsey and S. W. Feldberg, *Science*, 2001, **291**, 1519.
106. J. F. Smalley, H. O. Finklea, C. E. D. Chidsey, M. R. Linford, S. E. Creager, J. P. Ferraris, K. Chalfant, T. Zawodzinsk, S. W. Feldberg and M. D. Newton, *J. Am. Chem. Soc.*, 2003, **125**, 2004.
107. J. F. Smalley, S. B. Sachs, C. E. D. Chidsey, S. P. Dudek, H. D. Sikes, S. E. Creager, C. J. Yu, S. W. Feldberg and M. D. Newton, *J. Am. Chem. Soc.*, 2004, **126**, 14620.
108. A. S. Blum, T. Ren, D. A. Parish, S. A. Trammell, M. H. Moore, J. G. Kushmerick, G. L. Xu, J. R. Deschamps, S. K. Pollack and R. Shashidhar, *J. Am. Chem. Soc.*, 2005, **127**, 10010.

109. J. F. Smalley, M. D. Newton and S. W. Feldberg, *Electrochem. Commun.*, 2000, **2**, 832.
110. J. K. N. Mbindyo, T. E. Mallouk, J. B. Mattzela, I. Kratochvilova, B. Razavi, T. N. Jackson and T. S. Mayer, *J. Am. Chem. Soc.*, 2002, **124**, 4020.
111. L. T. Cai, H. Skulason, J. G. Kushmerick, S. K. Pollack, J. Naciri, R. Shashidhar, D. L. Allara, T. E. Mallouk and T. S. Mayer, *J. Phys. Chem. B*, 2004, **108**, 2827.
112. G. V. Nazin, X. H. Qiu and W. Ho, *Science*, 2003, **302**, 77.
113. X. Y. Xiao, B. Q. Xu and N. J. Tao, *Nano Letters*, 2004, **4**, 267.
114. L. A. Bumm, J. J. Arnold, T. D. Dunbar, D. L. Allara and P. S. Weiss, *J. Phys. Chem. B*, 1999, **103**, 8122.
115. K. Moth-Poulsen, L. Patrone, N. Stuhr-Hansen, J. B. Christensen, J. P. Bourgoin and T. Bjornholm, *Nano Letters*, 2005, **5**, 783.
116. R. Huber, M. T. Gonzalez, S. Wu, M. Langer, S. Grunder, V. Horhoiu, M. Mayor, M. R. Bryce, C. S. Wang, R. Jitchati, C. Schonenberger and M. Calame, *J. Am. Chem. Soc.*, 2008, **130**, 1080.
117. *Handbook of Chemistry and Physics*, CRC Press, Taylor and Francis Group, Boca Raton, 2008, Section 9.
118. B. Q. Xu and N. J. J. Tao, *Science*, 2003, **301**, 1221.
119. X. L. Li, J. He, J. Hihath, B. Q. Xu, S. M. Lindsay and N. J. Tao, *J. Am. Chem. Soc.*, 2006, **128**, 2135.
120. M. Magoga and C. Joachim, *Phys. Rev. B*, 1997, **56**, 4722.
121. X. Yao, J. X. Wang, F. M. Zhou, J. Wang and N. J. Tao, *J. Phys. Chem. B*, 2004, **108**, 7206.
122. D. Barrick, *Curr. Opin. Biotechnol.*, 1995, **6**, 411.
123. T. Denblaauwen, M. Vandekamp and G. W. Canters, *J. Am. Chem. Soc.*, 1991, **113**, 5050.
124. M. Vandekamp, R. Floris, F. C. Hali and G. W. Canters, *J. Am. Chem. Soc.*, 1990, **112**, 907.
125. A. C. F. Gorren, T. Denblaauwen, G. W. Canters, D. J. Hopper and J. A. Duine, *FEBS Letters*, 1996, **381**, 140.
126. G. VanPouderoyen, T. Denblaauwen, J. Reedijk and G. W. Canters, *Biochemistry*, 1996, **35**, 13205.
127. T. E. de Jongh, A. M. M. van Roon, M. Prudencio, M. Ubbink and G. W. Canters, *Eur. J. Inorg. Chem.*, 2006, 3861.
128. H. J. Wijma, M. J. Boulanger, A. Molon, M. Fittipaldi, M. Huber, M. E. P. Murphy, M. P. Verbeet and G. W. Canters, *Biochemistry*, 2003, **42**, 4075.
129. I. Willner, E. Katz and B. Willner, *Electroanalysis*, 1997, **9**, 965.
130. I. Willner and E. Katz, *Angew. Chem. Int. Ed.*, 2000, **39**, 1180.
131. B. Willner, E. Katz and I. Willner, *Curr. Opin. Biotechnol.*, 2006, **17**, 589.
132. Y. Xiao, F. Patolsky, E. Katz, J. F. Hainfeld and I. Willner, *Science*, 2003, **299**, 1877.

133. C. R. Hess, G. A. Juda, D. M. Dooley, R. N. Amii, M. G. Hill, J. R. Winkler and H. B. Gray, *J. Am. Chem. Soc.*, 2003, **125**, 7156.
134. J. J. Wei, H. Y. Liu, A. R. Dick, H. Yamamoto, Y. F. He and D. H. Waldeck, *J. Am. Chem. Soc.*, 2002, **124**, 9591.
135. H. Yamamoto, H. Y. Liu and D. H. Waldeck, *Chem. Commun.*, 2001, 1032.
136. D. H. Murgida, P. Hildebrandt, J. Wei, Y. F. He, H. Y. Liu and D. H. Waldeck, *J. Phys. Chem. B*, 2004, **108**, 2261.

CHAPTER 6

Cytochromes P450: Tailoring a Class of Enzymes for Biosensing

VIKASH R. DODHIA AND GIANFRANCO GILARDI

Imperial College London, Biochemistry Building, South Kensington, London, SW7 2AZ, UK and University of Turin, Turin, Italy

6.1 Introduction

The cytochrome P450 enzymes (P450s) are a family of heme-thiolate proteins that currently contains over 7000 members in species ranging from bacteria through to plants and mammals. The increased volume of the genome projects has made available a large amount of DNA sequences from which one can extract interesting information on this class of enzymes and infer sequence and structural similarities between different P450s from different species, with the implications that this has in designing protein immobilisation strategies. From the sequence database (http://pfam.sanger.ac.uk/) it is also possible to generate phylogenetic trees, an example of which is shown in Figure 6.1, from which interesting conclusions can be made: 1) all P450s evolved from a single ancestral gene probably of prokaryotic origin; 2) P450 divergence and evolution of life on earth are evolutionarily linked; 3) divergence between eukaryotic and prokaryotic P450s occurred around 1.4 billion years ago; 4) bacterial P450 102A1 (also called *B. megaterium* P450$_{BM-3}$) is more closely related to eukaryotic P450s than to prokaryotic P450s.

These enzymes were first recognised by Martin Klingenberg[1] who was studying the spectrophotometric properties of pigments present in the microsomal fraction from rat livers. Upon addition of the reducing agent sodium dithionite and bubbling carbon monoxide, a unique spectral absorbance band

---

Engineering the Bioelectronic Interface: Applications to Analyte Biosensing and Protein Detection
Edited by Jason Davis
© 2009 Royal Society of Chemistry
Published by the Royal Society of Chemistry, www.rsc.org

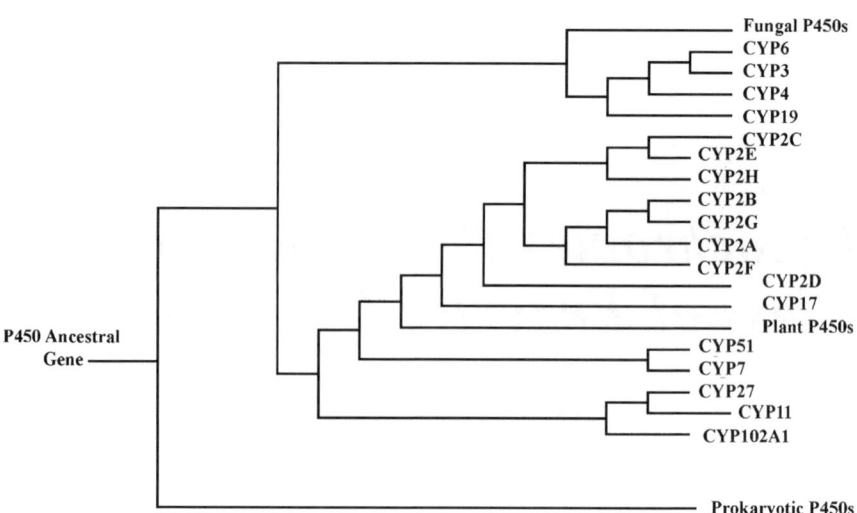

Figure 6.1 Summarised cytochrome P450 phylogenetic tree.

with a maximum at 450 nm was observed. Hence the name P450 was derived, *i.e.* "pigment" 450. The absorbance band at 450 nm is unique amongst heme proteins and serves as a signature for active P450 proteins. The heme domain of cytochrome P450 enzymes generally consists of approximately 460–500 amino acids with a cysteine molecule situated near the carboxy-teminus that provides the essential thiol ligand for the heme iron.

These enzymes play a critical role in the human body by catalysing reactions involved in xenobiotic metabolism, biosynthesis of steroid hormones, oxidation of unsaturated fatty acids and stereo-specific and regio-specific metabolism of fat-soluble vitamins and are located in various tissues ranging from the brain to the kidneys. For a more in-depth and complete description of the tissue expression levels, regulation, polymorphism, substrates, inhibitors and active site structure of the human cytochrome P450 enzymes, a recent review by Guengerich is recommended.[2]

This versatility in the recognition of different substrates, all of technological relevance either for the development of new drugs or for the detection or degradation of environmentally relevant molecules, has started an enormous amount of protein engineering work aimed at improving or expanding their catalytic performance and tailoring their surface for immobilisation for the creation of amperometric devices.[3] Indeed electrochemistry has been widely used as a tool to characterise both the electron transfer and the catalytic properties of these enzymes with the scope of developing biosensors and bioreactors for applications in environmental monitoring and pharmaceutical industry. However, in order to be catalytically active after immobilisation, the cytochrome P450 has to maintain the active conformational state guaranteed by the presence of a specific thiolate bond between the heme iron and a highly conserved cysteine residue, identified as a reduced CO complex with a maximum of absorbance at 450 nm.[1]

This chapter will start by covering fundamental aspects of P450 structure and function, leading to protein engineering by both rational design and directed evolution approaches to improve the performance of these enzymes. We will discuss the progress made on their immobilisation on electrode surfaces. Here progression of P450-electrode systems into amperometric biosensors or bioelectrocatalysts is considerably dependent on the efficiency of the electrode-driven P450 activity. In the few cases where electrocatalysis with immobilised human P450s has been reported, low product formation and a significant contribution from aspecific reactions between reactive oxygen species and substrate has been observed.[4,5] Improved coupling efficiency of immobilised P450s has only recently been reported thanks to the use of protein engineering methods.[6] These studies indicated the importance of relatively slow heterogeneous electron transfer rate constants in achieving efficient electron delivery and lower uncoupling in electrode-driven P450 enzymes.

6.2 Structure-function of Bacterial and Human Cytochromes P450

In general cytochromes P450 act as mono-oxygenases, catalysing the addition of oxygen to the substrate RH using the reducing power of NADPH as shown below:

$$RH + O_2 + H^+ + NADPH \longrightarrow ROH + H_2O + NADP^+$$

Two electrons originating from NADPH are transferred to the P450 *via* a redox partner flavoprotein such as cytochrome P450 reductase (or *via* a flavoprotein/iron sulfur protein) in the presence of a substrate and molecular oxygen. The substrate is oxidised and one atom of molecular oxygen is incorporated into the product. The reaction shown above is a gross generalisation of the diversity of reactions catalysed by cytochrome P450 enzymes as at least 40 different types of reactions catalysed by P450s have been identified.[7] These include hydroxylation, dealkylation, heteroatom oxygenation and epoxidations.

Depending on the nature of the redox partner, cytochromes P450 can be broadly divided into four different classes.[8] Class I P450s are generally mitochondrial or bacterial P450s and are three-component systems. They consist of a flavin adenine dinucleotide (FAD) containing reductase that oxidises NADH to NAD^+, an iron-sulfur protein (*e.g.* ferredoxin) that acts as an electron carrier and the cytochrome P450. Examples of this class include the mitochondrial P450 17A1 and the bacterial P450 101 ($P450_{cam}$). Class II P450s receive their electrons from NADPH *via* the membrane-bound cytochrome P450 reductase (CPR). CPR contains FAD and flavin mononucleotide (FMN) binding domains that directly oxidise NADPH and shuttle these electrons to the P450. This class includes microsomal P450s, of which examples are the human cytochromes P450 1A2, 2C9, 2C19, 2D6, 3A4 and 2C8. This classification system is not perfect as microsomal P450s can also receive their reducing equivalents from intermediates like cytochrome b_5.[9] Cytochrome b_5 can in turn be reduced by either CPR or cytochrome b_5 reductase, a NADH-utilising FAD-containing protein. Class III P450s are enzymes such as the bacterial P450 102A1 (P450$_{BM-3}$), which are catalytically self-sufficient single-polypeptide enzymes containing binding sites for heme, FAD and FMN. These enzymes are catalytically self-sufficient because they do not require an accessory reductase for activity since the reductase and P450 components are fused in the same polypeptide chain. Recently a new class of P450s (P450 116B2) has been identified from the bacterial species *Rhodococcus*.[8] In this new class of P450, an FMN-containing reductase is fused in a single polypeptide chain with a ferredoxin-like centre and a P450. The electrons are relayed to the P450 through an FMN centre and a [2Fe2S] ferredoxin-like component. This represents a new class of P450 and may indicate that more novel arrangements of P450s and redox partners are yet to be discovered.

Nebert and colleagues[10] published a recommended nomenclature system where P450s are divided into families and sub-families based on the principle that "a P450 protein sequence from one gene family is defined as usually having <40% resemblance to that from another family". Sub-families grouped together proteins of >60% sequence similarity. Genes encoding cytochromes P450 are named with the CYP prefix followed by the family, sub-family and isoenzyme member. For example cytochrome P450 2D6, encoded by the gene CYP2D6, is a cytochrome P450 from the 2 family, D sub-family and isoenzyme 6. Despite making the naming of enzymes clearer, it should, however, be noted that the naming system has no reflection on the function of the enzymes. Therefore members within the same family may have completely different functions and enzymes with similar roles may belong to different families.

Cytochromes P450 from different species are members of the same gene superfamily and they generally carry out oxidation reactions. Therefore it is not surprising that the solved X-ray crystal structures of these proteins from different species reveal a highly conserved fold. These structures include the P450 101 (P450$_{cam}$),[11] heme domain of P450 102A1 (P450$_{BM-3}$),[12] P450 107A1 (P450$_{eryF}$),[13] P450 108 (P450$_{terp}$),[14] P450 51,[15] P450 119[16] and recombinant rabbit P450 2C5.[17] The first six of these structures are all bacterial and 2C5 was the first mammalian P450 crystal structure to be solved. More recently, the crystal structures of human P450s including 2C9,[18,19] 2C8,[20] 3A4,[21,22] 1A2,[23] 2D6[24] and 2A6[25] have been solved. These P450s are all found to have triangular shape with one corner rich in β-sheets with the rest of the molecule mainly α-helical as shown in Figure 6.2. They have a high similarity in their secondary structure content and in the spatial location of these secondary structures within the tertiary structure.[26] Furthermore no other proteins apart from cytochromes P450 are found to have the P450-like fold.

Other common features seen in the cytochromes P450 crystallised to date are that the I helix is located diagonally across the heme cofactor and the L helix

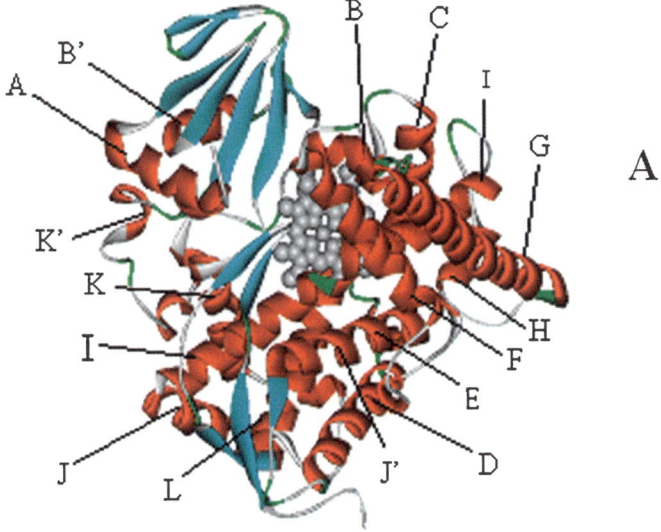

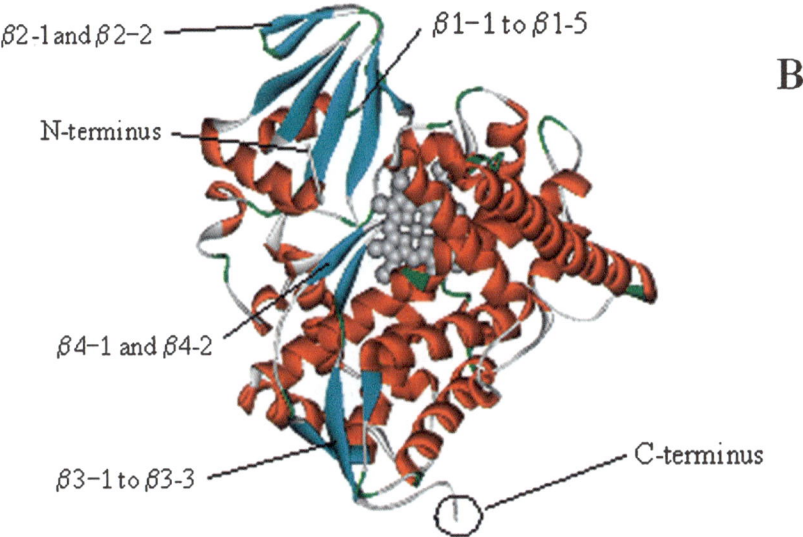

Figure 6.2 Structure of a typical P450 showing the position of the α-helices (red) and of the β-sheets (cyan) in relation to the heme cofactor (grey) and the N- and C-termini. Here the heme domain of bacterial P450 102A1 (P450$_{BM-3}$) is shown (PDB:2HPD).

runs behind the cofactor. Additionally cytochromes P450 have regions of highly conserved amino acid motifs that centre on the heme thiolate ligand and oxygen activation chemistry as shown in Table 6.1. The heme thiolate ligand motif located just prior to the L helix has a rigid architecture that affords

Table 6.1 Motifs found in the amino acid sequence of P450s.

Motif	Role
(A/G)Gx(E/D)T	In centre of the I helix and lies directly over the pyrrole ring B of the heme.
(P/E/K)(K/R)(G/N)	Follows β2-2 strand.
xPcxFxPE+a	Forms meander region in microsomal P450s and may have role in redox partner interactions.
F(G/S)xGx(H/R) xCxGxx(I/L/F)A	Contains the cysteine known to act as the ligand for the heme group.
Gxx(I/L/F)A	Found in the last few residues of the heme binding region and forms the beginning of the L helix.
(A/G)Gx(D/E)T(T/S)	Contains the threonine thought to be involved in oxygen activation.

Note: "x" signifies any amino acid, "+" a positively charged amino acid, "a" an aromatic amino acid and "c" a charged amino acid.

protection to the Cys ligand and maintains its position such that it can accept hydrogen bonds from peptide NH groups.[27] The other highly conserved region, the oxygen activation motif, is situated in the I helix near the heme iron and contains a Thr, which is situated in a position that allows the side-chain OH to form a hydrogen bond to the water molecule involved in oxygen activation.

Although the structures of the crystallised P450s display almost all of the secondary structures shown in Figure 6.2, there are some distinct differences between P450s and these differences are primarily in the regions involved in interaction with the substrate molecule. For instance the loop between the F and G helices, considered to have an important role in substrate interaction, is much shorter in P450 101 (P450$_{cam}$) than in either P450 2C5 or P450 102A1 (P450$_{BM-3}$).[26] Furthermore in the recently crystallised human P450 2C8, this F-G loop region was found to be more structured, giving rise to previously unseen F' and G' helices.[20]

In addition to retaining the ability to carry out the general P450 monooxygenation reaction, the other reason for cytochromes P450 adopting the general tertiary structure is to allow optimal interactions between the P450s and their redox partners such as cytochrome P450 reductase.[28] The distribution of positively and negatively charged amino acids in bacterial P450s such as 101 (P450$_{cam}$), the heme domain of 102A1 (P450$_{BM-3}$) and 108 (P450$_{terp}$) is such that there is an unequal distribution of charge across the protein producing a molecular dipole. This has a positive effect on the alignment between the redox-partner and the P450 and on the flow of electrons from the redox partner to the active site.[28]

6.3 The Need for Electrons: the Cytochrome P450 Catalytic Cycle

The utilisation of molecular oxygen and reducing equivalents (provided either by NAD(P)H or by an electrode in amperometric devices) by the P450 enzyme to carry out substrate oxidation is better understood through their catalytic cycle shown in Figure 6.3. The solution of the intermediates in the catalytic cycle of P450 101 (P450$_{cam}$) to atomic resolution by Schlichting and coworkers has aided the knowledge of previously unknown parts of the cycle.[29] This, with many previous and subsequent studies using a wide variety of techniques on various P450s, suggests all P450s activate molecular oxygen through a similar if not identical mechanism.[30,31]

The chemical changes undergone by the heme iron in the porphyrin ring are an important part of the cytochromes P450 in terms of substrate activation and subsequent mono-oxygenation. In the resting state, the iron is in the ferric low spin state Fe^{3+} and sits within the plane of the porphyrin ring. The sixth distal ligand is thought to be water.[32] In this state (**1**) the absorbance spectrum exhibits a maximum at around 420 nm. Upon substrate binding (RH) and displacement of the water molecule (**2**), the iron spin state equilibrium changes from one of low spin to high spin resulting in a change in the absorbance spectrum of the enzyme. Three specific types of substrate binding spectra have been identified, namely

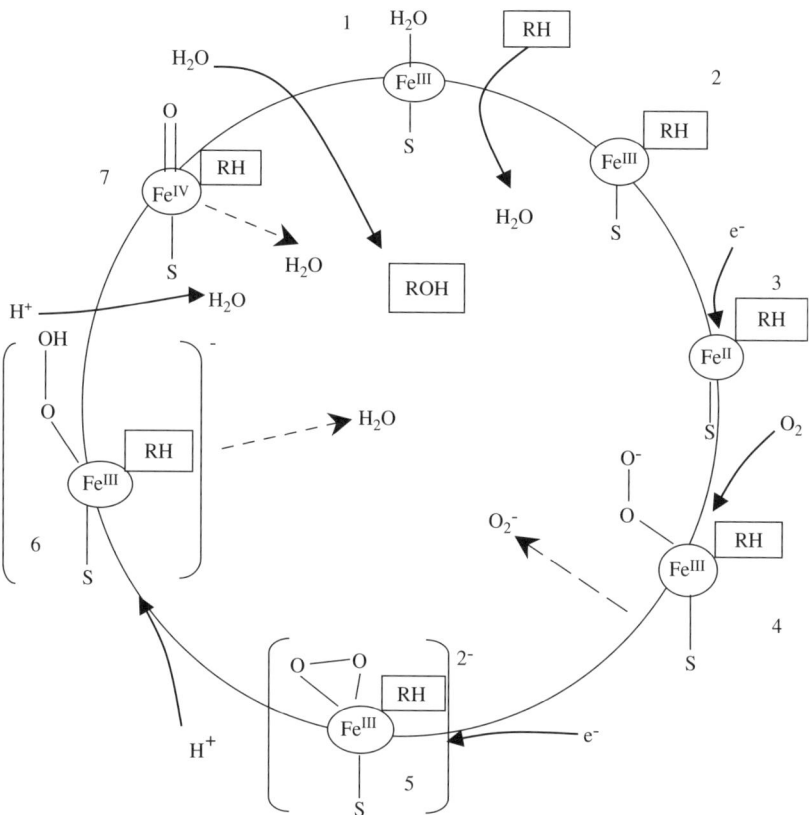

Figure 6.3 The cytochrome P450 catalytic cycle. The substrate is designated as RH and the oxidised product as ROH.

Type I, Type II and Reverse Type I.[33] Type I binding spectra are characterised by a decrease in the absorbance of the Soret peak at 420 nm coupled with an increase in the absorbance at around 390 nm (low spin to high spin), with an isosbestic point at around 407 nm. Most P450 substrates exhibit Type I binding spectra, and often the size of the spectral change indicates both the extent and the rate of metabolism, since there can be tight coupling between the spin and redox equilibria in microsomal and bacterial systems.[34] Examples of Type I substrates include arachidonic acid, camphor, debrisoquine, testosterone, ibuprofen and omeprazole. Type II binding spectra, caused by substrates such as aniline, cyanide and imidazole, result in the increase of the absorbance at 425–435 nm and a decrease at 390–405 nm, with the isosbestic point at 419 nm. Type II substrates typically tend to be inhibitors that contain non-bonded electrons such as a nitrogen lone pair and can ligate to the heme iron. Reverse Type I spectra are the mirror images of Type I spectra and can resemble Type II spectra since there is an increase in absorbance at 420 nm and a decrease at 390 nm. However, unlike Type II substrates, no ligation to the heme iron occurs and the spectral change is caused by the displacement of the water molecule by substrates such as caffeine, ethanol and theophylline.

Following substrate binding (**2**), the iron receives the first electron/reducing equivalent from its redox partner such as cytochrome P450 reductase or an iron-sulfur protein (**3**). The iron becomes reduced to the ferric Fe^{2+} state while still remaining high spin (**3**). This favours the binding of molecular oxygen (**4**), evidence for which was found when the P450 101 ($P450_{cam}$) system was examined in the electron-deficient di-oxygen form and in the electron-rich Fe^{2+} high spin state.[29] Binding of molecular oxygen leads to the formation of a di-oxygen complex (**4**). Upon transfer of a second electron (**5**), an Fe^{3+} peroxo complex is formed. This complex or its protonated hydroperoxo form, commonly referred to as Compound 0 (**6**), accepts a further proton resulting in the formation of water and an iron with a single oxygen ligand (**7**). This reactive intermediate, commonly referred to as Compound I, with its electrophilic oxygen donates its oxygen to the substrate. Upon completion, the substrate leaves the active site and water enters the site allowing further cycles to continue.

As well as leading to the turnover of substrate, the iron oxygen complexes in the P450 catalytic cycle shown in Figure 6.3 can lead to other products, where the consumption of reducing equivalents is uncoupled from the formation of product. Coupling between NADPH oxidation and substrate turnover in P450s varies widely. High levels of coupling have been reported for the bacterial P450 102A1 ($P450_{BM-3}$), which has been found to be close to 100% coupled,[35] while a mere 1.2% coupling has been reported for human P450 1A2 in a reconstituted system.[36]

Uncoupling can occur at three points within the cycle, leading to the formation of reactive oxygen species or water. This happens when the di-oxygen iron complex dissociates to give a super-oxide anion. Dissociation of the hydroperoxo iron complex giving hydrogen peroxide is the second point at which uncoupling can occur. In theory this so-called "peroxide shunt" reaction is reversible for all P450s and can be used to drive the P450 catalytic cycle. The

P450 152A1 enzyme is able to utilise this peroxide shunt mechanism to catalyse hydroxylation reactions, where an oxygen atom from hydrogen peroxide is efficiently introduced into fatty acid substrates such as myristic acid.[37] The final point at which uncoupling can occur is when the iron oxygen complex (Compound I) reacts with two electrons and two protons to give a water molecule.

These uncoupling events tend to be rare in bacterial P450s in comparison to microsomal P450s, where reactive oxygen species account for greater than 50% of the reducing equivalents consumed.[38] The level of coupling in microsomal P450 reconstituted systems appears to be highly dependent on the buffer components, the quantity of cytochrome P450 reductase added, the presence and quantity of cytochrome b_5, the identity and quantity of phospholipids, the salt concentration, the pH of the system and other factors such as the presence of magnesium chloride and detergents. A factor that does seem to be very important for some P450s, including human 3A4[38] is the addition of cytochrome b_5 to the reaction mixture. The addition of this redox protein has been demonstrated to greatly increase the level of coupling of NADPH consumption to product formation. The general effects of cytochrome b_5 have been reported to be either the reduction in the rate of NADPH consumption without a significant concurrent slowing of the product formation rate, or the stimulation of the activity of the protein.[39,40] Investigators have found that P450s such as 2B4, 1A2, 2E1 and 3A4 all show a marked increase in the level of coupling upon the addition of cytochrome b_5 to a reconstituted system.[36,38,39,41] However, this increase appears to be substrate dependent as illustrated by P450 3A4, where no increase in the level of coupling upon the addition of cytochrome b_5 was seen for the substrate cyclosporine, but a 5-fold increase was seen on the level of coupling for the hydroxylation of testosterone.[38] The mechanism by which cytochrome b_5 works is not one that is fully understood and there are several theories on how cytochrome b_5 reduces the level of cytochrome P450 uncoupling.[38,42–45]

6.4 Human Cytochromes P450 and Drug Metabolism

In humans, up to 57 different CYPs have been identified to date (http://drnelson.utmem.edu/CytochromeP450.html) and these have been subdivided into 18 families and 43 sub-families as summarised in Table 6.2. The majority of xenobiotics metabolised by P450s are lipophilic compounds, which include most drugs.[40] The human body will metabolise xenobiotics (drugs) in order to render them more soluble, with the overall aim being excretion primarily *via* urine. This results in reduced exposure of the body to the compound and a decrease in toxicity or side effects. Therefore the process of metabolism is a crucial aspect in the toxic potential, disposition and excretion of foreign compounds. The primary results of a metabolic process can therefore be:

  i. Transformation of the parent compound into a more polar metabolite often by the addition of hydrophilic or removal of hydrophobic groups;
  ii. An increase in molecular size and weight of the compound;
  iii. Facilitation of excretion of the compound.

Table 6.2 Families of human cytochrome P450 enzymes and their functions.

Family	Function	Sub-families	Members
CYP1	xenobiotic and steroid metabolism	3	CYP1A1, CYP1A2, CYP1B1
CYP2	xenobiotic and steroid metabolism	13	CYP2A6, CYP2A7, CYP2A13, CYP2B6, CYP2C8, CYP2C9, CYP2C18, CYP2C19, CYP2D6, CYP2E1, CYP2F1, CYP2J2, CYP2R1, CYP2S1, CYP2U1, CYP2W1
CYP3	xenobiotic and steroid metabolism	1	CYP3A4, CYP3A5, CYP3A7, CYP3A43
CYP4	arachidonic acid or fatty acid metabolism	6	CYP4A11, CYP4A22, CYP4B1, CYP4F2, CYP4F3, CYP4F8, CYP4F11, CYP4F12, CYP4F22, CYP4V2, CYP4X1, CYP4Z1
CYP5	thromboxane A_2 synthase	1	CYP5A1
CYP7	bile acid biosynthesis and steroid biosynthesis	2	CYP7A1, CYP7B1
CYP8	varied	2	CYP8A1 (prostacyclin synthase), CYP8B1 (bile acid biosynthesis)
CYP11	steroid biosynthesis	2	CYP11A1, CYP11B1, CYP11B2
CYP17	steroid biosynthesis	1	CYP17A1
CYP19	steroid biosynthesis	1	CYP19A1
CYP20	unknown function	1	CYP20A1
CYP21	steroid biosynthesis	2	CYP21A2
CYP24	vitamin D degradation	1	CYP24A1
CYP26	retinoic acid hydroxylation	3	CYP26A1, CYP26B1, CYP26C1
CYP27	varied	3	CYP27A1 (bile acid biosynthesis), CYP27B1 (vitamin D_3 1-alpha hydroxylase, activates vitamin D_3), CYP27C1 (unknown function)
CYP39	7-alpha hydroxylation of 24-hydroxycholesterol	1	CYP39A1
CYP46	cholesterol 24-hydroxylation	1	CYP46A1
CYP51	cholesterol biosynthesis	1	CYP51A1 (lanosterol 14-alpha demethylase)

However, the converse can also occur, where metabolism will either potentiate the activity of or activate a compound. For example the human P450 1A2 has been shown to play an important role in the activation of procarcinogens such as polyaromatic hydrocarbons, amines and heterocyclic amines.[46] The human P450 2E1 enzyme has been demonstrated to oxidise the common painkiller paracetamol to its hepatotoxic metabolite N-acetly-p-benzoquinoneamine.[47]

Drug or xenobiotic metabolism can be conveniently divided into two phases. Phase I is the alteration of the original molecule in order to add a functional group that makes the compound more soluble and/or more susceptible to conjugation in Phase II. The majority of Phase I metabolism is carried out by the P450 enzymes with other enzymes such as the flavin-containing monooxygenases (FMO) playing a minor role. Phase II metabolism reactions generally involve conjugation reactions where a highly soluble polar group such as a glucuronic acid is added to the xenobiotic in order to render the molecule more soluble and therefore facilitating excretion. These groups are added to a suitable functional group present on the foreign compound or one that has been introduced by Phase I metabolism.

It has been established that the human hepatic P450 isoforms that metabolise the majority of pharmaceuticals in clinical use are P450s 1A2, 2C8, 2C9, 2C19, 2D6 and 3A4.[48] These six enzymes account for nearly 90% of the cytochrome P450 mediated drug metabolism carried out in the human body. The pharmaceutical industry is becoming increasingly aware of the key roles of the P450 enzymes in the drug research and development process, which include: *in silico* methods, biosensor technology, dosage setting, influence of P450 polymorphism, metabolite and drug interaction prediction, and the future end to the assumption that "one drug dose fits all".

With six enzymes being involved in the transformation of such a large number of compounds it is immediately clear that any one enzyme will be involved in the metabolism of a large number of compounds of diverse structure and size, if the level of expression in the human liver for each enzyme were equal. However this is not the case and the liver content of these enzymes does not necessarily reflect the number of compounds metabolised by an enzyme.[49] This point is better illustrated in Figure 6.4. This opens the possibility of two compounds competing for the same enzyme or a compound inhibiting the metabolism of another compound leading to potentially fatal or disastrous consequences. This phenomenon has been demonstrated throughout history with a recent case being that of the anti-histamine drug terfenadine. Terfenadine was sold in the UK over the counter for treatment of allergic conditions such as hay fever but was later withdrawn from the market for safety reasons. Terfenadine, a human P450 3A4 substrate, when co-administered with other 3A4 substrates such as erythromycin (an antibiotic) or 3A4 inhibitors such as ketoconazole (an anti-fungal), caused potentially fatal arrhythmias.[50] Matters were further complicated when it was also discovered that flavonoids present in grapefruits also inhibited the metabolism of terfenadine.[51]

Moreover, certain compounds such as rifampicin (an antibiotic) can act as inducers of cytochromes P450[52] and potentiate the metabolism of other drugs such as steroid hormones used in oral contraceptive pills. Since in clinical practice it is extremely common for patients to be administered more than one drug for the same condition or an unrelated condition, the possibility of drug–drug interactions is extremely high.[53] For the pharmaceutical industry and

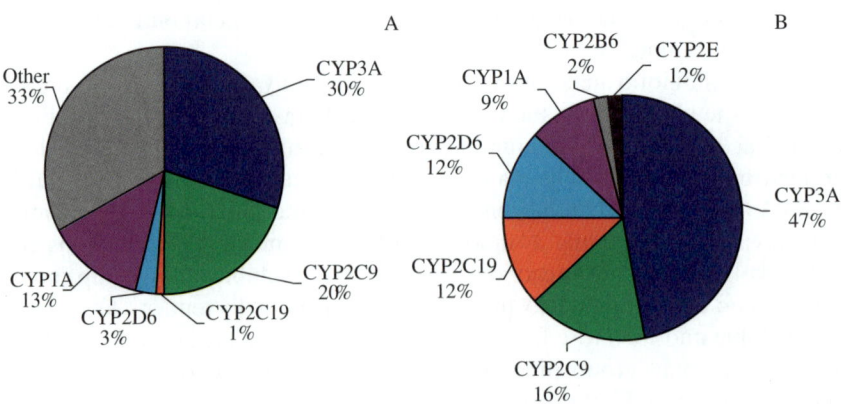

Figure 6.4 The hepatic content of human cytochrome P450 isoenzymes (**A**) and the involvement of each isoenzyme in CYP-mediated drug metabolism (**B**).[48]

clinicians this is a serious problem since these interactions have a huge implication on drug efficacy and safety.

Another factor that also has to be considered and is the subject of significant focus at present is the genetic difference between individuals that results in significant variations in P450-mediated drug metabolism.[54] Variations in the genes encoding a particular P450 enzyme within a population result in the expression of different phenotypes. This polymorphic nature of human cytochrome P450 enzymes has been known for nearly 30 years and numerous allelic variants of drug-metabolising human cytochromes P450 have been identified and cloned;[55] knowledge of the subject and its importance is still emerging. Polymorphic P450 enzymes, in particular human P450s 2C9, 2C19 and 2D6, are responsible for nearly 40% of cytochrome-P450-mediated drug metabolism and lead to the development of a significant number of adverse drug reactions. Below are some examples of the clinical consequences of polymorphic cytochrome-P450-mediated drug metabolism.

The allelic variants of the human P450 2C9 are becoming increasingly important in clinical practice since it is now recognized that toxicity associated with widely used drugs such as phenytoin and warfarin could in some part be due to these variations in 2C9 enzyme activity.[56,57] Of the 30 alleles that have been identified (http://www.cypalleles.ki.se/cyp2c9.htm), the most common are the *CYP2C9*2* and *CYP2C9*3* alleles.[56] Both alleles result in decreased metabolic activity with patients being homozygous for the *CYP2C9*3* having the lowest activity. These patients have an enzyme with catalytic efficiency 20-fold lower compared to the wild-type enzyme, resulting in poor clearance of 2C9 substrates and therefore increased drug levels in the body. This can often lead to bleeding complications with drugs like warfarin (an anti-clotting agent) and toxicity with drugs like phenytoin (an anti-epileptic).[56] The other allelic variant that is of interest is the *CYP2C9*6* allele. This allele has been found in the African-American population, resulting in the patients having no 2C9 activity and therefore when administered phenytoin, the patients exhibited severe toxicity.[58] These variations once again question both the notion of "one dose is suitable for all" and the general suitability of a drug across a range of ethnic groups. This emphasises the need to be able to predict the metabolic activity of these enzymes in a particular patient. However, current techniques do not allow routine testing of enzyme activity in individuals.

The human P450 2D6 enzyme is highly polymorphic and, to date, more than 60 allelic variants of this enzyme have been identified (http://www.imm.ki.se/CYPalleles/cyp2d6.htm), many of which involve splicing defects or frame shift mutations resulting in individuals with non-functional 2D6. Based on their ability to metabolise marker 2D6 substrates such as debrisoquine, the human population can be divided into three main categories. These categories are "poor metabolisers" (PMs), "extensive metabolisers" (EMs) and "ultra-rapid metabolisers" (UMs). PMs are individuals with lower capacity of metabolising 2D6 substrates, whereas UMs are able to rapidly clear the substrates. Phenotype investigations have shown that 5–10% of the Caucasian population are PMs, whilst in Orientals this figure is much lower (around 1%) and in African populations a range between 0 and 19% has been reported.[59] In general, there is a 2–5-fold difference between PMs and EMs in their capacity to metabolise 2D6 substrates. Therefore PMs obtain the same steady state serum levels as EMs at doses 50–80% lower.[60] In the case of drugs with narrow therapeutic windows, these differences can play a huge role in determining toxicity. Hence the ability to identify compounds that are 2D6 substrates early in drug development is of great interest to the pharmaceutical industry.

6.5 Protein Engineering of P450s to Improve or Expand their Catalytic Properties

The considerable substrate diversity, catalytic diversity and large number of sequences available for cytochrome P450 enzymes has led to a significant amount of interest in their engineering. The first studies on the engineering of cytochromes P450 were mainly focused on successful recombinant expression of these enzymes on a large scale in bacterial, yeast and insect cell systems. These studies led to attempts to increase the solubility of these enzymes for crystallisation, structure determination and other biophysical analyses. Furthermore, large-scale recombinant expression has allowed the production of significant quantities of cytochrome P450 enzymes for studies on catalytic specificity, the catalytic cycle and interactions with redox partners.

In addition to these studies, protein engineering of cytochromes P450 has also focused on engineering these enzymes for biotechnological applications. These studies can be divided into either directed evolution strategies or rational design depending on the engineering approach adopted and in the following sections, these two approaches will be reviewed in greater detail.

6.5.1 Directed Evolution of Cytochrome P450 Enzymes

Rational design approaches are sometimes limited due to a restricted knowledge of the relationship between sequence, structure and function and directed evolution of proteins is a powerful strategy that can partly overcome these limitations. The principle behind this approach is the generation of a library of randomly generated mutants and the screening of these mutants for particular characteristics *e.g.* improved catalytic properties. Improved mutants are isolated and used for the generation of more mutant libraries either by further random mutagenesis or gene recombination methods. The advantage of directed evolution is that even unpredictable mutations without an obvious relevance to the active site or structure might be found.[61] The potential limitations of this approach are that the screening of a large library is time consuming and a powerful and discriminate screening method is required. To this end, in our laboratory a rapid screening method based on the detection of a degradation product of NAD(P)H oxidation during substrate turnover has been developed and successfully applied to assess the substrate specificity of wild-type bacterial P450 102A1 towards non-natural substrates.[62,63]

Directed evolution has been applied to the bacterial 102A1 enzyme by several groups. Using five generations of random mutagenesis, Glieder and coworkers isolated the variant 139-3 that was highly efficient for alkane hydroxylation (C3-C8) and exhibited high activity towards fatty acids.[64] This variant had the highest turnover rate reported for an alkane hydroxylase and further mutagenesis of this variant generated highly regioselective and enantioselective alkane hydroxylases[65] such as the mutant 9-10A-A328V that hydroxylated octane at the 2-position to form *S*-2-octanol.[65] Another variant 1-12G hydroxylated long-chain alkanes primarily at the 2-position but formed *R*-2-alcohols. These variants led to the evolution of further mutants that were particularly active towards propane.[66,67] Furthermore, mutants 139-3, 1-12G and B could hydroxylate achiral cyclopentanecarboxylic acid derivatives stereoselectively and this was of importance since this molecule is an intermediate in the development of anti-HIV drugs.[68] The directed evolution of 102A1 has also been applied towards substrates of biotechnological importance. For example, a combination of site-directed and random mutagenesis has been used to generate mutants with enantioselectivity towards β-ionone.[69] Similarly the heme domain of the P450 102A1 has been evolved to produce authentic human metabolites of propranolol using the peroxide shunt pathway thereby removing the requirement for the costly NADPH cofactor.[70]

Random mutagenesis of P450 102A1 has also been applied to improve the stability of the enzyme in organic solvents. Variants with mutations at residues 235, 471 and 494 were found to be more resistant to acetone, acetonitrile, ethanol and dimethylformamide.[71] Additionally, directed evolution of the P450 102A1 enzyme has led to isolation of thermostable variants of this enzyme which maintain their activity at high temperatures.[72]

The other versatile bacterial P450 enzyme that has been the focus of various studies is the P450 101 enzyme. Directed evolution was applied to generate mutants of P450 101 that hydroxylated naphthalene with a range of regiospecificities in the absence of cofactors and additional redox proteins through the peroxide shunt pathway.[73,74] Similar studies have been used to isolate mutants that exhibited high activity towards indole oxidation and ethane oxidation.[75]

There has been little work on the directed evolution of mammalian cytochromes P450 for their exploitation in biotechnology. Random mutagenesis of human P450 1A2 in the putative substrate recognition sequences (SRS) and screening for improved activity towards natural substrates of 1A2, such as phenacetin and 7-ethoxyresorufin, resulted in the identification of 27 improved variants which displayed 30-fold variation in k_{cat} towards either or both of these substrates.[76] These approaches resulted in P450 1A2 variants with improved activity towards 7-methoxyresorufin and 2-amino-3,5-dimethylimidazo[4,5-f]quinoline (MeIQ).[77,78]

Semi-random mutagenesis of the six substrate recognition sequences in P450 2A6 resulted in mutants with novel catalytic activities towards the oxidation of indole and substituted indoles[79] and a double mutant isolated from this study was used in subsequent studies to identify novel indigoid inhibitors of cyclic dependent kinases.[80]

Random mutagenesis of P450 2B1 by Halpert and coworkers resulted in the isolation of variants with enhanced catalytic efficiency towards a panel of substrates, such as testosterone and several anticancer pro-drugs and also variants that exhibited enhanced thermostability and DMSO tolerance.[81-83] The same group applied this approach to the human P450 3A4 enzyme to generate variants that exhibited enhanced peroxide-mediated substrate oxidation towards a panel of substrates.[84]

Recently Gillam and coworkers created a library of chimeric mutants by DNA shuffling of human P450s 2C8, 2C9, 2C18 and 2C19. The diversity of the library was assessed using three luminogenic substrates, diclofenac and indole as probe substrates and variants with enhanced activity towards luciferin ME, luciferin H and diclofenac 4'-hydroxylation were identified.[85]

6.5.2 Rational Design of Cytochrome P450 Enzymes

Rational design has been applied to both bacterial and mammalian cytochrome P450 enzymes in order to exploit these enzymes for biotechnological purposes. Relevant to biosensor construction, the knowledge obtained from these studies such as the catalytic properties of the enzyme, the K_m and k_{cat} are all applicable to the construction of biosensors consisting of immobilised enzymes. The primary limitation of rational design is that ideally it requires the availability of the crystal structure or a homology model of the enzyme to guide the rational design. The primary biotechnological focus of rational design approaches to date has been:

- Site-directed mutagenesis studies to engineer substrate specificity;
- Engineering to create improved redox partner interactions.

The majority of site-directed mutagenesis studies on cytochrome P450 enzymes have been for the purpose of elucidating structure–function relationships of these enzymes. However, there have been a few studies directed towards altering the substrate specificity of these enzymes for potential use in biotechnological application and these are reviewed below.

The P450 101 enzyme has been engineered using site-directed mutagenesis for potential use in bioremediation. On the basis of the 3D structure,[11] substitution mutants T101M, T185F and V247M have been produced to alter the stereoselectivity and coupling of ethylbenzene hydroxylation.[86] Various mutants of Y96 have been engineered to create active sites that improve oxidation of phenyl derivatives such as diphenylmethane.[87] Furthermore, mutations of F87 and Y96 have been shown to enhance the activity of P450 101 towards polyaromatic hydrocarbons and polychlorinated benzenes[88,89] and the mutation T87F resulted in a 25-fold increase in the styrene oxidation activity of this enzyme.[90] The versatility of altering substrate specificity of the P450 101 enzyme was further demonstrated by the ability to engineer the enzyme to hydroxylate short-chain alkanes[91,92] and inert molecules such as pentachlorobenzene and hexachlorobenzene.[93]

The P450 102A1 enzyme has also been the focus of site-directed mutagenesis studies to improve its activity towards non-natural substrates. Substitution of F87 with various residues led to mutants with enhanced substrate specificity and regiospecificity in fatty acid hydroxylation[94] and in enhanced activity towards a variety of aromatic and phenolic substrates.[95] These initial mutants provided the basis for the further engineering of the P450 102A1 enzyme to create the mutant F87V/L188Q/A74G that was active towards indole, alkanes, cycloalkanes, arenes, heteroarenes, polyaromatic hydrocarbons and polychlorinated dibenzo-p-dioxins.[96,97]

Examples of site-directed mutagenesis studies for biotechnological application on cytochromes P450 other than the CYP101 and CYP102A1 are relatively scarce and this may be due to the fact that these enzymes show low initial activities and expression levels in heterologous hosts. However, the possibility of obtaining products that would otherwise be difficult through standard organic synthesis has meant that site-directed mutagenesis studies on other cytochromes have not escaped the attention of researchers. For example site-directed mutagenesis studies on the human steroid hydroxylating P450 11B1 and P450 11B2 have been used to rationally engineer the bacterial steroid hydroxylating enzyme P450 106A2 to produce specific hydroxylated steroids.[98–100] The regioselectivity for progesterone hydroxylation by P450 2B1 has also been re-engineered based on the X-ray crystal structure of P450 2C5. P450 2B1, a high-K_m progesterone 16α-hydroxylase, was converted to a low-K_m progesterone 21-hydroxylase by mutagenesis.[101] Furthermore, one of the mutants showed a very high k_{cat} and 80% regioselectivity for progesterone 21-hydroxylation.

In recent years recombinant DNA techniques have allowed the creation of various cytochrome P450 fusion proteins between natural and artificial redox partners in order to allow the expression and purification of high levels of catalytically self-sufficient protein. These catalytically self-sufficient cytochromes P450 have a wide range of exciting possible applications. Bacteria containing heterogously expressed cytochrome P450 fusion proteins would aid in the understanding of human drug metabolism, the consequences of drug–drug interactions and the activation of carcinogens. There is also the possibility of using these enzymes for bio-catalysis[102] or using hosts containing these proteins for bioremediation purposes. Therefore cytochrome P450 fusion proteins present a wide range of applications in drug metabolism studies, biocatalysis, bioremediation and, with careful selection of the redox partners within the fusion protein, these fusion proteins could also be used for biosensing purposes.[103,104] Furthermore, the careful selection of fusion partners could allow the creation of cytochrome P450 fusion proteins that have improved properties in comparison to the parent enzyme such as improved solubility, improved activity and enhanced coupling of cofactor consumption. This would prove to be particularly useful for microsomal-membrane-bound cytochromes P450 such as the drug-metabolising human cytochromes P450.

The earliest reported microsomal fusion protein was that of rat P450 1A1/rat NAPDH-cytochrome P450 reductase fusion protein constructed by Murakami and coworkers.[105] The enzyme was expressed in the holo form and it was active against the substrate 7-ethoxycoumarin. This led to constructions of further fusions between bovine P450 17A1 (P450c17) and yeast CPR and bovine P450 21 (P450c21) and yeast CPR that were active towards their respective substrates.[106,107] These studies initiated the construction of a number of microsomal P450/reductase fusion proteins to yield catalytically self-sufficient enzymes with improved properties that could be used for a range of studies and these are summarised in Table 6.3.

Fisher and coworkers constructed catalytically active fusions between bovine P450 17A and rat CPR and between rat P450 4A1 and rat CPR.[108] The group applied the same fusion protein principle to human P450 3A4 to create a P450 3A4/rat CPR fusion protein.[109,110]

A similar approach was used to engineer a human P450 1A1/rat CPR fusion protein that was active against 7-ethoxyresofurin, benzo[a]pyrene and

Table 6.3 Summary of the microsomal P450/microsomal NADPH-P450 reductase fusion proteins constructed to date.

P450 domain	Reductase domain	Substrate	k_{cat} (min^{-1})
Bovine CYP17A1	Rat CPR	Progesterone	30
Rat CYP4A1	Rat CPR	Lauric acid	30
Human CYP3A4	Rat CPR	Testosterone	7
Human CYP1A1	Rat CPR	Benzo[a]pyrene	1.4
Human CYP1A2	Rat CPR	7-ethoxyresofurin	0.24
Rat CYP1A1	Yeast CPR	7-ethoxyrcoumarin	0.4
Human CYP3A4	Yeast CPR	Testosterone	21
Rat CYP2C11	Rat CPR	Arachidonic acid	1.0
Rat CYP2C11	BMR	Arachidonic acid	0.2
Human CYP2D6	Human CPR	Bufuralol	4.1

zoxazolamine.[111] The group later extended the approach to human P450 1A2 and created a human P450 1A2/rat CPR fusion protein that, when purified, catalysed 7-ethoxyresofurin O-deethylation and phenacetin O-deethylation in the presence of NADPH and phospholipids.[112] Kondo and coworkers engineered a fusion between rat P450 1A1 and yeast CPR that catalysed 7-ethoxyresofurin O-deethylation.[113] In the study both mixed and fused systems had comparable activities that were found to decrease with increasing ionic strength suggesting an ionic interaction between the P450 and reductase domain, which decreased upon increasing ionic strength of the buffer. A similar construct between human P450 3A4 and yeast CPR was found to be more active at 6β-hydroxylation of testosterone than a mixed system.[114] However, coupling efficiency between NADPH utilisation and substrate hydroxylation in the absence of cytochrome b_5 was only 10% and increased to 50% in the presence of cytochrome b_5 for both fused and mixed systems.

Recently a fusion between human P450 2D6 and its redox partner human CPR was created by Deeni and coworkers.[115] The enzyme was able to catalyse bufuralol hydroxylation and dextromethorphan O-demethylation. However, activity of the enzyme was dependent on intermolecular electron transfer.

The primary objective behind the construction of P450/CPR fusions has been to create a system that would allow the convenient study of P450s without the problems associated with reconstituted systems. Therefore the activity of these fusions in comparison to the reconstituted systems is an important consideration. Whilst most fusion proteins have exhibited rates comparable to reconstituted systems, there have been exceptions reported.[112] However, it should be noted that the activity of mammalian P450s in both artificial reconstituted systems and fusion protein systems have been found to be highly dependent on the lipid nature,[108,116] salt composition[113,117] and the presence of detergents.[110] Therefore, comparison between systems is not always straightforward.

The construction of fusion proteins has not been limited to class II P450 systems and chimeras have also been made using components of class I and class III P450s. Microsomal cytochromes P450 have also been fused to domains from the bacterial enzymes P450 102A1 (P450$_{BM-3}$) and P450 101 (P450$_{cam}$). A fusion protein comprising the rat P450 2C11 fused to BMR (the reductase domain of P450 102A1) via a Pro-Ser-Arg linker was found to catalyse the metabolism of arachidonic acid.[118] However, its activity was lower than both a reconstituted system containing P450 2C11 and rat CPR and a fused system between P450 2C11 and rat CPR. Furthermore, activity of the 2C11/BMR fusion was greatly enhanced upon addition of purified reductase either of mammalian or bacterial origin indicating poor intra-molecular electron transfer.

Using the molecular Lego approach,[104] our laboratory has created fusion proteins consisting of the human P450s 2E1, 2C9, 2C19 and 3A4 fused to the reductase domain of bacterial P450 102A1.[119,120] The CYP2E1/BMR, CYP2C9/BMR, CYP2C19/BMR and CYP3A4/BMR chimeras were both active and correctly folded in the absence of detergent and, in comparison with the parent P450 enzyme, these chimeras showed greatly improved solubility properties. The chimeras were catalytically self-sufficient and presented turnover rates similar to those reported for the native enzymes in reconstituted systems, unlike previously reported mammalian cytochrome P450 fusion proteins. Furthermore, the specific activities of these chimeras were not dependent on the enzyme concentration present in the reaction and did not require the addition of accessory proteins, detergents or phospholipids to be fully active. The solubility, catalytic self-sufficiency and wild-type like activities of these chimeras would greatly simplify the studies of cytochrome-P450-mediated drug metabolism in solution.

The focus of plant P450 fusion protein research has been to engineer herbicide tolerance into plants and to harness their incredible biosynthetic P450 activities for human purposes. Lamb and coworkers constructed a fusion protein between P450 71B1 from *Thlaspi arvensae* and the CPR domain of *Catharanthus roseus* to study the effect of glycophosphate (Roundup®) on P450 71B1.[121] The study showed that glycophosphate inhibited the turnover of the polycyclic aromatic hydrocarbon benzo[a]pyrene (another herbicide) by the enzyme.

The expression of plant P450 fusion proteins in bacteria allows the production of medically important compounds that plants produce naturally at low levels. Schroder and coworkers constructed a fusion protein between *Catharanthus roseus* P450 71D12 and its reductase.[122] The fusion enzyme was successfully expressed in bacteria and exhibited tabersonine 16-hydroxylase activity. Tabersonine 16-hydroxylase is the first enzyme in a pathway that produces the medically important bisindoles vinblastine and vincristine, both of which have uses in the treatment of leukaemia. These compounds are produced naturally in plants at very low levels (0.0005% of plant mass) and chemical synthesis due to their complex structure is costly and difficult. Harnessing the biosynthetic capability of P450s to create an artificial synthetic pathway allows the mass production of these compounds.

The concept of engineering herbicide resistance into plants using P450s has not been limited to plant P450 fusions. Shiota and coworkers engineered a rat P450 1A1/yeast CPR fusion protein in tobacco plants that was shown to provide the plants with resistance to the herbicide chlortoluron.[123] Similarly the group constructed a fusion protein consisting of rat P450 1A1 fused to the iron sulfur protein maize ferredoxin (Fd) and the FAD-containing pea ferredoxin $NADP^+$ reductase (FNR).[124] The most active fusion, P450 1A1-FNR-Fd, when expressed in yeast was active towards 7-ethoxycoumarin and the herbicide chlortoluron.

Finally protein engineering studies using site-directed mutagenesis and rational design approaches have opened up a variety of possibilities that will allow the large substrate diversity and catalytic versatility of cytochrome P450 enzymes to be exploited for biotechnological applications such as biosynthesis, drug metabolism research, pro-drug activation, toxicology research, bioremediation, plant herbicide tolerance, bioelectrocatalysis and biosensing.

6.6 Interfacing Cytochromes P450 to Electrodes

Over the years it has been shown that the direct electrochemistry of wild type P450 enzymes in their catalytically active status (able to convert substrate into product) on unmodified electrodes is very difficult if not altogether impossible.[125–129] This has been ascribed to a number of reasons including the intrinsic instability of the enzymes, the presence of a deeply buried heme cofactor, the difficulty of sustaining the catalytic cycle without autoxidation and inactivation of the protein amongst others. A further complication is the possibility of these enzymes to carry out uncoupled reactions discussed in Section 6.3, and this makes it even more difficult to show the generation of a product from catalysis confirming the presence of an active enzyme on the electrode surface. These problems have only recently been successfully addressed by different combinations of electrode modification and protein engineering.[6]

One important parameter that bioelectrochemistry can measure is the midpoint potential (E_m) of cytochromes P450. This has proven to be a very tricky parameter to extract, as the values obtained by different authors are rather scattered depending on the method and electrode surface used. A recent interesting approach taken by Sligar and coworkers makes use of membrane-bound cytochromes P450 embedded in phospholipid bilayer nanodiscs. These consist of a palmitolyloleolylphosphatidylcholine bilayer kept together around the hydrophobic edges by two monomers of a membrane scaffold protein that forms an amphipathic helical ring defining the perimeter of a 10-nm-diameter and 4-nm-thick disc.[130] The approach has been used to study the midpoint potential of human cytochrome P450 3A4 and the ratio nanodisc: 3A4 was found to be 1 : 1. Spectro-potentiometric titrations carried out in the presence of redox mediators led to midpoint potentials of -220 ± 10 mV for the substrate-free P450 3A4 and -140 ± 5 mV for its testosterone-bound form.[131] This ~ 80 mV shift is similar to that observed for the bacterial P450 and has not been previously reported for the mammalian enzymes.

In order to obtain catalysis in a viable amperometric device, P450s must be stably immobilised onto electrode surfaces. Different strategies have been adopted, ranging from the direct adsorption to the entrapment within surfactants or polymers, with the use of gold nanoparticles and, in rare examples to date, with the covalent linkage *via* self-assembled monolayers on gold. These approaches will be discussed in the following sections.

6.6.1 Immobilisation on Unmodified Electrodes

As much of the knowledge available on P450 enzymology is based on the best characterised P450$_{cam}$ from *Pseudomonas putida*, many of the efforts in bioelectrochemistry have been dedicated to this enzyme. In a pioneering work, Hill and coworkers used recombinant P450$_{cam}$ on edge-plane graphite electrodes to measure midpoint potentials of -279 ± 11 and -143 ± 10 mV for the camphor-free and -bound respectively, under anaerobic conditions and pH 7.4.[132] As in these conditions the overall charge of the protein is negative, the interaction of the enzyme with the negatively charged electrodes was explained with the intervention of the patch of basic surface residues present on the proximal side of the protein, Arg-72, Arg-112, Lys344 and Arg-364, known to interact with the physiological redox partner putidaredoxin. These residues were used to insert unique cysteine mutants, Arg72Cys, Arg112Cys, Lys344Cys and Arg364Cys; the Lys344Cys mutant was also modified with N-Ferrocenylmaleimide. The mutants were studied on edge-plane graphite electrodes at high scan rates (10 V s^{-1}) where Lys344Cys gave a midpoint potential of -293 mV. At slow scan rates (<0.5 V s^{-1}) the cyclic voltammograms were dominated by oxygen reduction. No details on the catalytic oxidation were reported.[132,133]

Different data were found by Sheller and coworkers who measured the electrochemistry of P450$_{cam}$ using a glassy carbon electrode modified with sodium montmorillonite. They found a higher E$_m$ of -139 mV with no shift of E$_m$ upon substrate binding and high values (up to 152 s^{-1}) for the heterogeneous electron transfer rate constant (k_s). No product formation was reported in the study.[134]

Fleming and coworkers also reported the electrochemistry of P450 199A2 from *Rhodopseudomonas palustris* directly observed on pyrolytic graphite. The authors measure an E$_m$ of -132 mV and a high k_s of 550 s^{-1}, but no further product was reported in this work.[135]

To date, the overall emerging picture for the direct adsorption of P450s on unmodified electrode surfaces is that of electrochemically responding proteins with relatively high k_s but not able to turnover substrates into products.

6.6.2 Immobilisation with Surfactants, Polymers and Gold Nanoparticles

Electrochemical sensitivity strongly depends on both the concentration of the electroactive species at the electrode surface and their specific orientation to achieve optimal heterogeneous electron transfer. Often bare electrode surfaces have been found to distort or denature the marginally stable folded state of the proteins. For these reasons, much effort has been directed over the years on the immobilisation of enzymes on modified electrodes, where either the electrode or the protein or both have been tailored to achieve optimal electrochemical response.[5,6,125,128,129,134,136–140]

Electrode modification for immobilisation of P450s has been achieved in a number of ways, as summarised in Table 6.4, and has ranged from the use of films of positively charged surfactants such as didodecylammonium bromide (DDAB),[125–129,136,139,140] poly-(dimethyldiallylammonium chloride) (PDDA),[4,6,136,140,141] polyethylenimmine (PEI),[142] to the negatively charged poly(sodium 4-styrenesulfonate) (PSS),[5,143] to neutral polymers such as polyethylene oxide (PEO),[142] colloidal gold and chitosans,[137] to ordered self-assembled thiol-terminated chains such as cystamine-maleimide.[6,136]

DDAB is by far the most widely used polycation that has been applied on a variety of electrodes for both prokaryotic and eukaryotic P450s, with various degrees of success. DDAB films formed on edge-plane pyrolytic graphite have been used to characterise P450 2C9, 2C18 and 2C19 by Bernhardt and

Table 6.4 Overview of the main immobilisation strategies used for P450 enzymes.

Modification	Formula	Charge	Electrode	P450	References
DDAB	Br⁻ H₃C—N⁺(CH₃)—CH₂(CH₂)₁₀CH₃ with CH₃	+	EPG	P450BM3 BMP	Fleming et al., 2003
			Glassy carbon	2E1	Fantuzzi et al., 2004
			EPG	2C9, 18, 19	Shukla et al., 2005
			EPG	2C9	Jhonson et al., 2005
			EPG	P450c17	Jhonson et al., 2006
			Pyrolytic graphite	P450cam	Udit et al., 2006
			Pyrolytic graphite	P450BM3	Udit et al., 2006
			Screen printed graphite–Au	2B4, 1A2, CYP51b1	Shumyantseva et al., 2007
PEO	H—[O—CH₂(CH₂)₁₀CH₃]ₙ—OH		Plastic formed carbon	P450st	Matsumura et al., 2006
PDDA	Cl⁻ pyrrolidinium polymer (H₃C-N⁺-CH₃)ₙ	+	Au-MPS	3A4	Joseph et al., 2003
			Glassy carbon	2E1	Fantuzzi et al., 2004
			Au-MPA	2E1	Fantuzzi et al., 2004
			Glassy carbon	BMP	Fantuzzi et al., 2006
			Pyrolytic graphite	Microsomes	Krishnan et al., 2007
			SnO₂	BMP	Panicco et al., 2008
			Glassy carbon	3A4	Dodhia et al., 2008

PEI branched	[structure]	+	SnO$_2$	BMP	Panicco et al., 2008
PSS	[structure]	−	Carbon cloth Carbon cloth	P450cam 1A2	Estavillo et al., 2002 Estavillo et al., 2002
Cystamine-maleimide	[structure]		Au Au	2E1 3A4	Fantuzzi et al., 2004 Dodhia et al., 2008

coworkers[127] but no catalytic activity towards substrates has been reported. Martin and coworkers used the same derivatised electrodes to study human 2C9, finding a dependence of k_s on pH, consistent with the coupled electron-proton transfer required for the P450 catalytic cycle (Figure 6.3). The higher the pH, the faster the k_s: 139 s^{-1} at pH 6.0, 150 s^{-1} at pH 7.4 and 165 s^{-1} at pH 8.2. Although these authors measured the K_m for diclofenac (6.8 μM) and torsemide (11.4 μM) for the enzyme in solution, no product formation was reported on the electrode.[129] The same group used edge-plane pyrolytic graphite to study human, bovine and porcine P450c17. Here fast k_s values of 164 s^{-1} (human), 157 s^{-1} (bovine) and 153 s^{-1} (porcine) were measured, but again product formation was not reported.[128]

Hill and coworkers successfully used DDAB derivatised edge-plane graphite (EPG) electrodes to study the electrochemistry of cytochrome P450$_{BM-3}$ and its reductase and heme domains.[125] The authors reported very good cyclic voltammetry responses with E_m of −405 mV for the reductase and −244 mV for the heme domain. Electron transfer rate constants of 138 s^{-1} for the P450$_{BM-3}$ and 221 s^{-1} for the separated heme domain were measured. Although these relatively fast rate constants are in good agreement with those found in solution by other methods, it was found that, once generated, the ferrous heme rapidly bound oxygen, prior to dissociation leading to hydrogen peroxide and no substrate turnover.

Our group found that using glassy carbon electrodes modified with DDAB did not improve the performance of human P450 2E1 when compared to the results obtained from bare GC. While immobilisation on bare GC gave a k_s of 5 s^{-1}, the quality of the data obtained from DDAB did not allow measurement of this parameter. Both bare glassy carbon and its modification with DDAB did not give product formation, but it is important to note that the same enzyme on other electrode surfaces was able to generate product (Section 6.6.3).

Most probably the explanation of the lack of catalysis found by several groups when using DDAB is in the detailed spectroscopic study on the effect of DDAB on the structure of P450$_{BM-3}$ carried out by Udit and coworkers.[126] Optical spectroscopy (absorption and infrared) and voltammetry showed how the DDAB has the capacity of leading to the inactive P420 state, caused by the strong interactions between the protein and the surfactant. These are likely to be more crucial for P450 enzymes with a flexible scaffold, such as P450$_{BM-3}$, and the mammalian P450s that are known to be highly homologous to P450$_{BM-3}$. Could this be the reason why the literature often cannot demonstrate the catalytic products of P450s on DDAB electrodes?

Further evidence of this has been seen in recent cyclic voltabsorptometry work.[142] This technique combines optical and electrochemical data during potential sweep measurements and enables the detection of active and inactive P450 forms during voltammetry. Measurements on P450 BMP immobilised on nanocrystalline mesoporous tin-oxide (SnO$_2$) electrodes with polyethylenimmine (PEI) and polydiallyldimethylammonium chloride (PDDA) showed how the two polycations behave quite differently. PEI was found to stabilise the P420 form while PDDA stabilised the P450 as shown in Figure 6.5. The different effects of

Cytochromes P450: Tailoring a Class of Enzymes for Biosensing 177

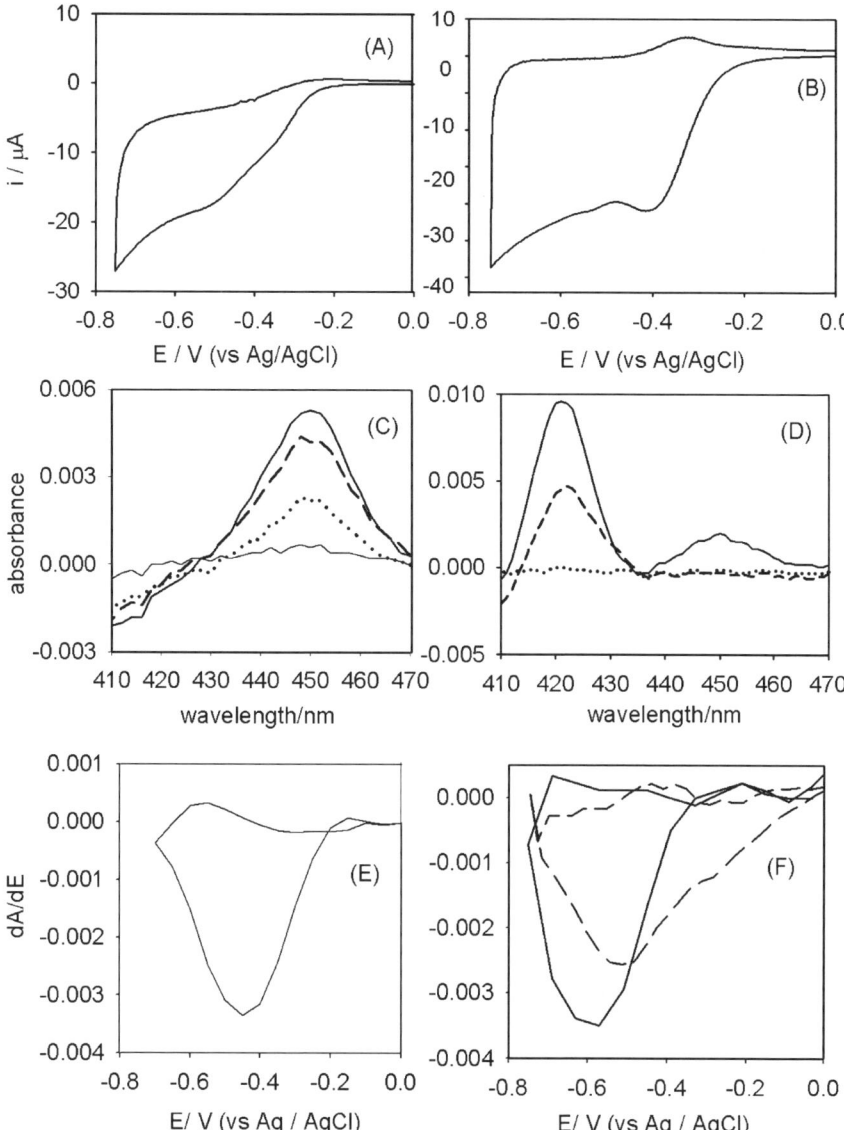

Figure 6.5 Spectroelectrochemical experiments recorded in presence of CO for BMP immobilised on SnO_2-PDDA and SnO_2-PEI electrodes.[142] (A) Cyclic voltammogram (CV) of the BMP/SnO_2-PDDA electrode; (B) Cyclic voltammogram (CV) of the BMP/SnO_2-PEI electrode; (C) Absorbance spectra collected during the cyclic scan on the BMP/SnO_2-PDDA. Different absorbance spectra were collected at different potentials: thin solid line at 0.2 V, round dots line at −0.42 V, dash line at −0.5 V and thick solid line at −0.75 V; (D) Absorbance spectra collected during the cyclic scan on the BMP/SnO_2-PEI. Different absorbance spectra were collected at different potentials: solid line at −0.75 V, dash line at −0.45 V and round dots line at −0.2 V; (E) Calculated DCVA at 450 nm for the BMP/SnO_2-PDDA; (F) Calculated DCVA at 420 nm (dash line) and 450 nm (solid line) for the BMP/SnO_2-PEI.

PEI and PDDA are ascribed to their different chemical characteristics. PEI contains primary and internal amino groups that can form H-bonds influencing the protein H-bonding network as suggested by Udit for DDAB.[126] Moreover, these amines could also react with the groups of the protein matrix responsible for the maintenance of the thiolate bond. On the contrary, the cyclic structure and the high pK_a of the quaternary ammonium group of PDDA keep it positively charged over a wide pH range and unable to form H-bonds with the protein matrix. One key observation made in this work is that when both the P450 and P420 forms coexist on the electrode, the P420 form dominates by far the cyclic voltammogram. In light of these findings, is it possible that there may be examples in the literature where the inactive P420 form was interpreted as the P450 form and therefore no product formation was observed?

Consistent with the better performance and stabilisation of PDDA on P450s, when human 2E1 and 3A4 were immobilised on GC or gold modified with mercaptopropionic acid (Au-MPA) modified with the above surfactant, the formation of the catalytic product was measured.[6,136]

In contrast with the observation made by Fleming *et al.* who reported a hydrogen peroxide dominance in the electrochemistry of $P450_{BM-3}$ when using DDAB,[125] Fuhr and coworkers used PDDA on gold modified with 3-mercapto-1-propenesulfonic acid (Au-MPS) to immobilise human P450 3A4 finding a minor contribution of hydrogen peroxide and modest but significant product formation from the dealkylation of verapamil and the hydroxylation of midazolam.[4]

Polystyrene sulfonate (PSS) on carbon cloth was used by Rusling and coworkers to study the epoxidation of styrene by human P450 1A2 *via* hydrogen peroxide.[5] Nakamura and coworkers instead described the electrochemistry of P450 from *Sulfolobus tokodaii* strain 7 (P450st) entrapped in plastic-formed carbon electrodes treated with P450st in polyethylene oxide (PEO), but without demonstrating catalytic activity.[144]

Gold nanoparticles have also been successfully used in conjunction with DDAB and chitosan to detect P450 electrochemistry. Liu and coworkers used films of colloidal gold nanoparticles and chitosan to encapsulate P450 2B6 on glassy carbon electrodes. An E_m of -207 mV and a k_s up to $16.4\,s^{-1}$ were measured. Here the catalytic activity was demonstrated with the C-hydroxylation and heteroatom release from bupronion, lidocaine and cyclophosphamide.[137]

Shumyantseva and coworkers used colloidal gold nanoparticles with DDAB on screen-printed graphite electrodes to study P450 2B4, 1A2 and 51B1. The later one showed an E_m of -26 mV and a k_s of $1.3\,s^{-1}$ with a demonstrated catalytic activity on the substrate lanosterol.[138,139]

6.6.3 Immobilisation by Covalent Linkage on Gold Electrodes: Use of Spacers

Although the use of surfactants or various polymers provide a relatively straightforward way to immobilise P450s on electrode surfaces, the covalent linkage of the enzymes either to bare or to derivatised gold electrodes has proven to lead to improved electron transfer rate constants and catalytic efficiency.

In general, the electrochemistry of redox enzymes on bare gold is very difficult to achieve. This has been repeatedly observed, starting from the early work on P450$_{cam}$ mutants[133] to the more recent work in our group on human P450 2E1.[136]

Non-covalent linkage of P450s on negatively charged electrode surfaces such as GC or Au-MPA is thought to be achieved thanks to the interaction of the patch of positively charged residues on the proximal side, opposite to the substrate binding region and optimal for electron transfer thanks to the shorter distance of the heme iron from the surface, as shown in Figure 6.6. This patch is present in virtually all P450s and it has been discussed for P450$_{cam}$,[132] P450 199A2[135] and human P450 2E1.[136]

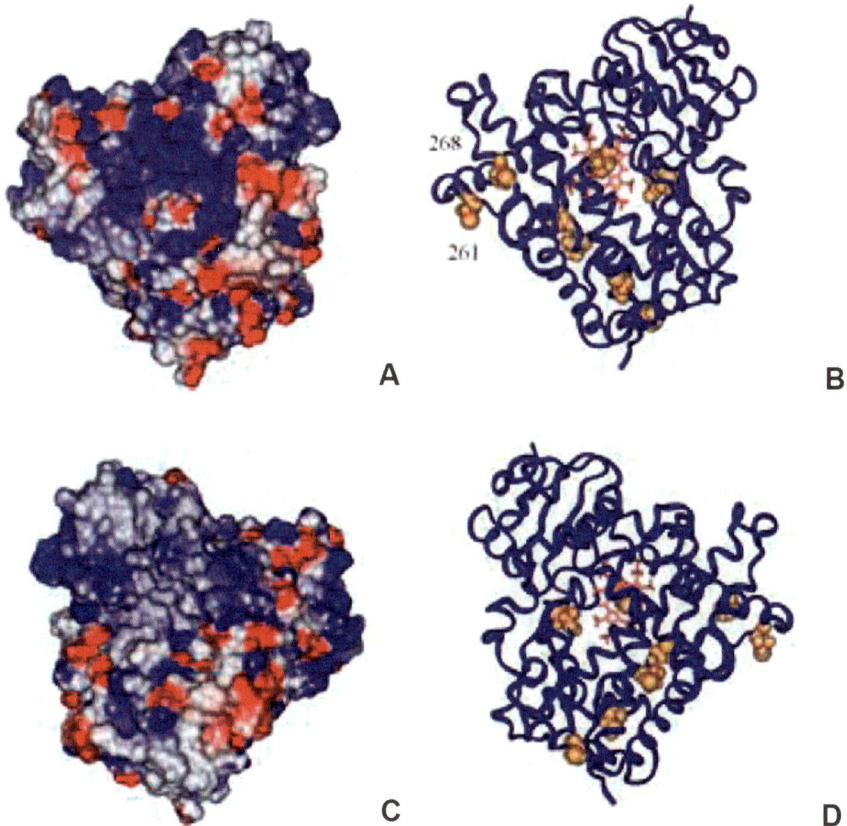

Figure 6.6 Delphi surface (**A, C**) and α-carbon ribbon (**B, D**) representation of P450 2E1.[136] (**A, B**) reductase binding face (proximal); (**C, D**) substrate entrance face (distal). Electrostatic potentials are indicated in blue (positive) and red (negative) and the cysteine residues are represented in yellow space filled.

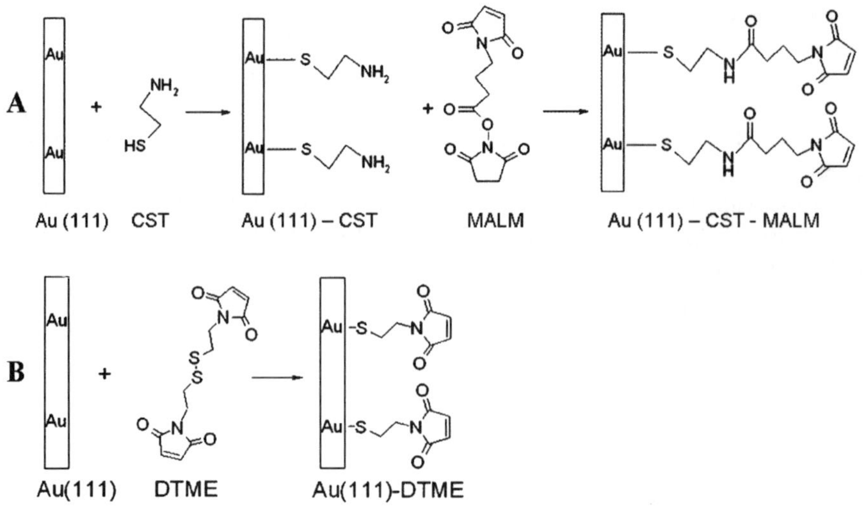

Figure 6.7 Schematic representation of strategies for the linkage of (A) cystamine-N-succinimidyl 3-maleimidopropionate (CST-MALM) and (B) dithio-bismaleimidoethane (DTME) spacers onto Au(111) surfaces.[148]

A higher level of order on the electrode surface is achieved when the protein[103] is covalently linked by exposed surface cysteine residues to spacers or linkers that in turn are covalently bound to the gold electrode. In this case several flexible spacer molecules can be used as self-assembling monolayers on gold exposing thiol reactive groups such as maleimido, thiol and disulfide groups able to mask the metal surface, as shown in Figure 6.7 and, at the same time, providing an anchor to the protein.[136,145–148]

This has been achieved for the first time with catalytic product formed in human P450 2E1, where the exposed Cys261 and 268 were used to link the enzyme to cystamine-maleimide on gold electrodes. A monolayer protein coverage of 1.1×10^{13} molecules cm^{-2} was achieved with a k_s of 10 s^{-1}, compared to a k_s of 5 s^{-1} for the same protein on GC. The catalytic ability to convert the substrate p-nitrophenol into the product p-nitrocatecol was quantified and a K$_m$ of 130 µM was calculated from a typical Michaelis–Menten curve determined for the immobilised protein as shown in Figure 6.8.

6.6.4 Protein Engineering to Control Protein Immobilisation and Catalytic Turnover on Electrode Surfaces

A further step in aiding not only the covalent-oriented immobilisation, but also the modulation of the catalytic performance of P450s involves the use of protein engineering strategies.

As early as in 2000, we proposed the construction of multi-domain P450 enzymes, where using the knowledge in genomics and proteomics as a

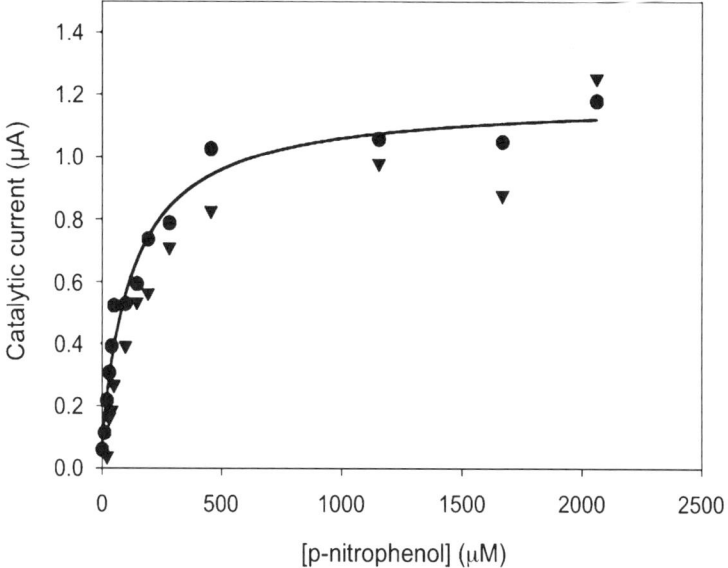

Figure 6.8 Michaelis–Menten curve determined for a p-nitrophenol titration on Au electrodes modified with cistamine-maleimide and covalently linked human cytochrome P450 2E1. Circles and triangles: p-nitrophenol in 500 mM potassium phosphate pH 7.0 for two different electrodes modified with Au/CYSAM/MALIM/P450 2E1.[136]

source, well-characterised electron transfer modules can be genetically fused to the P450 of interest.[103,104,119,120,140,149] The validity of this so-called molecular Lego approach was demonstrated with the chimera constructed with the fusion of the heme domain of *B. megaterium* cytochrome P450$_{BM-3}$ (BMP) and *D. vulgaris* flavodoxin (FLD) as shown in Figure 6.9. The choice of flavodoxin was motivated by its high sequence similarity to the FMN-binding domain of both P450$_{BM-3}$ and the cytochrome P450 reductase of class II P450s. The demonstration of the validity of this approach on the amenable bacterial system was key for its extension to more complex eukaryotic enzymes. Direct electrochemistry of BMP was reported on glassy carbon electrodes using poly-(dimethyldiallylammonium chloride) (PDDA) for film assembly. While BMP alone was found to be not very active toward the substrate even if a high k_s was measured, the catalytic activity is improved by almost six times when the non-physiological redox partner FLD is fused to it as shown in Figure 6.10A.[140] This was ascribed to a control role of the flavodoxin over the electron flow ensuring high coupling of the BMP. It is possible that in the case of BMP, electrons are wasted in uncoupled reactions with the production of reactive oxygen species. This is in keeping with the report by Gray and coworkers[150] where BMP was shown to convert oxygen to either water or hydrogen peroxide depending on the method of immobilisation on a carbon electrode.

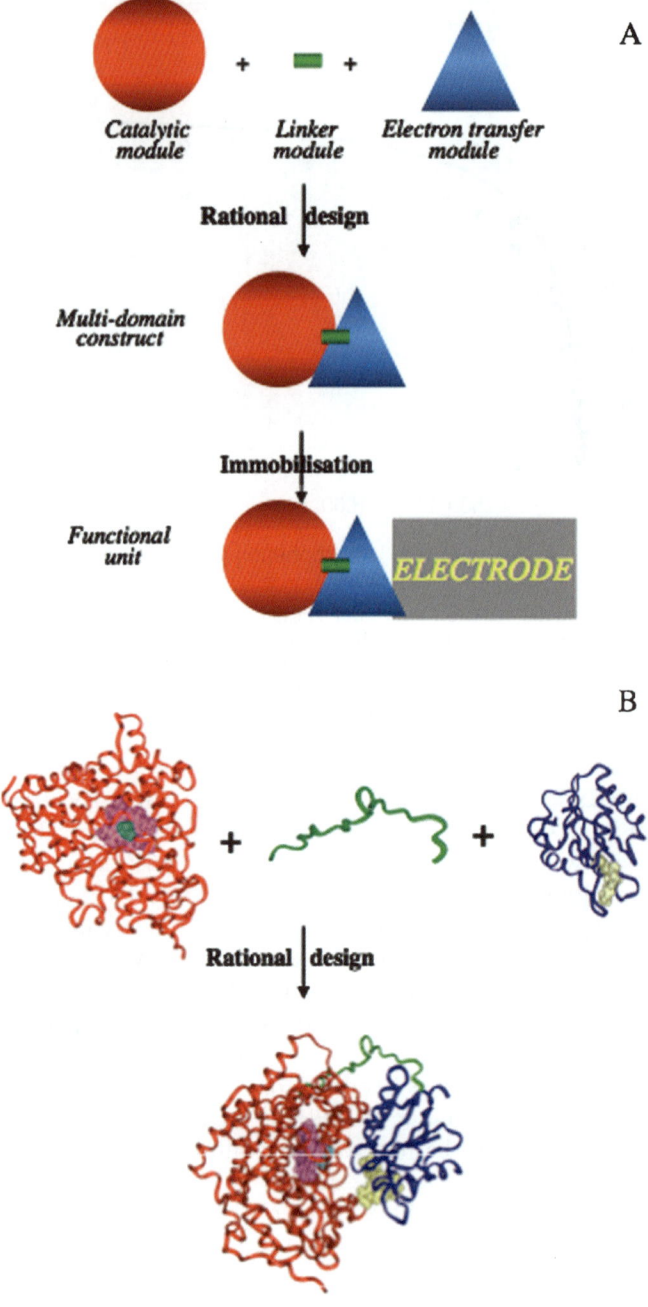

Figure 6.9 **A**. The *molecular Lego* approach.[149] Protein domains and loops are used to rationally link artificial redox chains enabling the electrochemically inaccessible catalytic module to communicate with the electrode. **B**. Application of the approach to the heme domain of cytochrome P450 BM3 using flavodoxin.

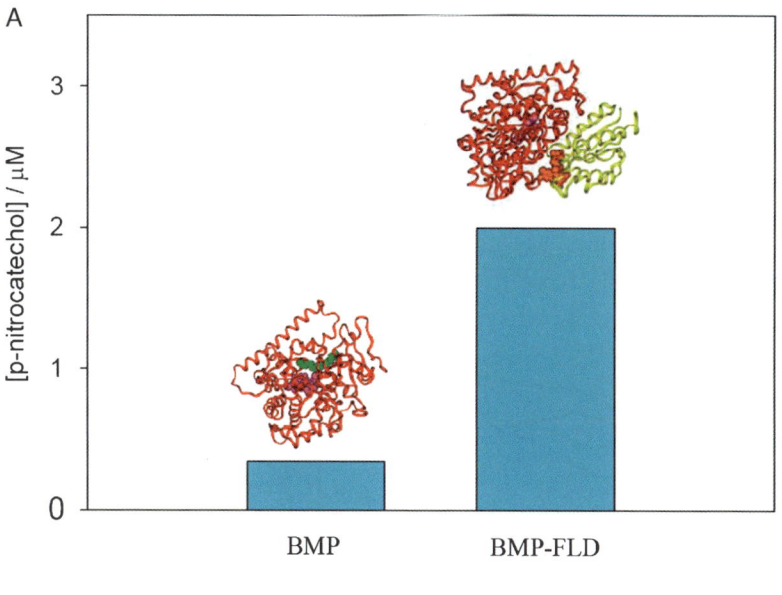
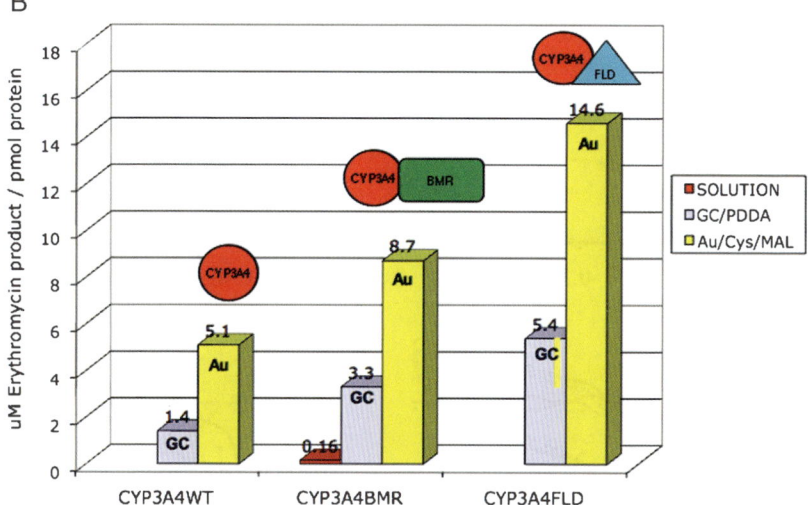

Figure 6.10 (**A**) Improvement in the product formation by BMP, the heme domain of P450 BM3 from *B. megaterium*, and its engineered chimera (BMP-FLD) with flavodoxin from *D. vulgaris*.[140] (**B**) Improvement in the product formation by human P450 3A4 from wild type (3A4WT), its engineered chimeras with the reductase domain of P450 BM3 from *B. megaterium* BMR (3A4BMR) and flavodoxin from *D. vulgaris* FLD (3A4FLD). All three proteins were immobilised on glassy carbon electrodes modified with PDDA (grey bars) and on gold electrodes modified with cystamine-maleimide self-assembled monolayers for covalent immobilisation of the proteins (yellow bars).

The molecular Lego was also used to modulate the coupling efficiency of the P450 most relevant to drug metabolism, human P450 3A4. The *N*-terminally modified human P450 3A4 was fused at the genetic level either to the reductase domain of P450 102A1 from *B. megaterium* (BMR) to create the CYP3A4/BMR or to flavodoxin from *D. vulgaris* (FLD) to create the CYP3A4/FLD. These proteins were chosen due to the significant sequence identity between these proteins and the human NAPDH-cytochrome P450 reductase (32.3% and 26.1% for the BMR and FLD, respectively). The CYP3A4, CYP3A4/BMR and CYP3A4/FLD were immobilised at poly-diallyldimethylammonium

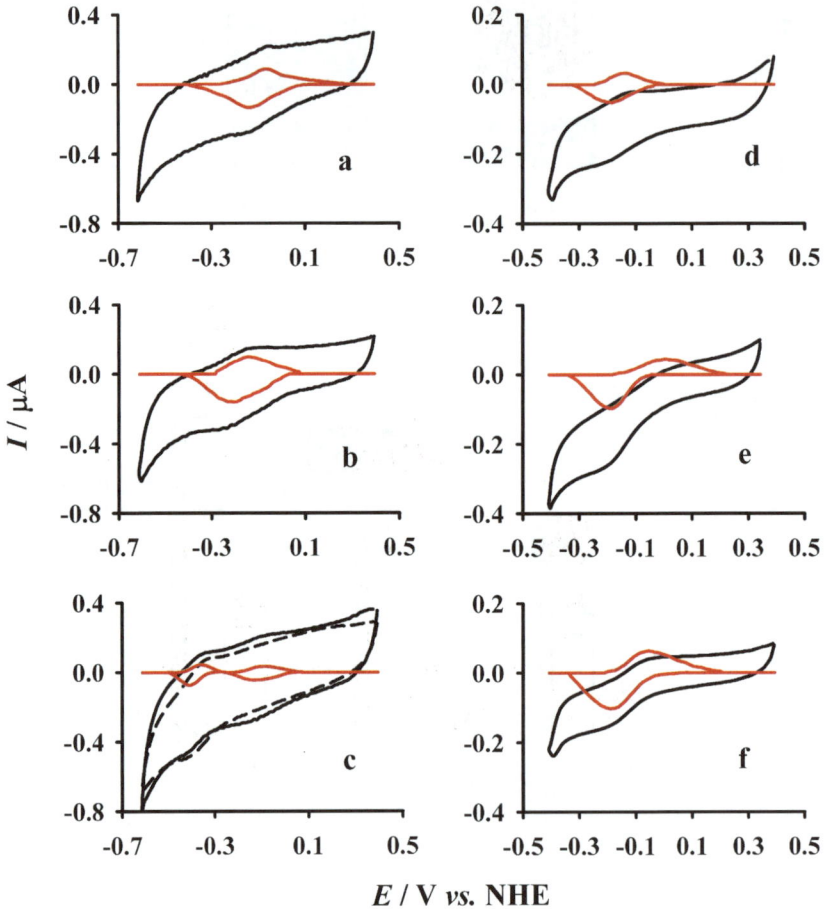

Figure 6.11 Anaerobic cyclic voltammograms (CV) of CYP3A4 (**a**), CYP3A4/BMR (**b**) and CYP3A4/FLD (**c**) on GC electrodes and of CYP3A4 (**d**), CYP3A4/BMR (**e**) and CYP3A4/FLD (**f**) on Au electrodes.[6] Scan rate 50 mV s^{-1} in 50 mM potassium phosphate with 100 mM KCl pH 7.4 at 25 °C. Shown are the original (black lines) and baseline corrected CVs (red lines intensity multiplied by 2 for clarity). The dashed line in (**c**) shows the CV of the FLD alone.

chloride (PDDA) modified glassy carbon electrodes and at thiol reactive cystamine-maleimide modified gold electrodes and cyclic voltammograms obtained for the three proteins on the two surfaces are shown in Figure 6.11.

The k_s on GC electrodes were found to vary from $378\,\text{s}^{-1}$ for 3A4, to $227\,\text{s}^{-1}$ for 3A4/FLD, to $281\,\text{s}^{-1}$ for 3A4/BMR. Interestingly product formation (quantity of product formed) and coupling efficiency (% of electron usage in the product formation) were found to vary as a function of the k_s, with the "slower" k_s leading to "higher" product formation and coupling. The data were confirmed when the same constructs were covalently linked to Au derivatised with cystamine-maleimide. Here slower k_s were found, $3.7\,\text{s}^{-1}$ for 3A4, $1.0\,\text{s}^{-1}$ for 3A4 BMR and $0.6\,\text{s}^{-1}$ for 3A4/FLD. Again the coupling efficiency was found to improve with the slower k_s: 1.9% for 3A4 and 19.9% for 3A4/FLD. The better performance for the slower k_s values was found to be consistent with a longer-lived iron-peroxy intermediate that leads to better controlled catalysis channelled in the formation of the product.[6] The data obtained in this work are shown in Figure 6.10B. This clearly demonstrates that a combined electrode and protein engineering approach may be necessary to achieve P450 enzymatic activity with turnover on electrode surfaces.

6.7 Conclusions

In conclusion the progress being made on many fronts to develop technologically viable cytochromes P450, spanning from the fine tuning of their catalytic parameters and substrate recognition, to the creation of new enzymes with the directed evolution approach, to the fast progress made in their bioelectrochemistry and applications in amperometric sensors, testifies the importance of this class of enzymes in tomorrow's biosensing technology.

The greatest progress made over the last five years is in the area of their immobilisation on electrode surfaces for environmental screening applications and drug metabolism toxicological tests. The measurements of parameters such as E_m, in some cases its shift upon substrate binding, heterogeneous electron transfer constants and potential pH dependence have been noted and a systematic scheme is now emerging. The demonstration of a catalytically active enzyme able to generate a product on the electrode surface has only become feasible during the past couple of years. Relatively straightforward immobilisation strategies, such as direct adsorption of the enzymes on electrode surfaces or the use of surfactants such as DDAB, although leading to high k_s values ($>100\,\text{s}^{-1}$), does not necessarily lead to catalytically active enzymes. Where colloidal gold or self-assembled monolayers have been used also to link the enzymes in a flexible conformation, slow k_s values have been obtained ($<20\,\text{s}^{-1}$) and catalytic activity/product formation demonstrated. This point has been reinforced by protein and electrode engineering that demonstrated the coupling efficiency can be modulated by controlling the electron transfer process from the electrode surface to the P450 enzyme.[6]

References

1. M. Klingenberg, *Arch. Biochem. Biophys.*, 1958, **75**, 376.
2. F. P. Guengerich, in *Cytochrome P450: Structure, Mechanism and Biochemistry*, ed. P. R. Ortiz de Montellano, Kluwer Academic/Plenum Publishers, New York, 2005.
3. E. M. Gillam, *Chem. Res. Toxicol.*, 2008, **21**, 220.
4. S. Joseph, J. F. Rusling, Y. M. Lvov, T. Friedberg and U. Fuhr, *Biochem. Pharmacol.*, 2003, **65**, 1817.
5. C. Estavillo, Z. Lu, I. Jansson, J. B. Schenkman and J. F. Rusling, *Biophys. Chem.*, 2003, **104**, 291.
6. V. R. Dodhia, C. Sassone, A. Fantuzzi, G. D. Nardo, S. J. Sadeghi and G. Gilardi, *Electrochem. Commun.*, 2008, *Electrochem. Commun.* 2008, **10**, 1744.
7. F. P. Guengerich, *Chem. Res. Toxicol.*, 2001, **14**, 611.
8. G. A. Roberts, G. Grogan, A. Greter, S. L. Flitsch and N. J. Turner, *J. Bacteriol.*, 2002, **184**, 3898.
9. A. Bridges, L. Gruenke, Y. T. Chang, I. A. Vasker, G. Loew and L. Waskell, *J. Biol. Chem.*, 1998, **273**, 17036.
10. D. W. Nebert, M. Adesnik, M. J. Coon, R. W. Estabrook, F. J. Gonzalez, F. P. Guengerich, I. C. Gunsalus, E. F. Johnson, B. Kemper, W. Levin, I. R. Phillips, R. Sato and M. Waterman, *DNA Cell Biol.*, 1987, **6**, 1.
11. T. L. Poulos, B. C. Finzel and A. J. Howard, *J. Mol. Biol.*, 1987, **195**, 687.
12. K. G. Ravichandran, S. S. Boddupalli, C. A. Hasermann, J. A. Peterson and J. Deisenhofer, *Science*, 1993, **261**, 731.
13. J. R. Cupp-Vickery and T. L. Poulos, *Nat. Struct. Biol.*, 1995, **2**, 144.
14. C. A. Hasemann, K. G. Ravichandran, J. A. Peterson and J. Deisenhofer, *J. Mol. Biol.*, 1994, **236**, 1169.
15. L. M. Podust, T. L. Poulos and M. R. Waterman, *Proc. Natl. Acad. Sci. USA*, 2001, **98**, 3068.
16. J. K. Yano, L. S. Koo, D. J. Schuller, H. Li, P. R. Ortiz de Montellano and T. L. Poulos, *J. Biol. Chem.*, 2000, **275**, 31086.
17. P. A. Williams, J. Cosme, V. Sridhar, E. F. Johnson and D. E. McRee, *Mol. Cell*, 2000, **5**, 121.
18. P. A. Williams, J. Cosme, A. Ward, H. C. Angove, D. Matak Vinkovic and H. Jhoti, *Nature*, 2003, **424**, 464.
19. M. R. Wester, J. K. Yano, G. A. Schoch, C. Yang, K. J. Griffin, C. D. Stout and E. F. Johnson, *J. Biol. Chem.*, 2004, **279**, 35630.
20. G. A. Schoch, J. K. Yano, M. R. Wester, K. J. Griffin, C. D. Stout and E. F. Johnson, *J. Biol. Chem.*, 2004, **279**, 9497.
21. P. A. Williams, J. Cosme, D. M. Vinkovic, A. Ward, H. C. Angove, P. J. Day, C. Vonrhein, I. J. Tickle and H. Jhoti, *Science*, 2004, **305**, 683.
22. J. K. Yano, M. R. Wester, G. A. Schoch, K. J. Griffin, C. D. Stout and E. F. Johnson, *J. Biol. Chem.*, 2004, **279**, 38091.
23. S. Sansen, J. K. Yano, R. L. Reynald, G. A. Schoch, K. J. Griffin, C. D. Stout and E. F. Johnson, *J. Biol. Chem.*, 2007, **282**, 14348.

24. P. Rowland, F. E. Blaney, M. G. Smyth, J. J. Jones, V. R. Leydon, A. K. Oxbrow, C. J. Lewis, M. G. Tennant, S. Modi, D. S. Eggleston, R. J. Chenery and A. M. Bridges, *J. Biol. Chem.*, 2006, **281**, 7614.
25. J. K. Yano, M. H. Hsu, K. J. Griffin, C. D. Stout and E. F. Johnson, *Nat. Struct. Mol. Biol.*, 2005, **12**, 822.
26. P. A. Williams, J. Cosme, V. Sridhar, E. F. Johnson and D. E. McRee, *J. Inorg. Biochem.*, 2000, **81**, 183.
27. T. L. Poulos and E. F. Johnson, in *Cytochrome P450: Structure, Mechanism, and Biochemistry*, ed. P. R. Ortiz de Montellano, Kluwer Academic/Plenum Publishers, New York, 3rd edn, 2005.
28. C. A. Hasemann, R. G. Kurumbail, S. S. Boddupalli, J. A. Peterson and J. Deisenhofer, *Structure*, 1995, **3**, 41.
29. I. Schlichting, J. Berendzen, K. Chu, A. M. Stock, S. A. Maves, D. E. Benson, R. M. Sweet, D. Ringe, G. A. Petsko and S. G. Sligar, *Science*, 2000, **287**, 1615.
30. P. R. Ortiz de Montellano and J. J. De Voss, *Nat. Prod. Rep.*, 2002, **19**, 477.
31. T. M. Makris, R. Davydov, I. G. Denisov, B. M. Hoffman and S. G. Sligar, *Drug Metab. Rev.*, 2002, **34**, 691.
32. R. Raag and T. L. Poulos, *Front. Biotransformation*, 1992, **7**, 1.
33. J. B. Schenkman, S. G. Sligar and D. L. Cinti, *Pharmacol. Ther.*, 1981, **12**, 43.
34. S. G. Sligar, D. L. Cinti, G. G. Gibson and J. B. Schenkman, *Biochem. Biophys. Res. Commun.*, 1979, **90**, 925.
35. S. S. Boddupalli, R. W. Estabrook and J. A. Peterson, *J. Biol. Chem.*, 1990, **265**, 4233.
36. H. Mayuzumi, C. Sambongi, K. Hiroya, T. Shimizu, T. Tateishi and M. Hatano, *Biochemistry*, 1993, **32**, 5622.
37. I. Matsunaga, A. Yamada, D. S. Lee, E. Obayashi, N. Fujiwara, K. Kobayashi, H. Ogura and Y. Shiro, *Biochemistry*, 2002, **41**, 1886.
38. A. Perret and D. Pompon, *Biochemistry*, 1998, **37**, 11412.
39. C. J. Patten and P. Koch, *Arch. Biochem. Biophys.*, 1995, **317**, 504.
40. W. Chen, R. M. Peter, S. McArdle, K. E. Thummel, R. O. Sigle and S. D. Nelson, *Arch. Biochem. Biophys.*, 1996, **335**, 123.
41. L. D. Gruenke, K. Konopka, M. Cadieu and L. Waskell, *J. Biol. Chem.*, 1995, **270**, 24707.
42. J. B. Schenkman and I. Jansson, *Pharmacol. Ther.*, 2003, **97**, 139.
43. T. D. Porter, *J. Biochem. Mol. Toxicol.*, 2002, **16**, 311.
44. H. Yamazaki, T. Shimada, M. V. Martin and F. P. Guengerich, *J. Biol. Chem.*, 2001, **276**, 30885.
45. H. Yamazaki, M. Nakajima, M. Nakamura, S. Asahi, N. Shimada, E. M. Gillam, F. P. Guengerich, T. Shimada and T. Yokoi, *Drug Metab. Dispos.*, 1999, **27**, 999.
46. F. P. Guengerich, T. Shimada, C. H. Yun, H. Yamazaki, K. D. Raney, R. Thier, B. Coles and T. M. Harris, *Environ. Health Perspect.*, 1994, **102**(Suppl 9), 49.

47. C. S. Lieber, *Physiol. Rev.*, 1997, **77**, 517.
48. D. A. Smith, S. M. Abel, R. Hyland and B. C. Jones, *Xenobiotica*, 1998, **28**, 1095.
49. J. A. Williams, R. Hyland, B. C. Jones, D. A. Smith, S. Hurst, T. C. Goosen, V. Peterkin, J. R. Koup and S. E. Ball, *Drug Metab. Dispos.*, 2004, **32**, 1201.
50. L. L. von Moltke, D. J. Greenblatt, S. X. Duan, J. S. Harmatz and R. I. Shader, *J. Clin. Pharmacol.*, 1994, **34**, 1222.
51. R. E. Benton, P. K. Honig, K. Zamani, L. R. Cantilena and R. L. Woosley, *Clin. Pharmacol. Ther.*, 1996, **59**, 383.
52. S. F. Zhou, *Curr. Drug Metab.*, 2008, **9**, 310.
53. T. Lynch and A. Price, *Am. Fam. Physician*, 2007, **76**, 391.
54. P. A. Wijnen, R. A. Op den Buijsch, M. Drent, P. M. Kuipers, C. Neef, A. Bast, O. Bekers and G. H. Koek, *Aliment. Pharmacol. Ther.*, 2007, **26**(Suppl 2), 211.
55. M. Ingelman-Sundberg, *Trends Pharmacol. Sci.*, 2004, **25**, 193.
56. R. L. Haining, J. P. Jones, K. R. Henne, M. B. Fisher, D. R. Koop, W. F. Trager and A. E. Rettie, *Biochemistry*, 1999, **38**, 3285.
57. A. R. Tabrizi, B. A. Zehnbauer, I. B. Borecki, S. D. McGrath, T. G. Buchman and B. D. Freeman, *J. Am. Coll. Surg.*, 2002, **194**, 267.
58. R. S. Kidd, T. B. Curry, S. Gallagher, T. Edeki, J. Blaisdell and J. A. Goldstein, *Pharmacogenetics*, 2001, **11**, 803.
59. T. E. Bapiro, J. A. Hasler, M. Ridderstrom and C. M. Masimirembwa, *Biochem. Pharmacol.*, 2002, **64**, 1387.
60. J. A. Hasler, R. W. Estabrook, M. Murray, I. Pikuleva, M. Waterman, J. Capdevila, V. R. Holla, C. Helvig, J. F. Falck, G. Farrell, L. S. Kaminsky, S. D. Spivack, E. Boitier and P. Beaune, *Mol. Aspects Med.*, 1999, **20**, 1.
61. F. H. Arnold, P. L. Wintrode, K. Miyazaki and A. Gershenson, *Trends. Biochem. Sci.*, 2001, **26**, 100.
62. G. E. Tsotsou, A. E. Cass and G. Gilardi, *Biosens. Bioelectron.*, 2002, **17**, 119.
63. G. Di Nardo, A. Fantuzzi, A. Sideri, P. Panicco, C. Sassone, C. Giunta and G. Gilardi, *J. Biol. Inorg. Chem.*, 2007, **12**, 313.
64. A. Glieder, E. T. Farinas and F. H. Arnold, *Nat. Biotechnol.*, 2002, **20**, 1135.
65. M. W. Peters, P. Meinhold, A. Glieder and F. H. Arnold, *J. Am. Chem. Soc.*, 2003, **125**, 13442.
66. R. Fasan, Y. T. Meharenna, C. D. Snow, T. L. Poulos and F. H. Arnold, *J. Mol. Biol.*, 2008, **383**, 1069.
67. R. Fasan, M. M. Chen, N. C. Crook and F. H. Arnold, *Angew. Chem. Int. Ed.*, 2007, **46**, 8414.
68. D. F. Munzer, P. Meinhold, M. W. Peters, S. Feichtenhofer, H. Griengl, F. H. Arnold, A. Glieder and A. de Raadt, *Chem. Commun. (Camb.)*, 2005, 2597.

69. V. B. Urlacher, A. Makhsumkhanov and R. D. Schmid, *Appl. Microbiol. Biotechnol.*, 2006, **70**, 53.
70. C. R. Otey, G. Bandara, J. Lalonde, K. Takahashi and F. H. Arnold, *Biotechnol. Bioeng.*, 2006, **93**, 494.
71. T. S. Wong, F. H. Arnold and U. Schwaneberg, *Biotechnol. Bioeng.*, 2004, **85**, 351.
72. O. Salazar, P. C. Cirino and F. H. Arnold, *Chembiochem*, 2003, **4**, 891.
73. H. Joo, A. Arisawa, Z. Lin and F. H. Arnold, *Chem. Biol.*, 1999, **6**, 699.
74. H. Joo, Z. Lin and F. H. Arnold, *Nature*, 1999, **399**, 670.
75. A. Celik, R. E. Speight and N. J. Turner, *Chem. Commun. (Camb.)*, 2005, **1**, 3652.
76. A. Parikh, P. D. Josephy and F. P. Guengerich, *Biochemistry*, 1999, **38**, 5283.
77. D. Kim and F. P. Guengerich, *Arch. Biochem. Biophys.*, 2004, **432**, 102.
78. D. Kim and F. P. Guengerich, *Biochemistry*, 2004, **43**, 981.
79. K. Nakamura, M. V. Martin and F. P. Guengerich, *Arch. Biochem. Biophys.*, 2001, **395**, 25.
80. Z. L. Wu, P. Aryal, O. Lozach, L. Meijer and F. P. Guengerich, *Chem. Biodivers.*, 2005, **2**, 51.
81. S. Kumar, C. S. Chen, D. J. Waxman and J. R. Halpert, *J. Biol. Chem.*, 2005, **280**, 19569.
82. S. Kumar, L. Sun, H. Liu, B. K. Muralidhara and J. R. Halpert, *Protein Eng. Des. Sel.*, 2006, **19**, 547.
83. L. Sun, C. S. Chen, D. J. Waxman, H. Liu, J. R. Halpert and S. Kumar, *Arch. Biochem. Biophys.*, 2007, **458**, 167.
84. S. Kumar, H. Liu and J. R. Halpert, *Drug Metab. Dispos.*, 2006, **34**, 1958.
85. W. Huang, W. A. Johnston, M. A. Hayes, J. J. De Voss and E. M. Gillam, *Arch. Biochem. Biophys.*, 2007, **467**, 193.
86. P. J. Loida and S. G. Sligar, *Biochemistry*, 1993, **32**, 11530.
87. S. G. Bell, C. F. Harford-Cross and L. L. Wong, *Protein Eng.*, 2001, **14**, 797.
88. C. F. Harford-Cross, A. B. Carmichael, F. K. Allan, P. A. England, D. A. Rouch and L. L. Wong, *Protein Eng.*, 2000, **13**, 121.
89. J. P. Jones, E. J. O'Hare and L. L. Wong, *Eur. J. Biochem.*, 2001, **268**, 1460.
90. D. P. Nickerson, C. F. Harford-Cross, S. R. Fulcher and L. L. Wong, *FEBS Lett.*, 1997, **405**, 153.
91. S. G. Bell, J. A. Stevenson, H. D. Boyd, S. Campbell, A. D. Riddle, E. L. Orton and L. L. Wong, *Chem. Commun. (Camb.)*, 2002, 490.
92. F. Xu, S. G. Bell, J. Lednik, A. Insley, Z. Rao and L. L. Wong, *Angew. Chem. Int. Ed. Engl.*, 2005, **44**, 4029.
93. X. Chen, A. Christopher, J. P. Jones, S. G. Bell, Q. Guo, F. Xu, Z. Rao and L. L. Wong, *J. Biol. Chem.*, 2002, **277**, 37519.
94. S. Graham-Lorence, G. Truan, J. A. Peterson, J. R. Falck, S. Wei, C. Helvig and J. H. Capdevila, *J. Biol. Chem.*, 1997, **272**, 1127.

95. W. T. Sulistyaningdyah, J. Ogawa, Q. S. Li, C. Maeda, Y. Yano, R. D. Schmid and S. Shimizu, *Appl. Microbiol. Biotechnol.*, 2005, **67**, 556.
96. D. Appel, S. Lutz-Wahl, P. Fischer, U. Schwaneberg and R. D. Schmid, *J. Biotechnol.*, 2001, **88**, 167.
97. W. T. Sulistyaningdyah, J. Ogawa, Q. S. Li, R. Shinkyo, T. Sakaki, K. Inouye, R. D. Schmid and S. Shimizu, *Biotechnol. Lett.*, 2004, **26**, 1857.
98. C. Virus, M. Lisurek, B. Simgen, F. Hannemann and R. Bernhardt, *Biochem. Soc. Trans.*, 2006, **34**, 1215.
99. M. Lisurek, B. Simgen, I. Antes and R. Bernhardt, *Chembiochem*, 2008, **9**, 1439.
100. B. Bottner, K. Denner and R. Bernhardt, *Eur. J. Biochem.*, 1998, **252**, 458.
101. S. Kumar, E. E. Scott, H. Liu and J. R. Halpert, *J. Biol. Chem.*, 2003, **278**, 17178.
102. E. M. Gillam, A. M. Aguinaldo, L. M. Notley, D. Kim, R. G. Mundkowski, A. A. Volkov, F. H. Arnold, P. Soucek, J. J. DeVoss and F. P. Guengerich, *Biochem. Biophys. Res. Commun.*, 1999, **265**, 469.
103. S. J. Sadeghi, Y. T. Meharenna, A. Fantuzzi, F. Valetti and G. Gilardi, *Faraday Discuss.*, 2000, **116**, 135.
104. G. Gilardi, Y. T. Meharenna, G. E. Tsotsou, S. J. Sadeghi, M. Fairhead and S. Giannini, *Biosens. Bioelectron.*, 2002, **17**, 133.
105. H. Murakami, Y. Yabusaki, T. Sakaki, M. Shibata and H. Ohkawa, *DNA*, 1987, **6**, 189.
106. T. Sakaki, M. Shibata, Y. Yabusaki, H. Murakami and H. Ohkawa, *DNA Cell Biol.*, 1990, **9**, 603.
107. M. Shibata, T. Sakaki, Y. Yabusaki, H. Murakami and H. Ohkawa, *DNA Cell Biol.*, 1990, **9**, 27.
108. C. W. Fisher, M. S. Shet, D. L. Caudle, C. A. Martin-Wixtrom and R. W. Estabrook, *Proc. Natl. Acad. Sci. USA*, 1992, **89**, 10817.
109. M. S. Shet, C. W. Fisher, P. L. Holmans and R. W. Estabrook, *Proc. Natl. Acad. Sci. USA*, 1993, **90**, 11748.
110. M. S. Shet, K. M. Faulkner, P. L. Holmans, C. W. Fisher and R. W. Estabrook, *Arch. Biochem. Biophys.*, 1995, **318**, 314.
111. Y. J. Chun, T. Shimada and F. P. Guengerich, *Arch. Biochem. Biophys.*, 1996, **330**, 48.
112. A. Parikh and F. P. Guengerich, *Protein Expr. Purif.*, 1997, **9**, 346.
113. S. Kondo, T. Sakaki, H. Ohkawa and K. Inouye, *Biochem. Biophys. Res. Commun.*, 1999, **257**, 273.
114. K. Hayashi, T. Sakaki, S. Kominami, K. Inouye and Y. Yabusaki, *Arch. Biochem. Biophys.*, 2000, **381**, 164.
115. Y. Y. Deeni, M. J. Paine, A. D. Ayrton, S. E. Clarke, R. Chenery and C. R. Wolf, *Arch. Biochem. Biophys.*, 2001, **396**, 16.
116. H. Yamazaki, E. M. Gillam, M. S. Dong, W. W. Johnson, F. P. Guengerich and T. Shimada, *Arch. Biochem. Biophys.*, 1997, **342**, 329.
117. H. Yamazaki, Y. F. Ueng, T. Shimada and F. P. Guengerich, *Biochemistry*, 1995, **34**, 8380.

118. C. Helvig and J. H. Capdevila, *Biochemistry*, 2000, **39**, 5196.
119. M. Fairhead, S. Giannini, E. M. Gillam and G. Gilardi, *J. Biol. Inorg. Chem.*, 2005, **10**, 842.
120. V. R. Dodhia, A. Fantuzzi and G. Gilardi, *J. Biol. Inorg. Chem.*, 2006, **11**, 903.
121. D. C. Lamb, D. E. Kelly, S. Z. Hanley, Z. Mehmood and S. L. Kelly, *Biochem. Biophys. Res. Commun.*, 1998, **244**, 110.
122. G. Schroder, E. Unterbusch, M. Kaltenbach, J. Schmidt, D. Strack, V. De Luca and J. Schroder, *FEBS Lett.*, 1999, **458**, 97.
123. N. Shiota, A. Nagasawa, T. Sakaki, Y. Yabusaki and H. Ohkawa, *Plant Physiol.*, 1994, **106**, 17.
124. T. Lacour and H. Ohkawa, *Biochim. Biophys. Acta*, 1999, **1433**, 87.
125. B. D. Fleming, Y. Tian, S. G. Bell, L. L. Wong, V. Urlacher and H. A. O. Hill, *Eur. J. Biochem.*, 2003, **270**, 4082.
126. A. K. Udit, K. D. Hagen, P. J. Goldman, A. Star, J. M. Gillan, H. B. Gray and M. G. Hill, *J. Am. Chem. Soc.*, 2006, **128**, 10320.
127. A. Shukla, E. M. Gillam, D. J. Mitchell and P. V. Bernhardt, *Electrochem. Commun.*, 2005, **7**, 437.
128. D. L. Johnson, A. J. Conley and L. L. Martin, *J. Mol. Endocrin.*, 2006, **36**, 349.
129. D. L. Johnson, B. C. Lewis, D. J. Elliot, J. O. Miners and L. L. Martin, *Biochem. Pharmacol.*, 2005, **69**, 1533.
130. A. Nath, Y. V. Grinkova, S. G. Sligar and W. M. Atkins, *J. Biol. Chem.*, 2007, **282**, 28309.
131. A. Das, Y. V. Grinkova and S. G. Sligar, *J. Am. Chem. Soc.*, 2007, **129**, 13778.
132. J. Kazlauskaite, A. C. G. Westlake, L. L. Wong and H. A. O. Hill, *Chem. Commun. (Camb.)*, 1996, **16**, 2189.
133. K. K. Lo, L. L. Wong and H. A. Hill, *FEBS Lett.*, 1999, **451**, 342.
134. C. Lei, U. Wollenberger, C. Jung and F. W. Scheller, *Biochem. Biophys. Res. Commun.*, 2000, **268**, 740.
135. B. D. Fleming, S. G. Bell, L. L. Wong and A. M. Bond, *J. Electroanal. Chem.*, 2007, **611**, 149.
136. A. Fantuzzi, M. Fairhead and G. Gilardi, *J. Am. Chem. Soc.*, 2004, **126**, 5040.
137. S. Liu, L. Peng, X. Yang, Y. Wu and L. He, *Anal. Biochem.*, 2008, **375**, 209.
138. V. V. Shumyantseva, T. V. Bulko, G. P. Kuznetsova, A. V. Lisitsa, E. A. Ponomarenko, I. I. Karuzina and A. I. Archakov, *Biochemistry (Mosc.)*, 2007, **72**, 658.
139. V. V. Shumyantseva, T. V. Bulko, Y. O. Rudakov, G. P. Kuznetsova, N. F. Samenkova, A. V. Lisitsa, Karuzina II and A. I. Archakov, *J. Inorg. Biochem.*, 2007, **101**, 859.
140. A. Fantuzzi, Y. T. Meharenna, P. B. Briscoe, C. Sassone, B. Borgia and G. Gilardi, *Chem. Commun. (Camb.)*, 2006, **12**, 1289.
141. S. Krishnan and J. F. Rusling, *Electrochem. Commun.*, 2007, **9**, 2359.

142. P. Panicco, Y. Astuti, A. Fantuzzi, J. R. Durrant and G. Gilardi, *J. Phys. Chem. B*, 2008, **112**, 14063.
143. B. Munge, C. Estavillo, J. B. Schenkman and J. F. Rusling, *Chembiochem*, 2003, **4**, 82.
144. H. Matsumura, N. Nakamura, M. Yohda and H. Ohno, *Electrochem. Commun.*, 2007, **9**, 361.
145. E. A. Smith, M. G. Erickson, A. T. Ulijasz, B. Weisblum and R. M. Corn, *Langmuir*, 2003, **19**, 1486.
146. G. J. Wegner, N. J. Lee, G. Marriott and R. M. Corn, *Anal. Chem.*, 2003, **75**, 4740.
147. I. Delfino, B. Bonanni, L. Andolfi, C. Baldacchini, A. R. Bizzarri and S. Cannistraro, *J. Phys.-Condens. Mat. 2007,* **19**.
148. V. E. V. Ferrero, L. Andolfi, G. Di Nardo, S. J. Sadeghi, A. Fantuzzi, S. Cannistraro and G. Gilardi, *Anal. Chem.*, 2008, **80**, 8438.
149. G. Gilardi, A. Fantuzzi and S. J. Sadeghi, *Curr. Opin. Struct. Biol.*, 2001, **11**, 491.
150. A. K. Udit, N. Hindoyan, M. G. Hill, F. H. Arnold and H. B. Gray, *Inorg. Chem.*, 2005, **44**, 4109.

CHAPTER 7
Label-free Field Effect Protein Sensing

JAN TKAC AND JASON J. DAVIS

Chemistry Research Laboratory, University of Oxford, South Parks Road, Oxford, OX1 3TA, UK

7.1 Interfacial Protein Detection

DNA microarray technology has generated valuable insight into various disease states, gene expression and co-regulated gene networks. An analysis of gene expression is not, however, sufficient to provide a knowledge of either protein abundance or protein function.[1] Though a DNA microarray highlights those genes which are turned on (expressed), the translation process itself, and its protein product, are further regulated by gene silencing (*via* RNAi).[2] Furthermore, the analysis of mRNA transcripts does not take into account post-translational modifications, such as proteolysis, phosphorylation, glycosylation or acetylation.[3] There is, therefore, a considerable requirement for the direct detection and quantification of specifically folded proteins in proteomics, drug design, disease prognosis and therapeutic development.[4,5] Mapping protein interaction and function, and the modulation of this by disease, is, though, a daunting task likely to evolve over decades. Compared with about 40 000 genes in the human genome, the human proteome is predicted to exceed 1 million.[6] Traditionally, the biochemical activities of proteins have been elucidated through the sequential study of individual proteins, a process which is both slow and labour intensive.[7] Classical, antibody-based detection methods (*e.g.* ELISA and immunospot assays) are, additionally, not currently capable of operation at the high throughputs accessible in a microarray format.[4]

Engineering the Bioelectronic Interface: Applications to Analyte Biosensing and Protein Detection
Edited by Jason Davis
© 2009 Royal Society of Chemistry
Published by the Royal Society of Chemistry, www.rsc.org

The development of protein microarrays, over the past decade,[8,9] has benefited from many of the lessons learned during the successful development of nucleic acid equivalents.[10] Proteins, however, demonstrate a staggering variety of chemistries, affinities and specificities and in many cases there is a requirement for multimerisation, partnership formation with other proteins or post-translational modification before activity or specific binding is significant.[11,12] There is, moreover, no simple process for the amplification of proteins (like PCR in the case of DNA) and expression and purification are neither facile nor guarantee functional integrity. Protein microarray challenges remain, then, considerable.

7.2 Protein Microarrays

The protein microarray consists (ideally) of a large number of capture agents that selectively and sensitively (with a low K_d) bind to protein targets at a solid/solution interface. Such surfaces should be treated in a way so as to minimise non-specific interaction and, thereby, enhance signal-to-noise ratio for target detection. These arrays may be formulated with two distinct objectives in mind. One may be to detect the abundance of proteins of interest in complex protein mixtures with highly specific capture agents for each target (analytical/diagnostic protein microarrays).[13,14] An alternative format is associated with resolving protein function, protein–protein interactions, receptor–ligand interactions, enzymatic activity and so on (functional protein microarrays).[15,16]

In general terms, the protein microarray detection platform is supported by the establishment and association of four distinct key elements: a suitable array substrate, chemistry providing highly specific and appropriately engineered surfaces, the development of high-throughput detection methods and the production of functional capture agents. Existing detection schemes can be, additionally, condensed into one of two types – those which are direct labelling/detection and those which are based on a sandwich format (Figure 7.1).[11] These will be discussed in more detail in Section 7.2.4.

7.2.1 Array Substrates

One key distinction emerges early in a survey of analysis methods: the selection of an array substrate (Table 7.1). Unlike oligonucleotides, proteins are broadly heterogeneous in size, shape and chemistry, and the diversity of applications (and detection methods) for protein arrays means that users need to "shop around" for an appropriate substrate. In many cases, several surfaces are tested prior to the one generating the best signal : noise being utilised in any given assay.[10]

Planar glass or silicon chips were the first surfaces to be used in protein arrays[17–19] and have associated low cost, low lot-to-lot variability, low fluorescent background and low r.m.s. roughness (a feature important for accurate array scanning). In many cases, however, these advantages are negated by

Label-free Field Effect Protein Sensing 195

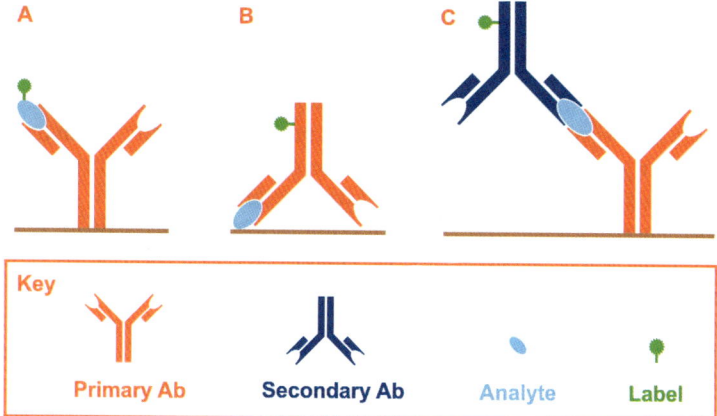

Figure 7.1 Protein microarrays can be generated by a number of means. A) Direct detection – a primary antibody (Ab) is surface immobilised prior to the injection of labelled protein; B) reverse-phase sample blot – proteins from the analyte sample attach non-specifically to a solid surface prior to the injection of labelled antibodies; C) sandwich detection – primary antibodies recognising one epitope on an analyte protein surface are initially immobilised. Following exposure of this surface to the sample solution, labelled secondary antibodies, which recognise a second epitope of the protein, are injected, and facilitate readout.

Table 7.1 Various substrates, which can be used for immobilisation of proteins with surface chemistry provided.

Type of immobilisation	Surface functionalised with	Targeted group
Non-specific/ non-covalent	PVDF, nitrocellulose, poly(L-lysine)	Group independent
Non-specific/ covalent	Aldehyde, epoxide, succinimidyl ester, isothiocyanate	Amine
	Epoxide	Thiol
Specific/ non-covalent	Avidin	Biotin tag
	Ni-NTA	His_6 tag
	Glutathione	GST tag
	Protein A/G	IgG Fc region
	Oligo DNA/PNA	DNA tag
Specific/covalent	Maleimide	Thiol
	Bromoacetyl	Thiol
	Thioester	Cys at N terminus
	Glyoxyl group	Aminooxy acetyl
	Diels–Alder reaction	Various
	"Click" chemistry	Azide OR alkyne
	Staudinger reaction (ligation)	Azide
3D surface	Agarose, PA gel, dextran, PDMS	Physical entrapment

Abbreviations: PVDF = poly(vinylidene difluoride), Ni-NTA = nickel(II) nitrilotriacetic acid, GST = glutathione-S-transferase, PNA = peptide nucleic acid, PDMS = polydimethylsiloxane.

rather undefined and variable (poorly reproducible) surface attachment chemistries. Nitrocellulose, which can be cheaply mass-produced, has a high binding capacity (surface area) and remains among the most popular substrates despite problems of high fluorescence background and non-specific binding.[19,20] PVDF (polyvinylidene fluoride) membranes offer a high protein binding capacity without modification but, consequently, no specific control of orientation.[21] Three-dimensional gels (agarose, polyacrylamide) have a low fluorescent background, high biocompatibility[22,23] and constitute a popular immobilisation platform by virtue of Biacore's CM5 (immobilised carboxymethyldextran layer) chips.[24] The surface area benefits of these translate into an approximately 100-fold increase in potential protein immobilisation than achievable on planar glass, this being somewhat negated by slow target diffusion (restricted access).[25] Gold surfaces are highly chemically tuneable and compatible with a range of detection methods, including surface plasmon resonance (SPR), fluorescence, scintillation counting and electrochemistry.[26–28] The generation of ultraclean surfaces prior to modification remains challenging but can be highly beneficial in maximising specific binding.[29,30]

7.2.2 Surface Chemistry and Immobilisation

A number of different approaches can be used to facilitate the immobilisation of recognition molecules on an array surface. They can be classified as covalent/specific, non-covalent/specific, covalent/non-specific and non-covalent/non-specific (Figure 7.2). A typical example of a non-covalent/non-specific immobilisation protocol is simple surface adsorption.[3,25] This can be achieved using electrostatic, hydrophobic or polar interactions. Though practically simple, this approach has many drawbacks including uncontrolled receptor orientation. Immobilisation may also be weak (reversible), denaturing and associated with high degrees of non-specific binding. The covalent coupling of amino acids to surface functionalities leads to robust immobilisation but again will, by default, lead to a heterogeneous population of an immobilised ligand and significant loss of tertiary structure.[8,24] Site-specific but non-covalent bioaffinity interactions can be utilised in establishing a biocompatible and orientated receptor immobilisation; *e.g.* avidin-biotin, His-tag, DNA directed, protein A/G (the latter used for oriented IgG immobilisation) and GST (glutathione S-transferase, which binds to glutathione) interactions on appropriately modified surfaces.[31–34] In all cases, care must be taken to minimise the impact of the introduced tag on ligand structure and stability. Engineered or native solution-exposed cysteine residues have been utilised in recent years to bind proteins to bare gold or maleimido/thiolated surfaces.[35,36]

Alternatively proteins can be immobilised on demand in being synthesised *in situ* on a chip surface. This approach utilises cell free protein expression directly on a chip with pre-arrayed DNA or RNA.[37] One method of doing this is the nucleic acid programmable protein array (NAPPA) (Figure 7.3), where expression occurs directly onto the microarray surface by utilising a

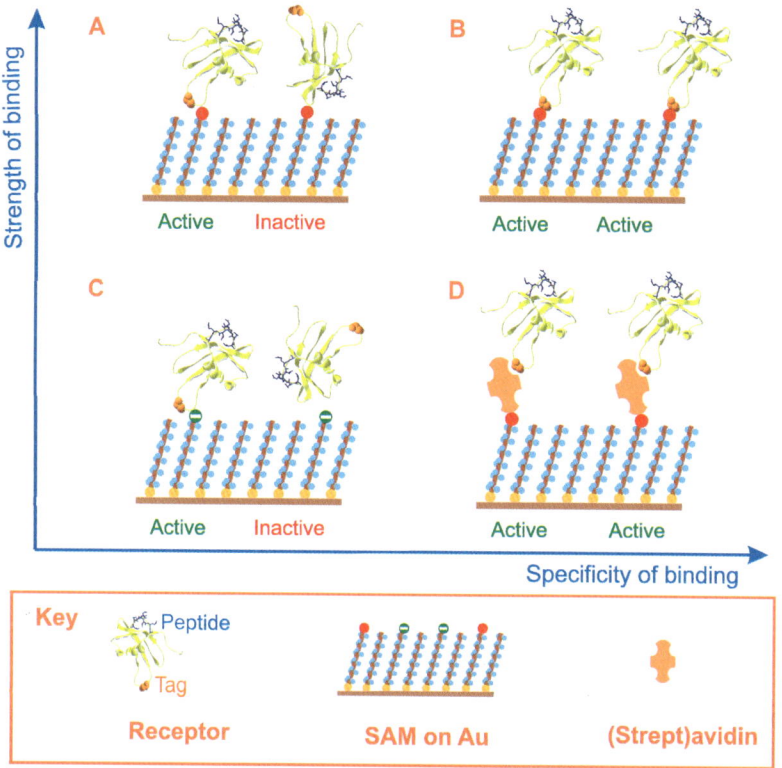

Figure 7.2 Schematic representation of protein immobilisation by implementation of A) covalent attachment (amine coupling shown as an example), B) covalent immobilisation *via* a surface-exposed protein cysteine residue, C) simple adsorption *via* electrostatic interactions and D) bioaffinity binding with biotin-streptavidin shown here as an example.

mammalian (cell-free) transcription/translation system. Plasmids, self-replicating (autonomous) loops of DNA, encoding recognition proteins such as GST fusions, are printed on a surface together with an anti-GST antibody, the latter being used to capture expressed protein. A second approach utilising print arrayed mRNA with a puromycin (an antibiotic which binds to nascent polypeptides) capture tag has been similarly and successfully used.[38] In a third form of *in-situ* protein synthesis called DAPA (DNA array to protein array), DNA arrays are utilised as a starting point for cell-free protein synthesis.[37] Specifically, a slide with covalently arrayed DNA molecules encoding a set of His_6-tagged proteins can be attached to a second slide functionalised with Ni-NTA (a tag capturing reagent). In such a configuration, synthesised proteins (up to 20 copies of each per strand of coding DNA) diffuse to the capture slide, where they are immobilised.[37] These *in-situ* protein synthesis methods appear to address several problems inherent in protein microarray fabrication; the nucleic

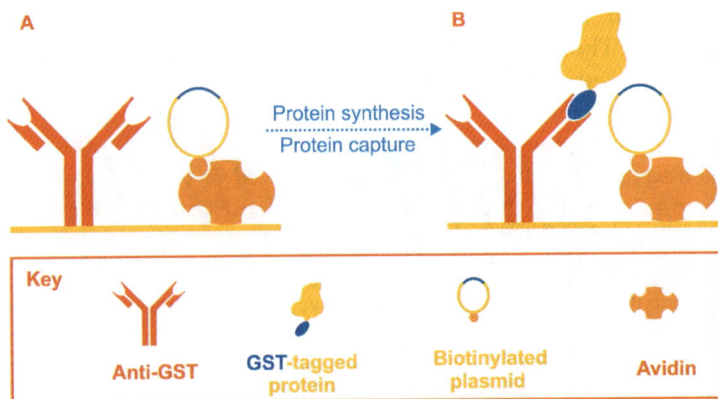

Figure 7.3 Self-assembled protein immobilisation by NAPPA (nucleic acid programmable protein array). A) Biotinylated plasmids encoding the proteins as GST fusion tags (glutathione S-transferase) are printed onto a surface pre-modified with avidin and anti-GST antibody. The array is exposed to a lysate for expression of the recombinant recognition proteins (having GST as a tag). B) The generated protein is subsequently surface captured by anti-GST antibody.

acid precursor arrays are stable at room temperature when dry prior to activation and derived capture protein immobilisation and purification requirements are met in one step.[39,40] One potentially significant drawback to this approach is, however, the possibility of protein misfolding outside native cellular environments (on synthesis).[41]

The use of nucleic acid aptamers as capture agents (see Section 7.2.3) greatly simplifies immobilisation issues because of their high levels of chemical and thermal stability compared to protein-based capture equivalents.[42] The amine coupling of suitably terminated aptamers on silane-treated glass surfaces[43,44] or the chemisorption of thiolated aptamers on gold,[45–48] for example, have been successfully used. Biotinylated aptamers have also been immobilised on streptavidin-coated glass or gold surfaces.[45,49] In many cases, immobilisation protocols are effectively characterised by surface imaging (far field fluorescence, scanning electron microscopy, near field atomic force microscopy) or spectroscopy (FTIR, ellipsometry) methods prior to assaying.

7.2.3 Capture Biomolecules

Capture agent selection is probably the most important issue in achieving high-performance protein detection and greatly influences the choice of array substrate, immobilisation protocol and detection method. A protein capture agent must be able to recognise its target among a potentially vast number of other protein species in a sample and then retain it throughout the measurement phase of the assay.[50] It should be highly specific for the protein of interest, with an affinity sufficient to capture protein target at very low (sub picomolar) concentration. The other important characteristic of a protein capture agent is its conformational stability at the surface during the pre-treatment, flowing or rinsing steps of the assay. Though the most frequently used capture agent remains the (immunoglobulin) antibody, as used in the "industry standard" immunoassay, antibodies do not function well in the microarray format, because, typically, only 20–30% of those which are commercially available can be used qualitatively.[11,51,52] For this reason, a number of groups have explored derivative recombinant single-chain antibodies.[53–55] Alternative peptide or nucleic-acid-based recognition entities have also been demonstrably powerful in recent years (Table 7.2).

Nucleic acid (DNA/RNA) aptamers, as mentioned, have high chemical/thermal stability, have little or no batch variation during production and are highly amenable to a number of labelling protocols.[56] Within these, "molecular beacons" (hairpin shaped DNA/RNA molecules internally modified by a fluorophore or redox probe) facilitate a real time monitoring of target binding.[57] These capture agents function *via* the interaction of target protein with specific three-dimensional loop structures adopted in solution (or at a surface under solution) and generally facilitate nanomolar target detection limits. A photoactivated form of nucleic acid aptamer, in which binding affinity is increased by a photoinitiated cross linking, is currently showing promise in

Table 7.2 Capture agents that would be applicable in the development of protein sensing devices.

Capture agent	Production	Target	Purpose of detection
Antibodies	Hybridoma cells	Antigens	Protein abundance
Antigen	Recombinant/extraction	Antibodies	Protein abundance
Fused proteins	Recombinant affinity	Proteins	Protein abundance and function
Phage-displayed proteins	DNA library	Proteins	Protein abundance and function
Antibody fragments	DNA library	Proteins	Protein abundance and function
DNA/RNA aptamers	SELEX	Proteins	Protein abundance
Peptide aptamers	DNA library	Proteins	Protein abundance and function
Peptides	Chemical synthesis	Antibodies/proteins	Protein abundance and function
Carbohydrates	C/E synthesis/extraction	Glycan binding proteins	Protein abundance and function
Small molecules	Chemical synthesis	Receptors/enzymes	Protein function
MIP	Chemical synthesis	Proteins	Protein abundance

MIP = molecule imprinted polymers, SELEX = systematic evolution of ligands by exponential enrichment process, C/E = chemical or enzymatic synthesis.

facilitating fluorescence assays to femtomolar limits.[58] Though these have been utilised in microarray format,[43,59] their application in complex mixtures currently suffers from inherent cation sensitivity[60,61] and the relatively low (by virtue of the limited chemical space spanned by 4–5 nucleotides) number of protein-recognising derivatives known to date.[59] Further application may require the development of novel nucleotides.[10] The vast majority of these nucleic acid assays are DNA based; RNA molecules are non-destructively attached to surfaces only with difficulty and, consequently, only a handful of RNA microarray reports have been published to date.[62–64] Synthetic peptides constitute interesting capture ligands: they can mimic protein biological activity, are easy to synthesise and manipulate, are thermally stable, are relatively cheap and can even be synthesised *in situ* by a well-established protocol.[65–67] Their downside lies with their low molecular mass (they are not easily accessible when non-specifically bound on a solid support) and lack of well-defined three-dimensional structure, both of which reduce specific targeting interactions.[50] Despite these difficulties, they have been successfully applied in microarrays composed of several thousand units (the assay being completed by the binding of a fluorophore-tagged secondary antibody).[68]

In recent years, new protein recognition moieties have emerged. Significantly, they can offer considerable flexibility and selectivity in recognition and considerable conformational stability.[69,70] These are the peptide aptamers and currently include anticalins® modelled on lipocalin structures,[71] trinectins® derived from a fibronectin III domain,[72] affibody® molecules, which are engineered Z domains of protein A[73] and others.[74] The number of peptide aptamers known already exceeds the number of functional DNA/RNA aptamers. Though affibodies have been successfully used in microarrays using different immobilisation protocols,[75] they exhibit rather large interaction interfaces and associated relatively low binding affinities.[76,77] Recently a newly developed peptide aptamer based on the use of a biologically inert human stefin A protein (called STM – Stefin Triple Mutant) as a stable scaffold has been utilised (Figure 7.4). The advantage of this approach lies within an ability to design recognition peptide sequences into the (otherwise inert) scaffold protein surface.[78] To date, this scaffold has been utilised successfully for both *in vivo* studies[79] and interfacial detection assays.[36,80–82]

7.2.4 Detection Tools

Detection techniques employed in microarray analysis can be broadly categorised into two classes: those based on sample labelling (with fluorophores, enzymes, nanostructures *e.g.* metal nanoparticles and carbon nanotubes) and those which are label-free techniques (*e.g.* surface plasmon resonance (SPR) and field effect methods including those based on nanowires/nanotubes). In direct labelling the entire sample is labelled prior to incubation with the arraying surface. The advantage of this is that it facilitates simultaneous multiple analyte analysis. An example of a commonly used label-free detection

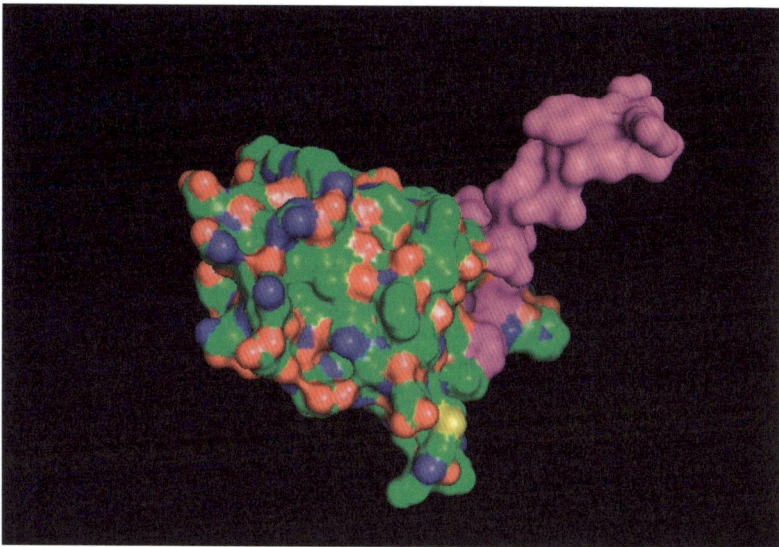

Figure 7.4 Schematic representation of the pep2 aptamer, derived from the STM scaffold protein by insertion of a peptide sequence, shown in purple, into the solution-exposed surface. Green represents hydrophobic patches, red and blue predicted negative and positive charges at pH 7.0 respectively. The anchoring cysteine residue is highlighted in yellow.

assay is the high-throughput reverse-dot blot assay in which a sample is spotted and then exposed to a range of labelled antibodies (Figure 7.1).[83] Though facile, this form of analysis is prone to high levels of non-specific signal generation.

7.2.4.1 Detection of Protein via Labelling

The majority of disease-related applications involving antibody microarrays reported to date, including those profiling the protein composition of serum in patients with prostate, pancreatic, bladder and lung cancers, are based on direct sample labelling.[84–87] Enzymes labels, traditionally used in ELISA (Enzyme Linked Immunosorbent Assay), can be utilised through either optical (fluorescent or chemiluminescent)[88–90] or electrochemical[91–94] detection of (specific substrate derived) product. The latter can be usefully amplified by redox cycling with a suitable solution phase redox partner to facilitate femtomolar levels of detection.[95] Fluorescent label detection methods, which can be readily interfaced with DNA microarray technologies, are fully compatible with microfluidic chips made of glass or polymers[96] and can also enable nM-fM detection limits comparable to ELISA.[54,97,98] Although organic dyes remain a popular, and chemically flexible, choice for fluorescent protein detection,[99,100] the equivalent application of highly emissive microparticles or nanoparticles has generated much recent interest.[98,101,102] Semiconducting quantum dots (QDs), which exhibit optical tuneability, photostability, brightness, broad excitation and sharp emission, have been, in particular, effectively utilised.[99,103,104] QD labels also facilitate detection by electrochemical means *e.g.* stripping voltammetry of component metals.[105,106]

The high surface area : volume characteristics of these particles, more generally, can also be utilised in non-optical assays when they are DNA loaded (the binding event subsequently being amplified by PCR).[98] Other nanoparticle tags utilised within protein microarrays include magnetic nanoparticles, which can be powerfully used in preconcentration/separation steps,[104,106] enzyme loaded carbon nanotubes[105,107,108] and gold nanoparticles.[104–106,109] Like quantum dots, the latter can be detected electrochemically by stripping voltammetry in an appropriate medium.[105] They can also be utilised in the catalytic reduction of silver ions, a process which can further amplify electrical or optical (scanometric) readout.[105,106,109]

Though potentially very sensitive (indeed detection down to the molecular level has been reported,[109,110]) labelling methods are not problem free. Firstly, labelling is time consuming, labour intensive and rarely equally efficient for all proteins in a given sample (this makes reliable quantification particularly difficult).[111] Of still greater concern is the potential influence of labels on the native protein fold and its binding properties.[15,111] These are significant problems which are by-passed in label-free detection protocols. Within these, those based on sandwich assays, which require no sample labelling, are particularly selective because two capture agents are used for every analyte. For the past two decades, the clinical need for low picomolar detection sensitivity has largely been met using enzyme tags in label-free ELISA assays, where signal amplification occurs through enzymic turnover of suitable substrate.[112] Though the inherent requirement (in any sandwich configuration) of having two sets of highly specific capture agents per analyte is demanding, a variety of successful label-free sandwich assays have now been reported.[97,113]

7.2.4.2 Label-free Detection Platforms

Although the desire for label-free protein sensing is clear, traditional platforms cannot generally offer suitably good detection limits (an issue particularly problematic for low-abundant protein quantification) or be readily multiplexed. [Sandwich assays, though potentially offering high degrees of signal amplification, are difficult to develop in a multiplexed (>30 antigens) configuration due to the requirement of two complete sets of antibodies for every target (and problems of cross-reactivity)]. The notable exceptions to this are the ultrasensitive capacitance/impedance protein detection protocols first introduced in the 1990s. These will be discussed in Section 7.3. Below is a general survey of traditional, new and emerging label-free detection platforms compatible with multiplexed assaying.

Mass spectrometry (MS) has played a key role in the advance of proteomics in the last decade. The integration of microarraying with MS has, in particular, generated a powerful new tool capable of providing chemical and structural information generally not available through other methods.[111,114] Most configurations operate by incident-laser-activated desorption of proteins from an array surface. These are then ionised, accelerated by an electric field, and finally separated according to mass/charge. Although operationally simple, fast and sensitive (picomolar levels of cancer markers are detectable fairly routinely),[115–117] MS methods remain problematic and significant challenges associated with normalising sample collection, experimental procedure and data interpretation remain.[111] Though small protein modifications (absent or oxidised amino acids, for example) can be detected,[115] low abundance proteins can only generally be sensed through initial pre-enrichment and/or separation.

Surface Plasmon Resonance (SPR) has been, since its introduction in 1983, the most frequently used label-free technique for detection of proteins.[118] It is an optical surface technique highly sensitive to change in the refractive index within the dielectric layer adjacent to a metal film. From a sensory perspective users are, of course, interested in adsorption/desorption triggered change and this technique enables such processes to be monitored in real time. The kinetics and thermodynamics of sample protein binding can be determined and the analyses have considerable use in the optimisation of bioreceptive layer formation. From an assaying perspective, the weakness of SPR lies in the relatively high equipment cost, high detection limits (nanomolar) and narrow linear range (1–2 orders of magnitude). With these limitations in mind, significant improvements have been made during the last two decades, including the development of spatially resolved sensing (imaging SPR)[119–122] and the use of enzyme[64,123,124] or nanoparticle[125–127] signal amplification methods. Commercial SPR imaging instruments currently enable the simultaneous monitoring of up to 1000 binding events.[128]

Atomic force microscopy (AFM) is one of the most important developments in surface based imaging and is almost unique in its spatial resolution. Developed in the early 1980s AFM is exquisitely sensitive to force interactions between a microfabricated nanometre scale tip and an underlying sample surface.[129] As the probe scans the surface, modulations in force triggered by the presence of any adsorbed entity are detected by cantilever deflection, recorded and translated into a three-dimensional image with Ångstrom-level spatial resolution. The technique has been successfully used not only for the preparation of protein nanoarrays,[130] but also for monitoring of protein–protein interactions.[131] Though the capturing of single protein molecules can be readily monitored, the high equipment costs, slow speed and experimental demands make imaging AFM configurations most useful in interfacial analyses and design rather than direct assaying. Derived technology, based on microcantilever deflection, can, however, be used in sensitive multianalyte (up to 50 000 cantilevers within a 1-cm^2 array have been generated[132,133]) detection.[134] These configurations are based on analyte binding to an appropriately functionalised (antibody[15,135] or antibody fragment[136]; recently peptide aptamers have also been used[137]) microcantilever through the detection of triggered changes in either surface stress or mechanical resonant frequency of the cantilever.[135,138,139] Detection limits down to low picomolar levels[136,140,141] in the presence of excess non-specific protein have been reported.[139,140] With an immobilised peptide aptamer receptor nM detection limits of a cancer marker have been reported in cell lysate.[137]

Electrochemical impedance/capacitance (EI/C) methods combine high sensitivity and low cost with potentially small equipment scale.[109,142–145] In comparison with alternative assays, EI/C analyses, which probe the interfacial properties of suitably modified electrodes through measuring target adsorption induced changes in electron transfer resistance or capacitance, are almost uniquely sensitive and multiplexable. Specific applications to protein detection will be discussed in Section 7.3.2.

Nanotube-/nanowire-transistor-based detection represents an early stage technology that has the potential to address many of the needs of label-free sensing in arrays.[146,147] Briefly (see Sections 7.3.4.1 and 7.3.4.2 for more detail), appropriately functionalised (most often with nucleic acid aptamers) nanowires demonstrate reliably detectable conductance change in the presence of protein targets. In some cases, these exhibit unprecedented levels of sensitivity.

In addition to the methods discussed above, label-free sensing techniques based on nanohole optical arrays,[148,149] Kelvin nanoprobes,[15,150] surface-enhanced Raman scattering,[106,151] localised surface plasmon resonance[106,149,152] and optical interference (ellipsometry),[15] including spatially resolved imaging derivatives,[153,154] are under development.

7.2.5 Ultrasensitive Protein Detection

Disease markers may be present only at sub-picomolar levels in biological fluid[110] and, in some cases, only a few molecules of a pathological protein may trigger disease development.[98] There exists, then, a need for ultrasensitive detection beyond the capabilities of current (for example) ELISA methodologies (particularly in early disease diagnosis). Several emerging ultrasensitive technologies exploit new labels such as metal/semiconducting nanoparticles and nanotubes or a combination of traditional tags (*e.g.* fluorescent dyes, electrochemical probes or enzyme labels) with new nanotechnological tools.[106,110,155] Using different nanoparticle/nanotube or DNA-barcode-based amplification strategies attomolar protein detection limits have been reported (Figure 7.5).[108,155–158] The compounding problem in many cases is the need (ideally) to perform such sensitive specific assays in the presence of the huge excesses (up to 10^7-fold) of non-specific protein present in complex biological fluid.

Ultrasensitive protein detection schemes may be based on any of a number of amplification schemes. The bio-barcode assay, for example, is based on tagging, in a sandwich format, a target protein with an antibody-labelled

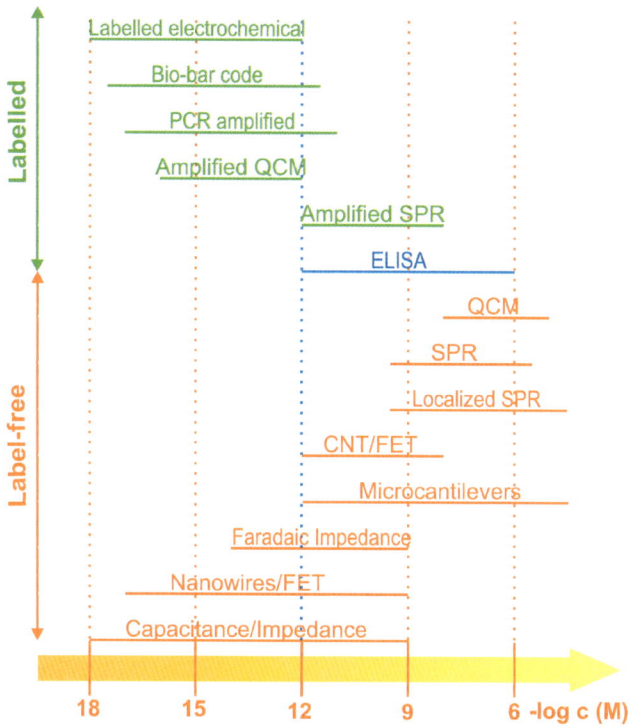

Figure 7.5 Typical protein concentration ranges accessible through various analytical techniques. Both labelled and label-free detection schemes are compared to ELISA.

nanoparticle pre-loaded with DNA. Amplification, in such cases, arises from the high loadings of nucleic acid and subsequent potential further amplification by standard PCR methods. With this approach attomolar detection limits have been reported.[156] Subsequent modifications of this scheme include derived colorimetric[159] and fluorescent detection assays[157] (of particle-released DNA) and the integration of the bio-barcode approach with a microfluidic format.[160] Though undoubtedly sensitive, one significant disadvantage of the bio-barcode assays is the requirement for both gold and (usually, for extraction purposes) magnetic nanoparticles specific for every target of interest.

Broadly equivalent amplification methodologies can be applied to electrochemical protein assaying. In tagging polystyrene microparticles with antibody and nucleic acid, for example, protein detection can be coupled with chronopotentiometry (guanine base oxidation) in producing a low femtomolar detection assay.[161] Enzymic amplification approaches to interfacial protein detection are now well established; the most common format for these is the use of an enzyme-tagged secondary antibody, the substrate turnover from which is associated with an electrochemical or optical signal detection.[91,92,162–164] A recently reported modification of this approach has utilised an enzyme-based precipitation reaction and SPR imaging analysis to facilitate pM target protein detection.[64] Even higher levels of signal amplification (fM protein detection) have been reported to be possible through the use of antibody-tagged, enzyme-loaded carbon nanotubes.[108] Interestingly, on utilising a gold nanoparticle as a catalytic entity instead of an enzyme the same (sandwich) format reportedly enables attomolar detection of prostate specific antigen.[155]

In shrinking the dimensions of a detection device one not only enables higher detection density but can also induce significant sensory benefits. It has been shown, for example, that mass action (transport) benefits can facilitate orders of magnitude increases in limits of detection.[165] In coupling dimensional advantages to electrokinetically enhanced transport 10^4-fold enhancements in sensitivity have been demonstrated in nanoscale transducers.[166] Field effect sensing based on nanowires, where the wire body itself is in direct contact with the analyte solution, seemingly offers detection limits comparable to the best achieved using labels (down to fM level) (Figure 7.5).[167] When appropriately modified, these nanowire devices have demonstrated, in some cases, a very large linear range and extraordinary levels of specificity (see below).

A number of ultrasensitive label-free assaying protocols currently exists in which neither the analyte nor the capture agent are labelled – these include those based on electroanalysis and nanoscale electronics (field effect transistor sensing – FET) (Figure 7.5). Electrochemical biosensors intimately coupling a biological recognition element to an electrode transducer have been a major focus in biosensor development. Impedance/capacitance assays are at least as sensitive as conventional ELISA[168,169] and in many cases can exceed this (detection limits down to 10^{-16} M[170] or 10^{-18} M[171] have been reported). These detection platforms additionally offer a wide linear range (3–4 orders of magnitude).[169,171] In some cases this can extend to an exceptional 7[170] or 9 orders[171] of magnitude. Bearing in mind the dynamic range of protein exhibited in human serum range (grams to tenths of picograms *per* ml) this is potentially highly significant. In the context of potentially important future developments, it is worth noting that these field effect methods not only enable ultrasensitive and label-free detection but also offer obvious high throughput possibilities (Figure 7.5, see Section 7.3).[172,173]

7.3 Label-free Field Effect Protein Detection

Electrical biosensors hold great promise because they are potentially cheap to fabricate, functionally flexible, highly sensitive and can be both readily miniaturised and multiplexed. This section will describe and expand upon the different approaches that can be utilised in label-free field effect protein detection, including the use of field effect transistors (FETs), semiconducting nanowires/nanotubes and impedance/capacitive assays operational in either Faradaic or non-Faradaic modalities.

7.3.1 Field Effect Transistor (FET) Based Protein Sensing

A classic field effect configuration consists of source and drain electrodes separated by a semiconducting channel. The physical diameter of the channel is fixed, but its effective electrical diameter can be perturbed by the electrical field applied to the third electrode – the gate.[174] A range of analogous configurations has been successfully utilised in ion sensing (ISFETs) since 1970.[175] Considerable developments since have enabled a range of biomolecular and even cellular applications.[176–180] If the default metallic gate structure is replaced by a biomolecular layer immobilised directly on the semiconductive channel, it is possible for specific recognition events to modulate detectable conductance change.[181] Though demonstrably effective, the precise mechanisms by which bio-FETs function remain, at least in some cases, highly debatable. In nucleic acid modified FET devices (which detect complementary nucleic acids), for example, conductance modulation is likely to be induced by both intrinsic analyte charge and recognition mediated changes in interfacial charge distribution.[182]

The ionic strength of the assay solution, a parameter which directly influences the double layer thickness (characterised by a Debye length λ_D) is important in all field effect protocols. Only binding which occurs within λ_D distances from the electrode surface is not screened by ions from the bulk solution and thus can generate a measurable signal. Theoretical calculations suggest, for example, that with a typical surface density of ssDNA of 1×10^{12} molecules cm^{-2}, sensor output will potentially be very low (~ 3–6 mV; dependent on the charge changes associated with recognition) in a physiological concentration of ions (~ 0.1 M).[182] Since Debye length varies as the inverse square root of the ionic strength (I) with a formula $\lambda_D \sim 0.32 * I^{-1/2}$ nm in aqueous solutions, operation in low ionic strength facilitates the detection of electrostatic change at greater distance from the electrode surface. [A typical λ_D in a buffer with physiological concentration of 0.1 M is ~ 1 nm and in 1 mM buffer it extends to 10 nm.] It is, then, important that the detection interface (and its application) is designed with this in mind. Successful DNA detection in an FET configuration is, for example, facilitated by both the high target charge and the fact that hybridisation occurs well within typical Debye lengths.[182–184]

The effects of Debye length on the viability of FET configurations have been nicely demonstrated in a number of recently published studies which have also suggested an upper ionic strength limit of 1 mM ($\lambda_D = 10$ nm).[182] Similar effects have been observed in antibody/semiconducting nanowire configurations[147,173,185] and in a systematic evaluation of biorecognition layer thickness distance (2.6–7.7 nm) in nucleic acid assays (Figure 7.6A).[186] In the latter case, device response was observed to fall exponentially as the DNA binding site (immobilised peptide nucleic acid) moved away from the transducing nanowire surface. In a streptavidin-based nanowire assay device (see Section 7.3.4) output signals measured in PBS buffer were observed to fall significantly with increasing ionic strength and to vanish altogether (total screening of binding induced charge changes) in 150 mM PBS buffer ($\lambda_D \sim 0.7$ nm).[187] Consistent

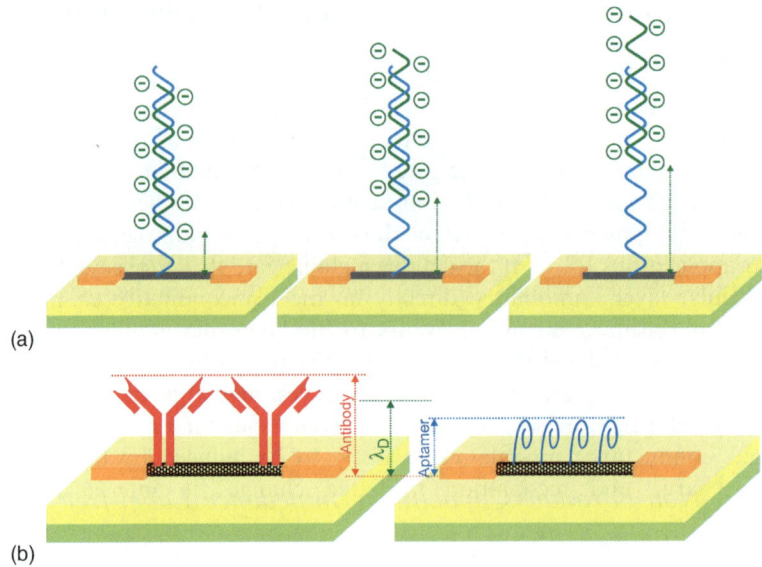

Figure 7.6 (a) Schematic representation of the varying electrostatic impact of DNA recognition at different localities of a surface-confined PNA. Though the target DNA length (and total charge) is constant, sequence differences modulate the binding position and, consequently, the degree of electrostatic gating observed at the underlying wire (figure redrawn with permission from reference 186). The greater the spatial separation between the biorecognition process and the transducer (here a nanowire), the lower the series sensitivity. (b) Schematic representation of a label-free protein biosensor based on CNT-FETs: antibody-modified CNT-FET (left) and nucleic acid aptamer-modified CNT-FET (right). (Figure redrawn with permission from reference 188.)

with the importance of localising charge changes close to the electrode surface, the size of the capture molecule has been observed to be important in mediating successful protein detection. The use of smaller biorecognition elements, of course, facilitates detection at higher ionic strength (10 mM phosphate buffer with $\lambda_D = 3$ nm).[80,182] It has been demonstrated, for example, that the detection of IgE (immunoglobulin E) can be achieved much more effectively using a DNA aptamer than with antibody;[188] the size of the latter relative to λ_D apparently compromising sensitivity (Figure 7.6B).

The possibilities and demands of field effect protein detection have been recently debated.[182,189] Though, in general terms, the same FET signal generation protocols used in DNA assays can be applied to detect protein analyte (through field changes induced by either intrinsic protein charge or by induced ionic redistribution after protein binding), proteins are bulky and have an associated lower charge density than nucleic acids. It is likely that the failure of early protein assays lies in the immobilisation protocols and high ionic strengths used.[189] In recent years, optimised immobilisation protocols have enabled low pM protein detection in an antibody-based FET configuration.[190] Most recently, a number of groups have reported the successful utilisation of nanometre-scale FET configuration in ultrasensitive protein detection (see Section 7.3.4).

7.3.2 Capacitance/Impedance Label-free Protein Sensing

In addition to the potential of utilising analyte and analyte-binding-based electrostatic charge changes directly, biorecognition can be detected and quantified through more subtle changes in interfacial electrical characteristics; capacitance/impedance assessments of charge transfer resistance or dielectric constant, in particular, can facilitate extremely sensitive detection in experimentally simple configurations.

7.3.2.1 Non-Faradaic Capacitance/Impedance Label-free Protein Sensing

An electrode immersed in an electrolyte solution can generally be described as a capacitor in its ability to store charge. Charged species and dipoles will be oriented at the electrode–solution interface giving the electrical double-layer a characteristic, potential tuneable, capacitance and impedance in any given medium. When a (bio)molecular film is deposited on the electrode surface solvated ions and water molecules are displaced from the electrode surface. It is a matter of debate, in such circumstances, as to whether the displacement of water, change in local dielectric (ε_r 2–5 for biomolecules vs. 80 for water) or electronic interactions associated with biorecognition are singularly or jointly responsible for biorecognition-triggered detectable changes in interfacial capacitance or impedance.[142,191] It has been suggested that, in non-Faradaic configurations (those without redox probe in the solution), detectable capacitance change arises primarily from the displacement of water and ions from the surface by the adsorbing analyte.[191] In such a format, a tightly packed, high-resistance biorecognition layer maximises the ability to detect small impedance/capacitance change. For Faradaic sensors the electrode surface needs to be simultaneously accessible to the redox species, in solution but not to the adsorption of other molecules.[142,191]

Capacitance/impedance changes are assessed through one of two experimental means; electrochemical impedance spectroscopy (EIS) or potential step amperometry. The former measures the current response generated through the application of a perturbative voltage on a sinusoidal signal.[145] Impedance measurements are typically run across a range of frequencies commonly centred at ~ 10 kHz.[142] The impedance or capacitance of the interface may, alternatively, be measured at a single frequency. Potential step amperometric methods sample the current response triggered by a potentiostatic step (typically 50 mV or so in amplitude). In running either assay in relatively low electrolyte concentration

($\sim$10 mM) the time constant ($\tau = C*R_s$) is relatively high, triggered current modulation relatively slow and measurement reliability high. In potentiostatic assays the current decay can be described by the following:

$$i(t) = \frac{U}{R_s} \exp\left(\frac{-t}{R_s C}\right) \qquad (1)$$

where $i(t)$ is the current in the circuit as a function of time, U is the pulse potential applied, R_s the resistance of the recognition layer, C the total capacitance measured at the working electrode/solution interface and t the time elapsed after the potential step application. By taking the logarithm of eqn (1), the linear relationship between $\ln(i)$ and t facilitates a straightforward least squares calculation of C_t and R_s.

A typical biorecognition surface used for capacitance detection consist of two layers – an insulation layer consisting of a double layer represented by C_i and a recognition layer represented by C_r. Analyte binding to the receptive generates a third layer represented by C_a. In general terms, the capacitance at such an electrode/solution interface is then described as being a composite of the capacitors in series (Figure 7.7) with the overall capacitance, C, expressed as:

$$\frac{1}{C} = \frac{1}{C_i} + \frac{1}{C_r} + \frac{1}{C_a} \qquad (2)$$

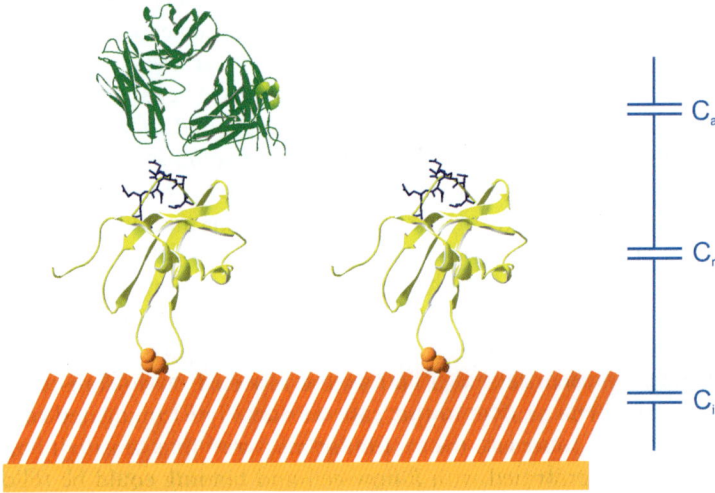

Figure 7.7 Schematic representation of a capacitive biosensor showing the overall capacitance of the layer as a composition of several capacitors in series. The first, C_i, represents the insulating layer e.g. SAM on gold and double layer, the second, C_r, the biorecognition layer and the last, C_a, being the contribution from the recognised analyte. All three layers must fall within the Debye length of the interface if analyte binding is to be detectable.

The lowest capacitance will dominate the total capacitance, meaning small capacitance changes (analyte induced) are best detected with a high-capacitance biorecognition layer.[142,192] It is, additionally, important to design the interface in such a way that it is not permeable to ions from solution (this will compromise detection performance).[142,191] The charge and permeability characteristics of a designed surface should, therefore, be a compromise between those two demands.[142] Although different transducer surfaces (*e.g.* semiconductor, metal oxides and metals) have been used in capacitative bioassays[142] modified gold films remain the most popular and chemically flexible (with longer saturation leading to more crystalline films of lower capacitance).[193–198] Importantly, these layers not only enable good levels of capacitance/impedance control but also facilitate controllable (*e.g.* orientated) bioimmobilisation in the generation of chemically or biologically reactive surface layers.[27,199,200]

To date, two different capacitative electrode configurations have been utilised in biosensing. One is based on the use of interdigitated electrodes (parallel metal plates), the second on analyses at a single working electrode. The former configuration, in which dielectric change between the plates is tracked by impedance, was first applied to a protein assay in a 1986 report. Within this, two copper conductors (25 μm high, 50 μm wide, 50 μm apart) were positioned on the surface of a parylene insulator. A 300-nm-thick layer of SiO_2 was then deposited on the polymer film and subsequently used for antibody immobilisation.[201] More recent work has suggested that increased (nM) sensitivity and responsivity can be achieved in impedance assays utilising interdigitated finger electrodes.[202] Detection sensitivity in these configurations should increase as plate separation is reduced; with this in mind, nanometre-scale interdigitated electrodes (with a separation gap of 200 nm) have been fabricated.[203]

Although antibodies have been the dominant biorecognition elements in capacitative protein sensing configurations, antibody fragments,[169,204] DNA/RNA aptamers,[205–209] a peptide epitope of the capsid protein[210] and peptide aptamers[81] have been recently used in these assays to detect proteins down to fM levels. The first report of a subnanomolar (~16 pM) detection limit by capacitance was published in 1988.[211] In this work, antibodies were immobilised on an oxidised silicon wafer by glutaraldehyde-mediated covalent coupling to an intervening silane SAM. Interestingly, performance was improved on lowering the fractional antibody surface coverage (presumably facilitating improved analyte access).[211] In 1992 another study was published in which comparable picomolar detection limits were reported on a silanised tantalum oxide surface.[192] The analyte in this case was anti IgG conjugated to β-galactosidase (in order to increase the molecular mass and thus sensitivity of assay). The assays were performed with a flow cell and binding could be followed in real time with a typical time of analysis being 10 min. Though the sensor surface could be effectively regenerated using glycine/HCl buffer its performance was ultimately compromised by non-specific protein binding.[192] A dramatic sensitivity increase was reported in 1997 using a potential step approach.[169] On antibody immobilisation through standard acid–amine coupling chemistry, the thiooctic acid modified gold surface was backfilled with hydrophobic 1-dodecanethiol and capacitative changes monitored by potential step chronoamperometry under flow. On optimisation, this assay facilitated a detection limit of ~10 fM for interleukine-2, interleukine-6 and HSA, and a linear range extending over three orders of magnitude. In extending this assay to one utilising polyclonal antibodies, the same group were able to demonstrate a highly impressive sub-femtomolar limit of detection and a linear range extending across more than seven orders of magnitude, though electrode reproducibility remained problematic ($\sim 30\%$).[169] In 2001 attomolar protein detection limits were reported. Interestingly, the authors not only noted a greater reliability of impedance assaying over capacitance assessments of the same configuration, but were also able to dramatically suppress non-specific binding events by injections of 100 mM KCl.[212] Though there remain problems with sample-to-sample and electrode reproducibility, these assays remain the state-of-the-art in highly sensitive label-free protein detection. These and related analyses have been summarised in two recent reviews.[142,191]

7.3.2.2 Faradaic Impedance Label-free Protein Sensing

A charged electrode surface will impart either an attractive or a repulsive force on ions near it. In Faradaic sensors the interaction of a charged redox species with this surface will be reflected in the charge transfer resistance (R_{ct}), a parameter representing the barrier associated with the redox species accessing the electrode. Interfacial binding events can be monitored by tracking changes in this on exposure of a suitable primed surface to analyte solution (most commonly using hexacyanoferrate as the diffusive probe).[213,214] This resistance (R_{ct}) has two origins: 1) the energy potential associated with the oxidation or reduction event at the electrode and 2) the electrostatic or steric energy barrier associated with the redox species reaching the electrode.[191] Ameur et al. have utilised this approach to establish an antibody-based assay sensitive to femtomolar levels of target across a four-orders-of-magnitude linear range.[204] In this case antibody immobilisation was through standard amine coupling on cysteamine-modified gold with non-specific protein interactions reported to be 0.1% of the signal.

Nucleic acid aptamers have been utilised in impedance detection assays on ITO[205] and gold[206] with nM detection limits but rather narrow linear ranges (typically two orders of magnitude). Related surface chemistry has also been extended to (gold) electrode arrays[207] and to pyrolysed carbon electrodes.[209] In most cases, protein binding at the aptamer recognition interface is associated with an increase in impedance (charge transfer resistance) of between 10 and reported 110%,[206,207,209] though, in one case, a decrease in charge transfer resistance was responsible for signal generation.[205] Femtomolar detection limits have been reported with nucleic acid aptamer modified surfaces using an "optimised immobilisation protocol" involving the chemisorption of thiolated DNA.[208] The aptamer surface was preferentially backfilled by mercaptohexanol rather than by mercaptoacetic acid to form a (reasonably) protein-resistive interface. Interestingly, detection limits were increased further by post-binding denaturation of the thrombin target protein. Very recently, a "molecular masking" method, in which protein resistive OEG films are selectively and sequentially desorbed from electrodes prior to exposure to the chemisorbing aptamer, has been applied to selective peptide aptamer functionalisation within an array of ten individually addressable gold microelectrodes.[81] Two different peptide aptamers were then used to detect cancer biomarkers with low pM detection limits.

7.3.3 Nanoscale Devices for Label-free Field Effect Protein Sensing

Nanostructures offer new and sometimes unique properties for the development of new and ultrasensitive sensors/biosensors. The combination of tuneability, device density and high sensitivity potentially attainable with these configurations (in which the transducer size is comparable to the analyte molecular dimensions) is formidable. During the past decade very considerable advances have been made in synthesising, characterising and manipulating metallic and semiconducting nanowires. With appropriate functionalisation, these structures offer much potential in the field of biosensing.

Semiconducting nanowires can be prepared as single-crystal structures either as p- or n-type materials. In an FET configuration they exhibit electrical performance characteristics comparable to or better than those achieved in the microelectronics industry with planar silicon devices. The synthetically tuneable electronic properties of these wires are particularly attractive. Under conditions of appropriate (low) ionic strength charge depletion/build-up associated with recognition events at their surfaces can be sufficiently sensitive to facilitate single molecule detection.[215]

7.3.3.1 Carbon Nanotube FET Protein Sensing

The utilisation of semiconducting carbon nanotubes in a FET sensing configuration was introduced in 1998.[216] By appropriate chemical design, nanotube side walls can be functionalised covalently or non-covalently (using π–π stacking interactions) in a way which facilitates a control of solubility and biofunctionalisation with retention of desirable electronic characteristics.[217–220] Though the mechanisms by which recognition mitigates a conductance change remains somewhat unclear[221] protein detection limits in the low nM or high pM range have now been reported[220] (though impressive, this is currently not competitive with sensitivities demonstrably achievable with semiconducting nanowires – see below).

In some configurations, biofunctionalisation may not be directly on the nanotube but on an electrostatically coupled gate (Figure 7.8).[222] In recent reports, this top gate method has been utilised in nM to pM immunoassaying and binding constant determination.[223,224] The direct exposure of the nanotube

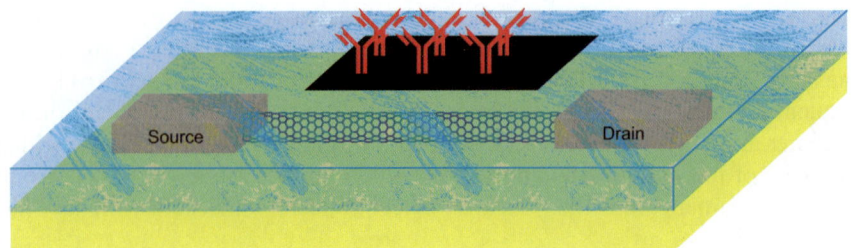

Figure 7.8 Schematic representation of a top-gated CNT-FET device. Biorecognition molecules can be immobilised on a relatively high surface area gate electrode remote from, but electrostatically coupled to, the current transducer. (Figure redrawn with permission from reference 223.)

itself to electrolyte brings with it problems brought on by the high degrees of non-specific protein adsorption typically observed on these surfaces. The use of PEI adlayers or surfactant has, however, been shown to considerably alleviate this.[225] These "direct modification" configurations have now been utilised in real time protein binding assays in devices that exhibit low electrical noise, good consistency, re-useability and nM detection limits.[223] Improved shadow mask methods have facilitated an increase in sensitivity of three orders of magnitude.[224] In very recent work, single protein molecule detection has been reported.[226] Amine coupling chemistry has been utilised in antibody based prostate specific antigen assays.[227] Covalently grafted nucleic acid aptamers have also been utilised in comparably sensitive FET configurations.[228] In a comparative study, both aptamer and monoclonal antibodies were utilised in an immunoglobulin E assay (Figure 7.6b).[188] Both receptors were immobilised covalently on a 1-pyrene butanoic acid succinimidyl ester modified nanotube (amine coupling). Interestingly, pM detection limits were achievable with the aptamer surface whereas the equivalent antibody functionalisation was both slower to respond to analyte (up to 60 minutes) and considerably less sensitive.

7.3.3.2 *Semiconducting Nanowire FET Protein Sensing*

The facile synthesis, high electron mobility and surface area : volume associated sensitivities of semiconducting nanowires make them potentially powerful transducers of chemical or biological recognition. Though the composition and characteristics of these are diverse, it is almost exclusively those of silicon that have been utilised in biosensing configurations.[229–231] Despite the fact that the potential of these constructs within "bottom-up" nanoelectronic fabrication and flexible surface chemical modification is, perhaps, unsurpassed, electronic tuning on this scale remains demanding.[232,233] In recent work, both boron and phosphorus (n and p respectively) doping has been demonstrated *via* monolayer annealing.

In 2001, the group of Lieber first demonstrated the chemical and real time protein detection capabilities of silane-modified boron-doped silicon nanowires.[146] These wires were grown *ex situ*, deposited on an oxidised silicon substrate and contacted by EBL defined leads. Several years later the same team established a multiplexed platform composed of 200 photolithographically fabricated individually addressable nanowires capable of low fM detection of cancer marker proteins in undiluted serum. Protein samples were delivered to the arrays using poly (dimethylsiloxane) polymer-defined fluidic channels. This work constitutes, arguably, the most impressive, scaleable and sensitive protein-detection platform to date. An interesting recent development is the integration of nanowire FET devices with electrokinetic control of transport.[166] Though preliminary work, this marriage appeared to facilitate attomolar detection limits and high levels of selectivity.

Though these constitute exciting and potentially powerful developments, the costs in time and equipment associated with the current "bottom-up" fabrication methods remain prohibitive to all but a few research groups worldwide. Related configurations established *via* "top-down" methods have, however, also proved to be problematic to fabricate and subject to material degradation during use. A CMOS-based immunosensor, in which device scale was reduced by anisotropic etching of a lithographically fabricated pattern, has recently been reported.[172] Though exceptionally sensitive (low fM) these devices exhibit considerable pH dependence and suffer from degradation problems.[172] A related top-down FET nanowire configuration has followed this initial study; here silane/anti-PSA antibody modification was successfully coupled to microfluidic delivery to generate attomolar detection limits and extraordinary levels of specificity (target detection in the presence of a 10^9-fold greater concentration of control protein), when the surface was blocked by ethanolamine.[234] A new FET sensing format, based on a vertical gap configuration, has also recently been demonstrated.[235] The device gap was fabricated by growing a 10-nm layer of silicon oxide on the active silicon area (to form the gate oxide), coating with chromium/gold to form the gate, then wet-etching a 15-nm vertical nanogap (100–400 nm long) between this and the silicon dioxide covering source and drain electrodes. Subsequent biotinylation of the gate surface generated a device capable of displaying a three-orders-of-magnitude current increase (gate voltage change 0.05 V) on exposure to nM levels of streptavidin.

7.4 Conclusions

The detection of protein by field effect methods has two decades of history. Only during the past five years or so, however, have characterisation and fabrication possibilities facilitated the ultrasensitive (sub-pM) label-free detection of protein in a multiplexed/multiplexable format. Though much remains to be done in reducing fabrication cost, increasing reproducibility and reducing non-specific response, these miniaturised, and potentially cheap, configurations have already demonstrated linear ranges and real time monitoring and detection limits that were entirely without precedent only a few years ago. One can suggest that advances during the next decade need "only" be developmental in nature and may lead to a robust collaboration between micro/nanoelectronics and biorecognition that is clinically exploitable on a significant scale.

References

1. S. P. Gygi, Y. Rochon, B. R. Franza and R. Aebersold, *Mol. Cell. Biol.*, 1999, **19**, 1720.
2. A. de Fougerolles, H. P. Vornlocher, J. Maraganore and J. Lieberman, *Nat. Rev. Drug Discov.*, 2007, **6**, 443.
3. K. Y. Tomizaki, K. Usui and H. Mihara, *Chembiochem*, 2005, **6**, 783.
4. S. F. Kingsmore, *Nat. Rev. Drug Discov.*, 2006, **5**, 310.
5. G. Mor, I. Visintin, Y. Lai, H. Zhao, P. Schwartz, T. Rutherford, L. Yue, P. Bray-Ward and D. C. Ward, *Proc. Natl. Acad. Sci. USA*, 2005, **102**, 7677.
6. L. Melton, *Nature*, 2004, **429**, 101.
7. D. A. Hall, J. Ptacek and M. Snyder, *Mech. Ageing Dev.*, 2007, **128**, 161.
8. G. MacBeath and S. L. Schreiber, *Science*, 2000, **289**, 1760.
9. H. Zhu, M. Bilgin, R. Bangham, D. Hall, A. Casamayor, P. Bertone, N. Lan, R. Jansen, S. Bidlingmaier, T. Houfek, T. Mitchell, P. Miller, R. A. Dean, M. Gerstein and M. Snyder, *Science*, 2001, **293**, 2101.
10. M. Eisenstein, *Nature*, 2006, **444**, 959.
11. J. LaBaer and N. Ramachandran, *Curr. Opin. Chem. Biol.*, 2005, **9**, 14.
12. B. B. Haab, *Curr. Opin. Biotechnol.*, 2006, **17**, 415.
13. M. Dufva and C. B. V. Christensen, *Expert Rev. Proteomic*, 2005, **2**, 41.
14. M. M. Ling, C. Ricks and P. Lea, *Expert Rev. Mol. Diagn.*, 2007, **7**, 87.
15. N. Ramachandran, D. N. Larson, P. R. H. Stark, E. Hainsworth and J. LaBaer, *FEBS J.*, 2005, **272**, 5412.
16. J. S. Merkel, G. A. Michaud, M. Salcius, B. Schweitzer and P. F. Predki, *Curr. Opin. Biotechnol.*, 2005, **16**, 447.
17. A. Wolter, R. Niessner and M. Seidel, *Anal. Chem.*, 2007, **79**, 4529.
18. K. Taniguchi, K. Nomura, Y. Hata, T. Nishimura, Y. Ksami and A. Kuroda, *Biotechnol. Bioeng.*, 2007, **96**, 1023.
19. A. J. Nijdam, M. M. C. Cheng, D. H. Geho, R. Fedele, P. Herrmann, K. Killian, V. Espina, E. F. Petricoin, L. A. Liotta and M. Ferrari, *Biomaterials*, 2007, **28**, 550.
20. C. Steinhauer, A. Ressine, G. Marko-Varga, T. Laurell, C. A. K. Borrebaeck and C. Wingren, *Anal. Biochem.*, 2005, **341**, 204.
21. H. Zhu and M. Snyder, *Curr. Opin. Chem. Biol.*, 2003, **7**, 55.
22. I. Saaem, V. Papasotiropoulos, T. Wang, P. Soteropoulos and M. Libera, *J. Nanosci. Nanotechnol.*, 2007, **7**, 2623.
23. L. L. Lv, B. C. Liu, C. X. Zhang, Z. M. Tang, L. Zhang and Z. H. Lu, *Electrophoresis*, 2007, **28**, 406.
24. I. Vostiar, J. Tkac and C. F. Mandenius, *Anal. Biochem.*, 2005, **342**, 152.
25. F. Rusmini, Z. Y. Zhong and J. Feijen, *Biomacromolecules*, 2007, **8**, 1775.

26. B. T. Houseman, J. H. Huh, S. J. Kron and M. Mrksich, *Nat. Biotechnol.*, 2002, **20**, 270.
27. J. C. Love, L. A. Estroff, J. K. Kriebel, R. G. Nuzzo and G. M. Whitesides, *Chem. Rev.*, 2005, **105**, 1103.
28. L. Yan, W. T. S. Huck and G. M. Whitesides, *J. Macromol. Sci.-Polym. Rev*, 2004, **C44**, 175.
29. R. T. Carvalhal, R. S. Freire and L. T. Kubota, *Electroanal*, 2005, **17**, 1251.
30. J. Tkac and J. J. Davis, *J. Electroanal. Chem.*, 2008, **621**, 117.
31. R. Y. P. Lue, G. Y. J. Chen, Y. Hu, Q. Zhu and S. Q. Yao, *J. Am. Chem. Soc.*, 2004, **126**, 1055.
32. K. Kato, H. Sato and H. Iwata, *Langmuir*, 2005, **21**, 7071.
33. A. Savchenko, E. Kashuba, V. Kashuba and B. Snopok, *Anal. Chem.*, 2007, **79**, 1349.
34. T. H. Ha, S. O. Jung, J. M. Lee, K. Y. Lee, Y. Lee, J. S. Park and B. H. Chung, *Anal. Chem.*, 2007, **79**, 546.
35. K. L. Brogan and M. H. Schoenfisch, *Langmuir*, 2005, **21**, 3054.
36. J. J. Davis, J. Tkac, S. Laurenson and P. K. Ferrigno, *Anal. Chem.*, 2007, **79**, 1089.
37. M. He, O. Stoevesandt, E. A. Palmer, F. Khan, O. Ericsson and M. J. Taussig, *Nat. Methods*, 2008, **5**, 175.
38. S. C. Tao and H. Zhu, *Nat. Biotechnol.*, 2006, **24**, 1253.
39. N. Ramachandran, E. Hainsworth, B. Bhullar, S. Eisenstein, B. Rosen, A. Y. Lau, J. C. Walter and J. LaBaer, *Science*, 2004, **305**, 86.
40. P. Angenendt, J. Kreutzberger, J. Glokler and J. D. Hoheisel, *Mol. Cell. Proteomics*, 2006, **5**, 1658.
41. P. Bertone and M. Snyder, *FEBS J.*, 2005, **272**, 5400.
42. T. Mairal, V. C. Ozalp, P. L. Sanchez, M. Mir, I. Katakis and C. K. O'sullivan, *Anal. Bioanal. Chem.*, 2008, **390**, 989.
43. K. Stadtherr, H. Wolf and P. Lindner, *Anal. Chem.*, 2005, **77**, 3437.
44. R. A. Potyrailo, R. C. Conrad, A. D. Ellington and G. M. Hieftje, *Anal. Chem.*, 1998, **70**, 3419.
45. A. Bini, M. Minunni, S. Tombelli, S. Centi and M. Mascini, *Anal. Chem.*, 2007, **79**, 3016.
46. R. Polsky, R. Gill, L. Kaganovsky and I. Willner, *Anal. Chem.*, 2006, **78**, 2268.
47. R. Y. Lai, K. W. Plaxco and A. J. Heeger, *Anal. Chem.*, 2007, **79**, 229.
48. A. E. Radi, J. L. A. Sanchez, E. Baldrich and C. K. O'sullivan, *J. Am. Chem. Soc.*, 2006, **128**, 117.
49. T. G. McCauley, N. Hamaguchi and M. Stanton, *Anal. Biochem.*, 2003, **319**, 244.
50. M. Cretich, F. Damin, G. Pirri and M. Chiari, *Biomol. Eng.*, 2006, **23**, 77.
51. B. B. Haab, M. J. Dunham and P. O. Brown, *Genome Biol. 2001*, **2**.
52. G. A. Michaud, M. Salcius, F. Zhou, R. Bangham, J. Bonin, H. Guo, M. Snyder, P. F. Predki and B. I. Schweitzer, *Nat. Biotechnol.*, 2003, **21**, 1509.

53. M. Hust and S. Dubel, *Trends Biotechnol.*, 2004, **22**, 8.
54. C. Wingren and C. A. K. Borrebaeck, *Omics*, 2006, **10**, 411.
55. C. Wingren, J. Ingvarsson, M. Lindstedt and C. A. K. Borrebaeck, *Nat. Biotechnol.*, 2003, **21**, 223.
56. C. K. O'sullivan, *Anal. Bioanal. Chem.*, 2002, **372**, 44.
57. W. H. Tan, K. M. Wang and T. J. Drake, *Curr. Opin. Chem. Biol.*, 2004, **8**, 547.
58. C. Bock, M. Coleman, B. Collins, J. Davis, G. Foulds, L. Gold, C. Greef, J. Heil, J. S. Heilig, B. Hicke, M. N. Hurst, G. M. Husar, D. Miller, R. Ostroff, H. Petach, D. Schneider, B. Vant-Hull, S. Waugh, A. Weiss, S. K. Wilcox and D. Zichi, *Proteomics*, 2004, **4**, 609.
59. J. R. Collett, E. J. Cho and A. D. Ellington, *Methods*, 2005, **37**, 4.
60. M. Gross, *Chem. World*, 2006, **3**, 48.
61. Y. X. Jiang, X. H. Fang and C. L. Bai, *Anal. Chem.*, 2004, **76**, 5230.
62. J. F. Lee, G. M. Stovall and A. D. Ellington, *Curr. Opin. Chem. Biol.*, 2006, **10**, 282.
63. E. J. Cho, J. R. Collett, A. E. Szafranska and A. D. Ellington, *Anal. Chim. Acta*, 2006, **564**, 82.
64. Y. Li, H. J. Lee and R. M. Corn, *Anal. Chem.*, 2007, **79**, 1082.
65. G. Henderson and M. Bradley, *Curr. Opin. Biotechnol.*, 2007, **18**, 326.
66. D. H. Min and M. Mrksich, *Curr. Opin. Chem. Biol.*, 2004, **8**, 554.
67. S. P. A. Fodor, J. L. Read, M. C. Pirrung, L. Stryer, A. T. Lu and D. Solas, *Science*, 1991, **251**, 767.
68. M. M. Reddy and T. Kodadek, *Proc. Natl. Acad. Sci. USA*, 2005, **102**, 12672.
69. R. C. Ladner and A. C. Ley, *Curr. Opin. Biotechnol.*, 2001, **12**, 406.
70. M. Crawford, R. Woodman and P. K. Ferrigno, *Brief. Function. Genom. Proteomics*, 2003, **2**, 72.
71. A. Skerra, *Rev. Mol. Biotechnol.*, 2001, **74**, 257.
72. L. H. Xu, P. Aha, K. Gu, R. G. Kuimelis, M. Kurz, T. Lam, A. C. Lim, H. X. Liu, P. A. Lohse, L. Sun, S. Weng, R. W. Wagner and D. Lipovsek, *Chem. Biol.*, 2002, **9**, 933.
73. K. Nord, E. Gunneriusson, J. Ringdahl, S. Stahl, M. Uhlen and P. A. Nygren, *Nat. Biotechnol.*, 1997, **15**, 772.
74. A. Skerra, *Curr. Opin. Biotechnol.*, 2007, **18**, 295.
75. B. Renberg, I. Shiroyama, T. Engfeldt, P. A. Nygren and A. E. Karlstrom, *Anal. Biochem.*, 2005, **341**, 334.
76. E. Wahlberg, C. Lendel, M. Helgstrand, P. Allard, V. Dincbas-Renqvist, A. Hedqvist, H. Berglund, P. A. Nygren and T. Hard, *Proc. Natl. Acad. Sci. USA*, 2003, **100**, 3185.
77. E. Wahlberg and T. Hard, *J. Am. Chem. Soc.*, 2006, **128**, 7651.
78. R. Woodman, J. T. H. Yeh, S. Laurenson and P. K. Ferrigno, *J. Mol. Biol.*, 2005, **352**, 1118.
79. A. Chattopadhyay, S. A. Tate, R. W. Beswick, S. D. Wagner and P. K. Ferrigno, *Oncogene*, 2006, **25**, 2223.

80. P. Estrela, D. Paul, P. Li, S. D. Keighley, P. Migliorato, S. Laurenson and P. K. Ferrigno, *Electrochim. Acta*, 2008, **53**, 6489.
81. D. Evans, S. Johnson, S. Laurenson, A. G. Davies, P. K. Ferrigno and C. Walti, *J. Biol.*, 2008, **7**.
82. S. Johnson, D. Evans, S. Laurenson, D. Paul, A. G. Davies, P. K. Ferrigno and C. Walti, *Anal. Chem.*, 2008, **80**, 978.
83. S. M. Chan, J. Ermann, L. Su, C. G. Fathman and P. J. Utz, *Nat. Med.*, 2004, **10**, 1390.
84. W. M. Gao, R. Kuick, R. P. Orchekowski, D. E. Misek, J. Qiu, A. K. Greenberg, W. N. Rom, D. E. Brenner, G. S. Omenn, B. B. Haab and S. M. Hanash, *BMC Canc.*, 2005, **5**.
85. J. C. Miller, H. P. Zhou, J. Kwekel, R. Cavallo, J. Burke, E. B. Butler, B. S. Teh and B. B. Haab, *Proteomics*, 2003, **3**, 56.
86. R. Orchekowski, D. Hamelinck, L. Li, E. Gliwa, M. VanBrocklin, J. A. Marrero, G. F. V. Woude, Z. D. Feng, R. Brand and B. B. Haab, *Cancer Res.*, 2005, **65**, 11193.
87. M. Sanchez-Carbayo, N. D. Socci, J. J. Lozano, B. B. Haab and C. Cordon-Cardo, *Am. J. Pathol.*, 2006, **168**, 93.
88. M. Herrmann, T. Veres and M. Tabrizian, *Lab Chip*, 2006, **6**, 555.
89. M. Tudorache, M. Co, H. Lifgren and J. Emneus, *Anal. Chem.*, 2005, **77**, 7156.
90. J. Yakovleva, R. Davidsson, A. Lobanova, M. Bengtsson, S. Eremin, T. Laurell and J. Emneus, *Anal. Chem.*, 2002, **74**, 2994.
91. M. Diaz-Gonzalez, M. B. Gonzalez-Garcia and A. Costa-Garcia, *Electroanal.*, 2005, **17**, 1901.
92. X.-M. Li, X.-Y. Yang and S.-S. Zhang, *TrAC Trends Anal. Chem.*, 2008, **27**, 543.
93. M. S. Wilson and W. Y. Nie, *Anal. Chem.*, 2006, **78**, 6476.
94. J. Wu, Y. Yan, F. Yan and H. Ju, *Anal. Chem.*, 2008, **80**, 6072.
95. J. Das, K. Jo, J. W. Lee and H. Yang, *Anal. Chem.*, 2007, **79**, 2790.
96. U. Bilitewski, *Anal. Chim. Acta*, 2006, **568**, 232.
97. K. Jaras, A. Ressine, E. Nilsson, J. Malm, G. Marko-Varga, H. Lilja and T. Laurell, *Anal. Chem.*, 2007, **79**, 5817.
98. H. Q. Zhang, Q. Zhao, X. F. Li and X. C. Le, *Analyst*, 2007, **132**, 724.
99. A. Waggoner, *Curr. Opin. Chem. Biol.*, 2006, **10**, 62.
100. C. Wingren and C. A. Borrebaeck, *Curr. Opin. Biotechnol.*, 2008, **19**, 55.
101. J. M. Nam, K. J. Jang and J. T. Groves, *Nat. Protoc.*, 2007, **2**, 1438.
102. R. De Palma, G. Reekmans, W. Laureyn, G. Borghs and G. Maes, *Anal. Chem.*, 2007, **79**, 7540.
103. H. M. E. Azzazy, M. M. H. Mansour and S. C. Kazmierczak, *Clin. Chem.*, 2006, **52**, 1238.
104. C. J. Johnson, N. Zhukovsky, A. E. G. Cass and J. M. Nagy, *Proteomics*, 2008, **8**, 715.
105. J. Wang, *Electroanal.*, 2007, **19**, 769.
106. N. L. Rosi and C. A. Mirkin, *Chem. Rev.*, 2005, **105**, 1547.

107. X. Yu, B. Munge, V. Patel, G. Jensen, A. Bhirde, J. D. Gong, S. N. Kim, J. Gillespie, J. S. Gutkind, F. Papadimitrakopoulos and J. F. Rusling, *J. Am. Chem. Soc.*, 2006, **128**, 11199.
108. J. Wang, G. D. Liu and M. R. Jan, *J. Am. Chem. Soc.*, 2004, **126**, 3010.
109. J. Wang, *Small*, 2005, **1**, 1036.
110. M. M. C. Cheng, G. Cuda, Y. L. Bunimovich, M. Gaspari, J. R. Heath, H. D. Hill, C. A. Mirkin, A. J. Nijdam, R. Terracciano, T. Thundat and M. Ferrari, *Curr. Opin. Chem. Biol.*, 2006, **10**, 11.
111. X. B. Yu, D. K. Xu and Q. Cheng, *Proteomics*, 2006, **6**, 5493.
112. L. Jantzie, V. Tanay and K. Todd, in *Handbook of Neurochemistry and Molecular Neurobiology*, Springer-Verlag, Berlin, 2007, Ch. 8.
113. A. Numnuam, K. Y. Chumbimuni-Torres, Y. Xiang, R. Bash, P. Thavarungkul, P. Kanatharana, E. Pretsch, J. Wang and E. Bakker, *Anal. Chem.*, 2008, **80**, 707.
114. K. R. Kozak, M. W. Amneus, S. M. Pusey, F. Su, M. N. Luong, S. A. Luong, S. T. Reddy and R. Farias-Eisner, *Proc. Natl. Acad. Sci. USA*, 2003, **100**, 12343.
115. L. Favre-Kontula, Z. Johnson, T. Steinhoff, A. Frauenschuh, F. Vilbois and A. E. I. Proudfoot, *J. Immunol. Methods*, 2006, **317**, 152.
116. S. M. Patrie and M. Mrksich, *Anal. Chem.*, 2007, **79**, 5878.
117. G. Malik, M. D. Ward, S. K. Gupta, M. W. Trosset, W. E. Grizzle, B. L. Adam, J. I. Diaz and O. J. Semmes, *Clin. Cancer Res.*, 2005, **11**, 1073.
118. B. Liedberg, C. Nylander and I. Lundstrom, *Sens. Actuat.*, 1983, **4**, 299.
119. M. Malmqvist, *Nature*, 1993, **361**, 186.
120. M. A. Cooper, *Nat. Rev. Drug Discov.*, 2002, **1**, 515.
121. J. O. Foley, K. E. Nelson, A. Mashadi-Hossein, B. A. Finlayson and P. Yager, *Anal. Chem.*, 2007, **79**, 3549.
122. S. O. Jung, H. S. Ro, B. H. Kho, Y. B. Shin, M. G. Kim and B. H. Chung, *Proteomics*, 2005, **5**, 4427.
123. B. E. Erickson, *Anal. Chem.*, 2004, **76**, 386A.
124. T. T. Goodrich, H. J. Lee and R. M. Corn, *J. Am. Chem. Soc.*, 2004, **126**, 4086.
125. S. P. Fang, H. J. Lee, A. W. Wark and R. M. Corn, *J. Am. Chem. Soc.*, 2006, **128**, 14044.
126. L. He, M. D. Musick, S. R. Nicewarner, F. G. Salinas, S. J. Benkovic, M. J. Natan and C. D. Keating, *J. Am. Chem. Soc.*, 2000, **122**, 9071.
127. L. A. Lyon, M. D. Musick and M. J. Natan, *Anal. Chem.*, 1998, **70**, 5177.
128. J. Homola, *Chem. Rev.*, 2008, **108**, 462.
129. G. Binnig and H. Rohrer, *Rev. Mod. Phys.*, 1987, **59**, 615.
130. K. B. Lee, S. J. Park, C. A. Mirkin, J. C. Smith and M. Mrksich, *Science*, 2002, **295**, 1702.
131. M. Lee, D. K. Kang, H. K. Yang, K. H. Park, S. Y. Choe, C. Kang, S. I. Chang, M. H. Han and I. C. Kang, *Proteomics*, 2006, **6**, 1094.
132. K. Salaita, Y. H. Wang, J. Fragala, R. A. Vega, C. Liu and C. A. Mirkin, *Angew. Chem.-Int. Ed.*, 2006, **45**, 7220.
133. C. N. LaFratta and D. R. Walt, *Chem. Rev.*, 2008, **108**, 614.

134. D. J. Muller and Y. F. Dufrene, *Nat. Nanotechnol.*, 2008, **3**, 261.
135. K. M. Goeders, J. S. Colton and L. A. Bottomley, *Chem. Rev.*, 2008, **108**, 522.
136. N. Backmann, C. Zahnd, F. Huber, A. Bietsch, A. Pluckthun, H. P. Lang, H. J. Guntherodt, M. Hegner and C. Gerber, *Proc. Natl. Acad. Sci. USA*, 2005, **102**, 14587.
137. W. Shu, S. Laurenson, T. P. J. Knowles, P. Ko Ferrigno and A. A. Seshia, *Biosens. Bioelectron.*, in press, corrected proof.
138. J. Fritz, M. K. Baller, H. P. Lang, H. Rothuizen, P. Vettiger, E. Meyer, H. J. Guntherodt, C. Gerber and J. K. Gimzewski, *Science*, 2000, **288**, 316.
139. G. H. Wu, R. H. Datar, K. M. Hansen, T. Thundat, R. J. Cote and A. Majumdar, *Nat. Biotechnol.*, 2001, **19**, 856.
140. M. Yue, J. C. Stachowiak, H. Lin, R. Datar, R. Cote and A. Majumdar, *Nano Lett.*, 2008, **8**, 520.
141. C. A. Savran, S. M. Knudsen, A. D. Ellington and S. R. Manalis, *Anal. Chem.*, 2004, **76**, 3194.
142. C. Berggren, B. Bjarnason and G. Johansson, *Electroanal.*, 2001, **13**, 173.
143. E. Bakker and Y. Qin, *Anal. Chem.*, 2006, **78**, 3965.
144. J. G. Guan, Y. Q. Miao and Q. J. Zhang, *J. Biosci. Bioeng.*, 2004, **97**, 219.
145. E. Katz and I. Willner, *Electroanal.*, 2003, **15**, 913.
146. Y. Cui, Q. Q. Wei, H. K. Park and C. M. Lieber, *Science*, 2001, **293**, 1289.
147. F. Patolsky, G. F. Zheng, O. Hayden, M. Lakadamyali, X. W. Zhuang and C. M. Lieber, *Proc. Natl. Acad. Sci. USA*, 2004, **101**, 14017.
148. J. Ji, J. G. O'Connell, D. J. D. Carter and D. N. Larson, *Anal. Chem.*, 2008, **80**, 2491.
149. J. N. Anker, W. P. Hall, O. Lyandres, N. C. Shah, J. Zhao and R. P. Van Duyne, *Nat. Mater.*, 2008, **7**, 442.
150. A. K. Sinensky and A. M. Belcher, *Nat. Nanotechnol.*, 2007, **2**, 653.
151. A. E. Grow, L. L. Wood, J. L. Claycomb and P. A. Thompson, *J. Microbiol. Methods*, 2003, **53**, 221.
152. E. Hutter and J. H. Fendler, *Adv. Mater.*, 2004, **16**, 1685.
153. Z. H. Wang, Y. H. Meng, P. Q. Ying, C. Qi and G. Jin, *Electrophoresis*, 2006, **27**, 4078.
154. M. Schaferling and S. Nagl, *Anal. Bioanal. Chem.*, 2006, **385**, 500.
155. J. Das, M. A. Aziz and H. Yang, *J. Am. Chem. Soc.*, 2006, **128**, 16022.
156. J. M. Nam, C. S. Thaxton and C. A. Mirkin, *Science*, 2003, **301**, 1884.
157. B. K. Oh, J. M. Nam, S. W. Lee and C. A. Mirkin, *Small*, 2006, **2**, 103.
158. S. Y. Hou, H. K. Chen, H. C. Cheng and C. Y. Huang, *Anal. Chem.*, 2007, **79**, 980.
159. J. M. Nam, A. R. Wise and J. T. Groves, *Anal. Chem.*, 2005, **77**, 6985.
160. K. A. Shaikh, K. S. Ryu, E. D. Goluch, J. M. Nam, J. W. Liu, S. Thaxton, T. N. Chiesl, A. E. Barron, Y. Lu, C. A. Mirkin and C. Liu, *Proc. Natl. Acad. Sci. USA*, 2005, **102**, 9745.
161. J. Wang, G. D. Liu, B. Munge, L. Lin and Q. Y. Zhu, *Angew. Chem.-Int. Ed.*, 2004, **43**, 2158.

162. M. A. Hayes, N. A. Polson, A. N. Phayre and A. A. Garcia, *Anal. Chem.*, 2001, **73**, 5896.
163. F. Lacharme, C. Vandevyver and M. A. M. Gijs, *Anal. Chem.*, 2008, **80**, 2905.
164. A. D. Wellman and M. J. Sepaniak, *Anal. Chem.*, 2006, **78**, 4450.
165. P. E. Sheehan and L. J. Whitman, *Nano Lett.*, 2005, **5**, 803.
166. J.-R. Gong, W. Lu, G. Zheng and C. M. Lieber, *Abstracts of Papers, 235th ACS National Meeting, New Orleans, LA, United States*, 2008, April 6–10.
167. F. Patolsky and C. M. Lieber, *Mater. Today*, 2005, **8**, 20.
168. X. B. Yu, R. Lv, Z. Q. Ma, Z. H. Liu, Y. H. Hao, Q. Z. Li and D. K. Xu, *Analyst*, 2006, **131**, 745.
169. C. Berggren and G. Johansson, *Anal. Chem.*, 1997, **69**, 3651.
170. C. Berggren, B. Bjarnason and G. Johansson, *Biosens. Bioelectron.*, 1998, **13**, 1061.
171. M. Bart, E. C. A. Stigter, H. R. Stapert, G. J. de Jong and W. P. van Bennekom, *Biosens. Bioelectron.*, 2005, **21**, 49.
172. E. Stern, J. F. Klemic, D. A. Routenberg, P. N. Wyrembak, D. B. Turner-Evans, A. D. Hamilton, D. A. LaVan, T. M. Fahmy and M. A. Reed, *Nature*, 2007, **445**, 519.
173. F. Patolsky, G. F. Zheng and C. M. Lieber, *Nat. Protoc.*, 2006, **1**, 1711.
174. J. V. Veetil and K. M. Ye, *Biotechnol. Prog.*, 2007, **23**, 517.
175. P. Bergveld, *IEEE Transactions on Biomedical Engineering*, 1970, **BM17**, 70.
176. A. Sibbald, *J. Mol. Electron.*, 1986, **2**, 51.
177. P. Bergveld, *Sens. Actuat. B-Chem.*, 2003, **88**, 1.
178. S. Caras and J. Janata, *Anal. Chem.*, 1980, **52**, 1935.
179. M. Gotoh, E. Tamiya and I. Karube, *J. Membr. Sci.*, 1989, **41**, 291.
180. L. Bousse, *Sens. Actuat. B-Chem.*, 1996, **34**, 270.
181. S. N. Kim, J. F. Rusling and F. Papadimitrakopoulos, *Adv. Mater.*, 2007, **19**, 3214.
182. A. Poghossian, A. Cherstvy, S. Ingebrandt, A. Offenhausser and M. J. Schoning, *Sens. Actuat. B-Chem.*, 2005, **111**, 470.
183. A. Poghossian, S. Ingebrandt, M. H. Abouzar and M. J. Schoning, *Appl. Phys. A-Mater. Sci. Process.*, 2007, **87**, 517.
184. P. Estrela and P. Migliorato, *J. Mater. Chem.*, 2007, **17**, 219.
185. G. F. Zheng, F. Patolsky, Y. Cui, W. U. Wang and C. M. Lieber, *Nat. Biotechnol.*, 2005, **23**, 1294.
186. G. J. Zhang, G. Zhang, J. H. Chua, R. E. Chee, E. H. Wong, A. Agarwal, K. D. Buddharaju, N. Singh, Z. Q. Gao and N. Balasubramanian, *Nano Lett.*, 2008, **8**, 1066.
187. E. Stern, R. Wagner, F. J. Sigworth, R. Breaker, T. M. Fahmy and M. A. Reed, *Nano Lett.*, 2007, **7**, 3405.
188. K. Maehashi, T. Katsura, K. Kerman, Y. Takamura, K. Matsumoto and E. Tamiya, *Anal. Chem.*, 2007, **79**, 782.
189. P. Bergveld, *Biosens. Bioelectron.*, 1991, **6**, 55.

190. H. T. Wang, B. S. Kang, F. Ren, S. J. Pearton, J. W. Johnson, P. Rajagopal, J. C. Roberts, E. L. Piner and K. J. Linthicum, *Appl. Phys. Lett.*, 2007, **91**, 222101.
191. J. S. Daniels and N. Pourmand, *Electroanal.*, 2007, **19**, 1239.
192. A. Gebbert, M. Alvarezicaza, W. Stocklein and R. D. Schmid, *Anal. Chem.*, 1992, **64**, 997.
193. J. J. Davis, *Chem. Commun.*, 2005, 3509.
194. W. T. S. Huck, *Angew. Chem.-Int. Ed.*, 2007, **46**, 2754.
195. S. Zhang, C. M. Cardona and L. Echegoyen, *Chem. Commun.*, 2006, 4461.
196. M. Mrksich, *Chem. Soc. Rev.*, 2000, **29**, 267.
197. I. S. Choi and Y. S. Chi, *Angew. Chem.-Int. Ed.*, 2006, **45**, 4894.
198. J. M. Seminario, *Nat. Materials*, 2005, **4**, 111.
199. D. Chen and J. H. Li, *Surf. Sci. Rep.*, 2006, **61**, 445.
200. V. M. Mirsky, M. Riepl and O. S. Wolfbeis, *Biosens. Bioelectron.*, 1997, **12**, 977.
201. A. L. Newman, K. W. Hunter and W. D. Stanbro, *Chemical Sensors: 2nd International Meeting, Proc.*, 1986, 596.
202. R. F. Taylor, I. G. Marenchic and R. H. Spencer, *Anal. Chim. Acta*, 1991, **249**, 67.
203. Z. W. Zou, J. H. Kai, M. J. Rust, J. Han and C. H. Ahn, *Sens. Actuat. A-Phys.*, 2007, **136**, 518.
204. S. Ameur, C. Martelet, N. Jaffrezic-Renault and J. M. Chovelon, *Appl. Biochem. Biotechnol.*, 2000, **89**, 161.
205. M. C. Rodriguez, A. N. Kawde and J. Wang, *Chem. Commun.*, 2005, 4267.
206. A. E. Radi, J. L. A. Sanchez, E. Baldrich and C. K. O'sullivan, *Anal. Chem.*, 2005, **77**, 6320.
207. D. K. Xu, D. W. Xu, X. B. Yu, Z. H. Liu, W. He and Z. Q. Ma, *Anal. Chem.*, 2005, **77**, 5107.
208. Y. Xu, L. Yang, X. Y. Ye, P. A. He and Y. Z. Fang, *Electroanal.*, 2006, **18**, 1449.
209. J. A. Lee, S. Hwang, J. Kwak, S. Il Park, S. S. Lee and K. C. Lee, *Sens. Actuat. B-Chem.*, 2008, **129**, 372.
210. J. Rickert, W. Gopel, W. Beck, G. Jung and P. Heiduschka, *Biosens. Bioelectron.*, 1996, **11**, 757.
211. P. Bataillard, F. Gardies, N. Jaffrezicrenault, C. Martelet, B. Colin and B. Mandrand, *Anal. Chem.*, 1988, **60**, 2374.
212. M. Dijksma, B. Kamp, J. C. Hoogvliet and W. P. van Bennekom, *Anal. Chem.*, 2001, **73**, 901.
213. M. Dijksma, B. Kamp, J. C. Hoogvliet and W. P. van Bennekom, *Langmuir*, 2000, **16**, 3852.
214. M. Dijksma, B. A. Boukamp, B. Kamp and W. P. van Bennekom, *Langmuir*, 2002, **18**, 3105.
215. F. Patolsky, G. F. Zheng and C. M. Lieber, *Anal. Chem.*, 2006, **78**, 4260.
216. S. J. Tans, A. R. M. Verschueren and C. Dekker, *Nature*, 1998, **393**, 49.

217. B. C. Satishkumar, L. O. Brown, Y. Gao, C. C. Wang, H. L. Wang and S. K. Doorn, *Nat. Nanotechnol.*, 2007, **2**, 560.
218. J. J. Davis, K. S. Coleman, B. R. Azamian, C. B. Bagshaw and M. L. H. Green, *Chem.- Eur. J.*, 2003, **9**, 3732.
219. B. R. Azamian, J. J. Davis, K. S. Coleman, C. B. Bagshaw and M. L. H. Green, *J. Am. Chem. Soc.*, 2002, **124**, 12664.
220. B. L. Allen, P. D. Kichambare and A. Star, *Adv. Mater.*, 2007, **19**, 1439.
221. I. Heller, A. M. Janssens, J. Mannik, E. D. Minot, S. G. Lemay and C. Dekker, *Nano Lett.*, 2008, **8**, 591.
222. M. Abe, K. Murata, T. Ataka and K. Matsumoto, *Nanotechnology*, 2008, 19, 045505.
223. M. Abe, K. Murata, A. Kojima, Y. Ifuku, M. Shimizu, T. Ataka and K. Matsumoto, *J. Phys. Chem. C*, 2007, **111**, 8667.
224. S. Takeda, A. Sbagyo, Y. Sakoda, A. Ishii, M. Sawamura, K. Sueoka, H. Kida, K. Mukasa and K. Matsumoto, *Biosens. Bioelectron.*, 2005, **21**, 201.
225. M. Shim, A. Javey, N. W. S. Kam and H. J. Dai, *J. Am. Chem. Soc.*, 2001, **123**, 11512.
226. A. Star, J. C. P. Gabriel, K. Bradley and G. Gruner, *Nano Lett.*, 2003, **3**, 459.
227. C. Li, M. Curreli, H. Lin, B. Lei, F. N. Ishikawa, R. Datar, R. J. Cote, M. E. Thompson and C. W. Zhou, *J. Am. Chem. Soc.*, 2005, **127**, 12484.
228. H. M. So, K. Won, Y. H. Kim, B. K. Kim, B. H. Ryu, P. S. Na, H. Kim and J. O. Lee, *J. Am. Chem. Soc.*, 2005, **127**, 11906.
229. W. Lu and C. M. Lieber, *J. Phys. D-Appl. Phys.*, 2006, **39**, R387.
230. N. Stutzmann, R. H. Friend and H. Sirringhaus, *Science*, 2003, **299**, 1881.
231. M. Muccini, *Nat. Materials*, 2006, **5**, 605.
232. J. C. Ho, R. Yerushalmi, Z. A. Jacobson, Z. Fan, R. L. Alley and A. Javey, *Nat. Materials*, 2008, **7**, 62.
233. Y. Li, F. Qian, J. Xiang and C. M. Lieber, *Mater. Today*, 2006, **9**, 18.
234. A. Kim, C. S. Ah, H. Y. Yu, J. H. Yang, I. B. Baek, C. G. Ahn, C. W. Park, M. S. Jun and S. Lee, *Appl. Phys. Lett.*, 2007, 91, 103901.
235. H. S. Im, X. J. Huang, B. Gu and Y. K. Choi, *Nat. Nanotechnol.*, 2007, **2**, 430.

CHAPTER 8
Biological and Clinical Applications of Biosensors

PAUL KO FERRIGNO

Section of Experimental Therapeutics, Leeds Institute of Molecular Medicine, St James's University Hospital, Beckett Street, Leeds LS9 7TF, UK

There is an increasing need for devices that will (1) allow the real-time, simultaneous monitoring of multiple analytes in biological samples, to serve the growing fields of proteomics and the emerging field of systems biology; (2) allow the early detection of disease biomarkers in patient samples, especially serum and urine, that will allow early therapeutic intervention so as to reduce harm to the patient and costs to the health care system; or (3) allow the sensitive detection of biological material, whether toxins or infectious organisms, in environmental and security monitoring. This chapter will focus on the first two applications, introducing each in turn, in terms that will be suitable to an audience of chemists, physicists and engineers as well as biologists. The emphasis throughout will be on providing a brief but authoritative summary of the numbers, types and concentrations of molecules that are likely to be of interest to the biosensor community.

8.1 Biosensing for Pure Biological Research

8.1.1 The Challenges of "Omics" and "Systems" Approaches

Biosensors are traditionally viewed as devices for the detection of one or at most a handful of biological molecules. However, the technologies and know-how

Engineering the Bioelectronic Interface: Applications to Analyte Biosensing and Protein Detection
Edited by Jason Davis
© 2009 Royal Society of Chemistry
Published by the Royal Society of Chemistry, www.rsc.org

that are taken for granted in this area would be of enormous value in the emerging fields of systems biology and synthetic biology, where the emphasis is on the measurement of all the products of an organism's genome.

Genome-wide sequence data and the attendant availability of gene expression microarrays have led to a catalogue of genes and some insight into their expression in different biological conditions. This is perhaps best illustrated in the model organism *Saccharomyces cerevisiae* (baker's yeast), which has been at the forefront of systems biology for over 15 years. The first complete chromosome to be sequenced was yeast Chromosome III[1] and the yeast genome was also the first complete genome to be sequenced.[2] However, the catalogue of parts has no value unless we understand how those parts interact functionally. The early sequencing of the *S. cerevisiae* genome thus led to the first rigorous microarray analysis of gene expression[3] and has also led to high-throughput analysis of metabolites.[4,5] Efforts have also been made to immobilise all of the proteins expressed from the yeast genome on microscope slides to allow for protein–protein interaction analysis,[6] which represents a direct cross-over into the traditional arena of the biosensor community – the detection of an organism's proteins. However, the data obtained from many of these experiments remain flawed, as is discussed below. All of these technologies, which have been and continue to be applied to an ever-increasing number of other organisms including *Homo sapiens*, are really no more than highly multiplexed biosensors, but they still lack the ability to integrate biological information into the detection process – so an enzyme may be predicted to be expressed, but there is no attempt to build a multiplexed sensor to monitor the activity of multiple enzymes. A key challenge for the future, then, will be the integration of biosensing expertise with the desire and need to capture the complexity of biology and disease.

8.1.2 Biological Complexity

Genome sizes vary widely between organisms, and there are many tables in standard texts that describe this. High-throughput genome sequencing is bringing this complexity out of textbooks and into every biological laboratory. On the date of this writing, the National Centre for Biological Information (NCBI) genome database contains the genomes of 4637 species, with multiple species described by more than one complete genome sequence (http://www.ncbi.nlm.nih.gov/sites/entrez?db=Genome&itool=toolbar). At first glance, the widespread availability of genome sequences should make the task of the biosensing community easier – one simply needs to make probes against the relevant gene products, and devise sensing mechanisms. In practice, however, a genomic overview vastly oversimplifies the biological reality where proteins exist in multiple chemically distinct isoforms, cover a huge range of copy numbers (single copy to millions of copies per cell) and are frequently found in sub-cellular compartments distinct from those that encode them (see sections on mitochondria and chloroplasts below). The following paragraphs attempt to capture these challenges.

8.1.2.1 The Biological Complexity of Model Organisms

Between kingdoms, genome size correlates very roughly with the size and complexity of an organism – viruses generally have smaller genomes than bacteria, unicellular organisms generally have smaller genomes than multicellular organisms and so on. However, when addressing organisms of similar evolutionary complexity (*e.g.* all multicellular organisms), genome size does not reliably predict biological complexity – plants have larger genomes than primates (including humans) and so on. Furthermore, some organisms possess genetic information encoded outside of the nuclear genome. This information can be carried on plasmids (circular fragments of single- or double-stranded DNA that are prevalent in microbes), or in the genomes of extra-nuclear organelles such as mitochondria or chloroplasts. In addition, knowledge of genome sequence leaves large gaps – which predicted genes are made into mRNA? How many splice variants are made from each gene? How many of the mRNAs are translated into protein, and under what conditions? How are the proteins modified post-translationally and how does this affect function? Post-translational modification of amino acid side chains alone represents a minefield of complexity, with over 300 potential modifications known, of which the most common are by glycosylation or phosphorylation of serine, threonine or tyrosine, ubiquitination of lysine, cysteine, serine or threonine, methylation or acetylation of lysine or arginine, hydroxylation of prolines, disulfide bond formation between cysteine residues, amidation of phenylalanine, sulfation of tyrosine residues or gamma-carboxylation of glutamic acid.

The goal of the following section is to attempt to summarise the hurdles that need to be overcome for successful integration of our knowledge of molecular biology (by which I mean the interactions between biological molecules) with the ability to catalogue all the bio-molecules that result from the expression of an organism's genome – not just the gene products of transcribed RNA, but the proteins translated from mRNA, and the metabolic products that are made by enzymes made from a proportion of those mRNAs.

8.1.2.2 *Escherichia coli*

The genome of K12 laboratory strains of *E. coli* comprises more than 4.6 million base pairs, rising to more than 5.5 million for pathogenic strains such as O157:H7. Added complexity derives from the fact that *E. coli* cells typically carry extra-genomic DNA in the form of self-replicating plasmids that carry information that may for example be useful to the carrier in times of stress. There is a large range of plasmids that cells may carry, and any given population may be expected to carry a multiplicity of plasmid types. For example, in one study a single strain of *E. coli* isolated from a patient was found to harbour five plasmids of which one, pKL1, comprised just 1549 base pairs and appears to encode only the genetic information it needs for its own replication,[7] which can be made to reach more than 2500 copies per cell.[8] In contrast, a plasmid of more than 240 000 base pairs (pAPEC01-R), which carries 224 protein-coding genes, was isolated from a pathogenic *E. coli* strain.[9] These numbers illustrate for a relatively simple organism first the huge dynamic range that multiplexed biosensors would need to be able to address, from single genes or proteins to many thousands of genes encoding proteins present in perhaps several million copies. They also illustrate the problems of ensuring biological specificity in the detection at each sensor feature of a single, usually rare biological species in the midst of a competing mixture of many multiply repetitive competing molecules. Finally, this analysis serves to make a point that will occur again and again – a knowledge of the sequence of a genome, even with robust predictions of which genes code for proteins, cannot provide a full picture of the organism's biology. In this case complexity derives from the biology encoded in plasmids, but there are many sources of confounding information.

8.1.2.3 Arabidopsis Thaliana

Another well-studied organism is the simple plant *A. thaliana* (thale cress). *Arabidopsis* cells contain a nuclear genome with 199 million base pairs as well as a mitochondrial genome of 366 924 base pairs and a chloroplast genome of 154 478 base pairs. Surprisingly, the nuclear genome of *A. thaliana* is predicted to encode more proteins (25 498; the Arabidopsis Genome Initiative, 2000) than the human nuclear genome (21 660; http://www2.bioinformatics.tll.org.sg/Homo_sapiens/index.html). Across the plant kingdom, chloroplast genomes are generally between 120 000 and 240 000 base pairs, and are predicted to code for over 100 proteins. Indeed, more than 250 spots can be visualised in 2D electrophoresis on plant chloroplasts, but not all of them emanate from the chloroplast genome. Once again a genetic analysis alone cannot address the biology of the chloroplast, because as many as 40% of these proteins are encoded in the nuclear, rather than the chloroplast, genome.[10]

8.1.2.4 Biological Complexity of Human Cells

The human genome comprises slightly over 3 billion base pairs encoding an estimated 21 660 genes. Each gene is regulated independently, and proteins made from these genes may be present within cells at between a single copy and as many as a million copies (for proteins involved in ribosomal structure, for example). If we estimate the volume of a cell with dimensions of the order of 10 μm to be 1 pL, the concentration range of proteins within that cell is therefore 1.7 pM to 1.7 μM. Assuming a biosensor with a sample volume of 100 μL, and a sample of 10 000 cells (such as might be isolated by laser capture microdissection from a tumour biopsy) whose total volume would be 10 000 pL, the dilution factor would be 1 : 10 000 and the sample protein concentration would range from 0.17 fM to 0.17 nM. These numbers set the target detection range for a multiplexed biosensor.

8.1.2.5 Within Cell Fractions

Eukaryotic cells are defined by a plasma membrane that contains multiple intra cellular membrane-bound compartments (organelles) including the nucleus, which contains the genome.

8.1.2.5.1 The Plasma Membrane. The plasma membrane is the double lipid bilayer, 7.5 nm wide, which surrounds cells. It comprises lipids and proteins, many of which are heavily glycosylated. The ratio of lipid molecules to protein molecules within the plasma membrane has been calculated to be 50 : 1. Because lipids are approximately 50 times smaller than the average protein, they each make up 50% of the plasma membrane by mass.

Double lipid bilayer membranes also define the many organelles found in eukaryotic (but not prokaryotic) cells, including the nucleus, the mitochondrion, the Golgi apparatus, the peroxisome and (in plants only) the chloroplast. In addition, there are many membranous vesicles within cells which mediate trafficking between these compartments. These include endosomes that originate at the plasma membrane and either return to the plasma membrane or fuse with each other, acidify and eventually become lysosomes, sites of protein degradation; vesicles from the ER carrying newly synthesised proteins and that are destined to fuse with the Golgi or directly with the plasma membrane; and Golgi vesicles that are destined to fuse with the plasma membrane to release secreted or transmembrane proteins at the cell surface. Because lipids will pose different challenges to biosensors than proteins, one school of thought is that one should ignore the membrane-associated proteins and focus on the tractable soluble fraction. However, many clinically relevant proteins are membrane-associated (including the vast majority of drug targets, which are transmembrane proteins) and an alternative view is that there is a pressing need for biosensors that can address the membrane-associated proteome. We will now focus on two major intra-cellular membrane-bound compartments.

8.1.2.5.2 Mitochondrial Genomes. Mitochondrial genomes range in size from 16 000 base pairs (encoding 13 proteins) in animal cells to as many as 2.4 million base pairs in plant cells (http://www.ncbi.nlm.nih.gov/genomes/genlist.cgi?taxid=2759&type=4&name=Eukaryotae%20Organelles). As with chloroplasts (see above), the genomic information available fails to capture the biological complexity of mitochondria, as 99% of the proteins essential for mitochondrial function in human cells are encoded in the nuclear rather than in the mitochondrial genome.

8.1.2.5.3 The Nucleus. Each nucleus of a human cell contains a complete genome (3 billion base pairs, corresponding to a total length of 1.5 m) organised into 22 autosomal and 2 sex chromosomes. DNA in each chromosome is packed into nucleosomes in which it winds around a core complex of highly basic proteins called histones. There are an estimated 25 million nucleosomes per nucleus, each comprising 2 molecules each of histones H2A, H2B, H3 and H4 (8 histone molecules in total per nucleosome) corresponding to 200 million molecules of the highly similar histone proteins per nucleus. These themselves represent both a challenge to biosensing, being highly similar proteins with unusually basic characters, but also an opportunity, as it has recently become clear that post-translational modification of the histones by acetylation or methylation of lysine residues in their tails is intimately associated with the regulation of gene expression in what has been termed "the histone code".[11,12] The nucleus comprises further functional (as opposed to membrane-confined) compartments, such as the nucleolus (700 proteins[13]) and the spliceosome (more than 110 proteins[14]), which can be isolated by affinity purification.

8.1.3 The Types of Device Required

The study of proteomics is currently dominated by the use of mass spectrometry, a technique that relies on the differential movement of ionised species in a vacuum, under the influence of an electric field, to determine specific mass/charge ratios that can be used to identify each species. This is particularly relevant to the study of proteins in organisms whose genomes have been sequenced, as bioinformatic approaches can be used to (i) predict the precise sequence of open reading frames (the sections of genes that code for protein, as opposed to intervening or regulatory DNA sequences) and the proteins coded by them, (ii) the likely peptides produced either enzymatically or by the ionisation process from each predicted open reading and (iii), with much less certainty, the effect that post-translational modification of each residue may have on migration in the mass spec. These predictions are then used to analyse the spectra produced in each experiment, and if sufficient numbers of peptides can be identified for a given reading frame, the conclusion will be that that open-reading frame is indeed expressed. Quantification is now possible in these experiments using techniques that rely on the isotopic labelling of specific amino acids in parallel or control experiments. However, the technique suffers from many limitations. For example, (i) it is easily swamped by abundant proteins, which must be removed from a sample if less-abundant species are to be detectable; (ii) it is possible to detect spectra that are signatures of a state, or to determine the identity of a protein that contributes to one peak in such a spectrum, but only very rarely is it possible to obtain both a global signature and insights into the biology that underlies that signature; and (iii) the technique is so sensitive to minor variations that results tend not to be reproducible between laboratories, with variables from ambient temperature to sample handling time having effects that outweigh the biological variation that is being explored. Given the widespread application of mass spectrometry, there have been many reviews of these issues and readers are referred here to just one, which discusses attempts to use MS to analyse the urinary proteome.[15]

The second most common approach involves the use of protein microarrays. These were first described by Haab et al.,[16] who showed that proteins spotted onto a range of chemically modified microscope slides could recruit specific antibodies from a model solution of 114 competing antibodies. In contrast, antibodies that had been immobilised on the slides largely failed to detect their target proteins when these were presented in a similar model mixture. For both systems, failure rates increased as the target protein was present at lower concentrations, and were unacceptable in the clinically relevant range of 10 s ng mL^{-1}. These failures are most likely due to catastrophic loss of structure of the proteins or antibodies when immobilised on the glass surface – even where attempts are made to keep proteins hydrated, interactions between amino acid side chains and the surface will inevitably lead to progressive unfolding. Nonetheless, a number of suppliers now sell protein or antibody microarrays, although few come with any guarantee of success.

A further drawback of the microarray approach described in the preceding paragraph is that the proteins to be detected need to be fluorescently labelled. The addition of one or more molecules of a hydrophobic dye molecule is likely to also lead to protein unfolding, at least locally and perhaps also globally. Because protein–protein interactions and enzyme activity are both intimately dependent on protein structure and conformations, protein labelling is likely to lead to the loss of relevant biological information. To avoid this drawback will require the development of so-called "label-free" methods for the detection of biomolecules. Such approaches rely on the effects that biomolecules can have on the physical properties of a sensor surface and include surface plasmon resonance, acoustic waveguides, resonant sensors including quartz crystal microbalances and microcantilevers, and electrical devices where biomolecules alter the surface charge or the current flow at an electrode in solution. Optical methods are the most highly developed, but will suffer from the restriction of the wavelength of light on pixel density. Resonant sensors perform poorly in liquids and also suffer greatly from thermal effects. Electrochemical techniques offer perhaps the best hope for highly multiplexed, high-density, sensitive and robust devices with the required sensitivities and dynamic ranges. However, all of the label-free techniques suffer from a major drawback – the need to show that signals detected are due to specific interactions rather than to interactions with non-specific but often highly abundant competing molecules. Clearly, the long experience of the biosensor community has much to offer here.

8.2 Biosensing for Clinical Applications

8.2.1 The Clinical Problems – Diagnosis, Prognosis, Personalised Medicine

The term diagnosis originates from Greek and can be translated as "seeing through", or "knowing through". A doctor's role is to determine the best course of treatment or advice for a patient based on what they know i.e. on past history and present symptoms – that is to say, on the basis of a diagnosis. In contrast, a patient's prognosis describes the future – how they are likely to fare on a course of treatment. The term prognostic assay describes one that seeks to identify patients who are likely to benefit from a given treatment. Currently, both diagnosis and prognosis are based on statistical analyses of observations made across populations of patients and "normal" individuals. This approach is becoming less and less acceptable to clinicians and the healthcare industry as increasingly sophisticated technologies promise to deliver personalised diagnosis and prognosis that are based on the patient's own physiology and pathology. To be effective, these technologies need to be able to assess not only the impact of the patient's own genetic make-up on the course of a disease, but also the interplay between the patient and their environment. This will call not only for increasingly sophisticated technologies, devices and assays, but also for a solid understanding of the clinical background to each assay: much of the increased power will come simply from combining existing markers into one assay. This will have several beneficial effects – simplifying decision-making, reducing variability, increasing robustness and most importantly reducing the time that patients and their doctors need to wait for test results. The goal of this section is to set the scene for such development, bearing in mind that new tests will allow better measurements of existing markers as well as enabling the measurement of markers for which no tests exist. The development of technologies that will enable the discovery of novel biomarkers is beyond the scope of this book, but the biosensor community should have much to contribute.

8.2.1.1 Body Fluids

The human body is made up, on average, of 20% (by mass) proteins, nucleic acids and complex sugars, 7% minerals and 15% fat, with the remaining mass being water. This water is divided between intra-cellular water (40% of total body mass) and extra-cellular water (20%). Extra-cellular water is either contained within the vasculature (*i.e.* in blood), or is referred to as interstitial water, which accounts for 15% of body mass and bathes organs, tissues and cells. The two major interstitial fluids are the lymph and cerebro-spinal fluid (CSF). Water within the vasculature makes up the vast majority of blood and accounts for 5% of body weight in a volume of 4–6 L in an average male weighing 70 kg. The key feature of these fluids is that they represent a continuum – fluid continuously leaches out of the vasculature into tissues, and is taken back up; cells take up water through pinocytosis and related events at the plasma membrane, and release it equally. Thus blood, urine, saliva, tears and all other fluids which are readily sampled from patients may be expected to produce a picture that reflects the health of that individual. The completeness of the picture will depend on the fluid sampled, with "specialised" fluids such as saliva or semen being less informative about the overall state than blood. Urine represents a mechanism for the elimination of waste and thus is not thought to contain much information. In fact, many cells, proteins and nucleic acids can be detected in urine, and it is expected that increased information will be derived from this source as more sensitive and specific methods for biomolecular detection become available.

8.2.1.2 The Composition of Body Fluids

8.2.1.2.1 Urine. The average individual produces approximately 1 L of urine a day, resulting from the filtration of as much as 180 L of fluid by the kidneys. The pH of urine (generally held to be acidic) can in fact vary from pH 4.5 to pH 8. Major components of urine (other than water) include ammonia (NH_3, also combining with H^+ to give ammonium ions NH^{4+}), urea ($9\,mg\,mL^{-1}$), formed from NH^{4+} in the liver, and creatinine ($1.5\,mg\,mL^{-1}$) that all result largely from protein breakdown. Uric acid is formed from the turnover of purines, and is found in blood (240 µM) as well as urine (25 µmoles per day). Urine also contains sulfate ions that result from the oxidation of cysteine. Elevated levels of Ca^{2+} in urine can be indicative of osteoporosis. Some hormones and signalling molecules such as dopamine can be found in urine, while epinephrine or nor-epinephrine are only present in the form of breakdown metabolites. Testosterone and other steroid hormones are present as 17-ketosteroids including androsterone. Surprisingly, urine also contains both intact proteins and nucleic acids. In a recent analysis, it was estimated that urine contains more than 1500 proteins[17] although their identification and application remain in their infancy and current approaches are not quantitative.[15] In a normal individual, protein secretion in the urine is approximately $50\,mg\,day^{-1}$. As in serum, a single protein (albumin) makes up the bulk of this – 60% in the case of urine. Protein levels in urine are held to be abnormal if they exceed $150\,mg\,day^{-1}$, and levels of $1\,g\,day^{-1}$ indicate nephritis (inflammation of the kidney) while levels greater than $3\,mg\,day^{-1}$ can be associated with diabetes. In the latter case, one would also find sugar (glucose) in the urine. High levels of bilirubin in the serum or the urine would indicate liver problems, ranging from viral or alcoholic hepatitis to liver failure or even right heart failure.

8.2.1.2.2 Serum/plasma. Blood is an extremely complex environment, comprising cells, proteins, sugars, lipids and a carefully regulated ionic environment. Wikipedia includes a comprehensive list of human blood components, many with concentrations, although the page does not comprise any references and information must therefore be verified (http://en.wikipedia.org/wiki/List_of_human_blood_components). The following section gives only information for which one or more authoritative reference(s) is available. If blood is spun in a centrifuge the cells can be removed, leaving behind plasma, representing between 50 and 60% of blood by volume and 5% of total body weight. The liquid portion of blood that remains following the formation of a semi-solid clot by the natural process of coagulation is serum which is essentially identical to plasma from which the clotting proteins (fibrinogen, factor II, factor V and factor VII have been removed) – although one would also expect loss of other constituents at random as they are caught up in the growing clot. As it flows around the body, blood will potentially collect information from every tissue once per cycle (exit from heart to re-entry to the heart) and is expected to carry a sample of proteins secreted from every organ, representing both the healthy and disease state. For example, in adults bone alone has a total blood flow of between 2 and 400 mL min^{-1}. Because of its quantity (4.5 litres per adult, of which as much as 30% can be lost without significant problems) ease of handling and relative simplicity, serum is a widely used diagnostic material, closely followed by plasma and serum produced from blood to which coagulation inhibitors such as heparin or the chelating agent EDTA have been added.

Normal blood comprises three major types of cells. Red blood cells, also called erythrocytes, whose function is to carry oxygen *via* haemoglobin, are present at 4–6 × 10^9 cells mL^{-1} of blood, with females having a slightly lower average value (4.8 × 10^6) than males (5.4 × 10^6). Proteins present on the surface of red blood cells give rise to the ABO antigens that are responsible for the famous incompatibility between blood donations from different individuals. White blood cells are the cells of the immune system and are present at between 4 and 11 × 10^6 cells mL^{-1}. The third type comprises platelets, which lack a nucleus and so are not strictly speaking cells; these play a role in the repair of injured blood vessels and are present at 1–300 × 10^6 mL^{-1}. White blood cells are further divided into neutrophils (3–6 × 10^6 mL^{-1}), lymphocytes (1.5–4 million cells mL^{-1}), eosinophils (150–300 000 cells mL^{-1}), basophils (0–100 000 cells mL^{-1}) and monocytes (300–600 000 cells mL^{-1}). Neutrophils, eosinophils and basophils may be collectively referred to as granulocytes or as polymorphonuclear leukocytes (PMNs) due to their distinctively shaped nuclei that stain with different dyes (eosinophils stain with acidic dyes, basophils with basic and neutrophils with neutral dyes).

In common with the rest of the extra-cellular fluids, the pH of blood is 7.4 ± 0.05. Blood contains roughly 300 equivalents per L (one electrical equivalent is 1 mol of an ionised substance divided by its valency), most of which are contributed by Na$^+$ (present at ~145–150 mM) and the counter-ions Cl$^-$ (~120–125 mM) and HCO$_3^-$ (~27 mM). Other important cations are K$^+$ (~4–5 mM), Ca^{2+} (100 mg L^{-1} or 2.5 mM, of which almost 50% is complexed with counter-ions *e.g.* HCO$_3^-$ and citrate, 35% is bound to albumin and 10% to globulin while the rest is ionised) and iron (1.3 mg L^{-1} or 23 µM in men, slightly lower in women), absorbed as Fe^{2+} and converted to the ferrous Fe^{3+} form in plasma, where it is mostly bound to transferrin, a plasma protein iron carrier. Important anions include phosphorus (120 mg L^{-1}, almost 70% of which is in organic compounds with the rest circulating as PO$_4^{3-}$, HPO$_4^{2-}$ and H$_2$PO$_4^-$) and I$^-$ (3 µg L^{-1}). Ca^{2+} levels are of particular importance in blood as it is required for coagulation at steps involving the activation of factor X by factor VII or factor VIII, the activation of thrombin by factor Xa and as a cofactor for factor XIII in clot stabilisation. It is worth noting that chelation of calcium may not be sufficient to prevent all the steps involved in clotting, as negatively charged surfaces including glass or collagen can directly activate factor XII.

Blood also contains high levels of glucose (between 70 and 100 mg L^{-1}, roughly 4 to 6 mM) as well as some acetone and ketones, in the form of ketone bodies (10 mg L^{-1}). Finally, blood carries lipids. The normal concentration of cholesterol in plasma is 1.2–2 g L^{-1}, although levels higher than 1.8 g L^{-1} are strongly indicative of ischaemic heart disease. Lipids including cholesterol are usually solubilised in plasma in the form of lipo-proteins such as the very-low-density lipoproteins (VLDL), low-density lipoproteins (LDL, which deliver cholesterol to cells) and high-density lipoproteins (HDL, which deliver cholesterol to the liver for elimination) that are important markers of cardiovascular risk, particularly atherosclerosis. The ratio of HDL to LDL, combined with plasma cholesterol levels, is thought to be a better marker of cardiovascular health than any one alone.

Proteins in blood have been classified into three groups: the albumins (carrier proteins), the globulins (mainly antibodies and steroid binding proteins) and the fibrinogens (proteins involved in clotting). The protein concentration of normal human serum is 30–80 mg mL^{-1} – and 90% of this protein by mass comprises a single protein species, albumin, while 99% of plasma protein by mass comprises just 22 proteins[18] (Table 8.1). The remaining 1% of plasma protein by mass may comprise a further 10 000 proteins.[19] Antibodies, whose role is to recognise potentially dangerous molecules such as those on the surface of, or secreted by, pathogens, are present at high levels. For example, immunoglobulin G (IgG) is 10 mg mL^{-1}, IgA is 2 mg mL^{-1}, IgM is 1.2 mg mL^{-1}, IgD is 30 μg mL^{-1} and IgE is 0.5 pg mL^{-1}.

The majority of non-antibody plasma proteins are synthesised in the liver. Other proteins of interest in serum include the acute phase proteins (synthesised in response to stress) such as α1-globulin and haptoglobin, whose concentrations are 2.5 mg mL^{-1} and 1.5 mg mL^{-1} respectively, increasing by up to four-fold during an acute phase response (cited in Ahlqvist[20]). Clotting factors that are abundant in plasma and serum include fibrinogen (>1 mg mL^{-1} in normal individuals), factor VIII (anti-hemophilic factor), factor XIII (fibrin stabilising factor) and von Willebrand factor. Steroid-hormone binding proteins are also abundant. Also of interest, although harder to detect, are polypeptide hormones, growth factors, interleukins and cytokines (a term that designates an interleukin whose sequence has not been determined) that mediate intercellular communication. Interleukins and cytokines play specialised, but still poorly understood, roles in inflammation and infection and in diseases including auto-immunity and cancer. In the case of hormones, precise quantification of levels is hampered not only by the lack of suitable technologies but also by the constant fluctuations imposed both by the regulation of their expression by feedback loops and by response to environmental and physiological changes in any individual. Accurate numbers are available for the best-studied protein hormones. For example, ACTH (adrenocorticotropic hormone, a protein just 39 amino acids long) levels are maximal in the morning at less than 50 pg mL^{-1} plasma (10 pM), dropping by more than 90% over the course of a normal day, although levels can increase to close to 500 pg mL^{-1} (100 pM) in response to stress. Para-thyroid hormone (PTH is in

Table 8.1 Some of the most abundant proteins in serum and plasma, with their concentrations. PSA, a widely studied marker of clinical relevance for screening and diagnosis of prostate cancer is given for comparison.

Protein	Plasma concentration
Albumin	45–50 g/mL
Immunoglobulin G (IgG)	10 mg/mL
Fibrinogen	2–4.5 mg/mL
Immunoglobulin A (IgA)	2 mg/mL
α2-macroglobulin	1.5–4.2 mg/mL
Immunoglobulin M (IgM)	1.2 mg/mL
Hemopexin	0.5–1 mg/mL
Haptoglobin	0.4–1.8 mg/mL
Transthyreitin	0.25 mg/mL
Coagulation factors II, VII, IX, X	0.2 mg/mL
anti-thrombin-III	170–300 µg/mL
ceruloplasmin	150–600 µg/mL
steroid hormone binding globulin (SHBG)	33 µg/mL
Immunoglobulin D (IgD)	30 µg/mL
Transferrin	30–65 µg/mL
Thyoxine-binding globulin	15 µg/mL
α1-antiprotease	13–14 µg/mL
Prostate Serum Antigen (PSA), in health	< 4 ng/mL

a similar range (10–55 pg mL^{-1}) although its levels do not fluctuate during the day. Insulin levels are 0–500 pM, with a total secretion (stimulated by food intake) of approximately 290 nmoles per day. Similarly, a normal thyroid gland will secrete 80 µg (103 nmol) of thyroxine (the thyroid hormone also called T4) and 4 µg (7 nmol) of triiodothyronine (T3) in any 24-hour period, which leads to plasma levels of 103 nM and 2.3 nM, respectively. Notably, the vast majority of thyroid hormones are bound to soluble proteins in plasma, and free T4 is inferred to be only 20 ng L^{-1} and free T3 only 3 ng L^{-1} plasma – roughly 1/1000th of the secreted amount of T4 and 1/300th the amount of T3. Some of the circulating thyroid hormones are known to be bound to serum albumin, an abundant protein that has been hypothesised to serve just such a carrier role. Although albumin is a huge problem for most measurement techniques, and is usually removed from samples before processing, there have been suggestions that biosensors may be able to use the albumin fraction itself as an enriched source of the biological molecules to be interrogated. It is also worth mentioning two hormones that may be of interest in coming years. Leptin, a 167-amino acid long protein hormone, is secreted from fat cells to the circulation and signals a requirement for decreased food uptake; its activity is opposed by other hormones including ghrelin, a 28-amino acid long protein secreted by the stomach which promotes both growth and food uptake.

In order to detect disease biomarkers in plasma, it is crucial that we understand the protein composition of normal plasma. This has led for example to worldwide collaborations between as many as 35 laboratories (The Human Proteome Organisation Plasma Proteome Project[21]) as well as intensive efforts by others.[18,22,23] The most recent attempt to accomplish this involved a mass spectrometric analysis of plasma and built upon earlier work to yield a list of 697 proteins (not including immunoglobulins) whose presence in plasma can be detected with 99% confidence.[23]

Useful non-proteinaceous markers in serum include potassium (may indicate fluid imbalance or problems with kidneys) and calcium (normally 85–105 mg L^{-1}; elevation indicative of thyroid problems or possibly bone metastases in cancer; decrease associated with *e.g.* vitamin D deficiency, kidney failure or pancreatitis). The latter two are more useful when assessed together with plasma alkaline phosphatase activity, which normally lies in the range of 39 to 117 units L^{-1} blood. A major problem with serum, however, is the vast dynamic range – proteins of clinical interest may be present in serum at pg mL^{-1} or less, while albumin is in the tens of mg mL^{-1}, giving a concentration range of at least 10 logs. As outlined by Horvatovich *et al.*,[24] if the limit of detection of a biosensor is of the order of 1 fmol, one would need 50 µL of serum to be able to detect a protein biomarker of average molecular weight (50 kDa in the example of Horavovitch *et al.*) present in the ng mL^{-1} range. However, 50 µL of serum equates to 4 mg of total protein, which saturates conventional proteomic tools such as mass spectrometers and is likely to prove a daunting problem for many biosensor platforms unless biological specificity can be built into the platform. In addition to ions and proteins, blood will also carry a sampling of the chemical messengers that mediate much of intercellular communication. These messengers include acetylcholine (600 000 molecules of acetylcholine may be released at a nerve ending to stimulate the corresponding muscle), catecholamines including adrenaline (epinephrine, nor-epinephrine) and dopamine, steroids (including testosterone and oestrogen and their derivatives) and retinoids. Plasma epinephrine levels vary from 60 to 300 pg mL^{-1} (1.8 nM) at rest to 40–600 pg mL^{-1} during activity. In contrast, nor-epinephrine levels approaching 5000 pg mL^{-1} may be achieved during heavy exercise.[25] Roughly 10% of plasma epinephrine is free *i.e.* not bound to either effector or carrier proteins. Plasma dopamine is ~35 pg mL^{-1} (0.23 nM). Testosterone in plasma is largely bound to proteins including the B-globulin sex-steroid-binding hormone (65% total testosterone) and albumin (33% total testosterone); the same proteins sequester similar proportions of total oestrogen in women. Total plasma testosterone is 3–10 ng mL^{-1} (10–35 nM) in men and less than a tenth of that in women. It should be noted that in women, testosterone, oestrogen and other hormone levels vary up to 10-fold with the menstrual cycle and vary further with pregnancy (Table 8.2).

Other steroid hormones include glucocorticoids involved in the regulation of metabolism as well as the suppression of inflammation, such as cortisol (130 ng mL^{-1} or 370 nM) and corticosterone (4 ng mL^{-1}) and mineralocorticiods that regulate ionic secretion such as aldosterone (60 pg mL^{-1} or 170 pM) and deoxycorticosterone (also 60 pg mL^{-1}). Cortisol is largely bound to protein (the α-globulin transcortin, in which form cortisol lacks biological

Table 8.2 Variation in selected hormones following and during pregnancy. hCG and hCS are two related forms of the protein hormones generically called human growth factor. Prolactin is also a protein hormone and is structurally related to hCS and hCG, while oxytocin is another polypeptide hormone unrelated to hCG/hCS. Estradiol and progesterone are steroid hormones.

Hormone	Normal	Pregnancy peak	Time of peak
hCG	<3 ng/mL	5 mg/mL	First trimester
hCS	<3 ng/mL	15 mg/mL	Term
Estradiol	5–35 pM	16 ng/mL	Term
progesterone	0.9–18 ng/mL	190 ng/mL	Term
prolactin	8 ng/mL	200 ng/mL	Term
oxytocin	25 pg/mL	?	?Late labour?

activity but is available as a reservoir of active hormone for release) whereas aldosterone appears to be largely free. Although signalling molecules such as acetylcholine that work at neuro-muscular junctions should not in theory be present in blood, the existence of auto-immune disease such as myasthenia gravis, where auto-antibodies are made against acetylcholine, demonstrates that they can be found there – at least in pathological states. Defects in catecholamine metabolism associated with phenylketonuria can lead to mental retardation, and are associated with increased levels of phenylalanine in the blood (as well as in tissues and urine). Mediators of neuronal communication including amino acids (glutamate and aspartate are excitatory amino acids for the nervous system, while glycine and gamma-amino butyrate [GABA] are inhibitory) may be present in blood, and it will be interesting to determine whether these can be used to monitor neural health and injury. Finally, the blood also carries gases, most notably NO released by endothelial cells that line blood vessels.

8.2.1.2.3 Cerebro-spinal Fluid. The brain and the spinal cord are bathed in a liquid called cerebro-spinal fluid (CSF) that is largely synthesised in the brain by the choroid plexus. Some CSF leaches into plasma, but CSF may also be directly accessed by lumbar puncture. Electrolyte composition is broadly similar to that of plasma, from which it only really differs in lacking bulk protein (CSF has just $200\,\mu g\,mL^{-1}$ total) and cholesterol ($2\,\mu g\,mL^{-1}$ compared to more than $1\,mg\,mL^{-1}$ in plasma).

8.2.1.2.4 Synovial Joint Fluid. Joints such as the knee, the shoulder and the elbow consist of interacting bone and cartilage surfaces that are subject to significant wear and tear and undergo constant repair. The synovial fluid, whose major role is thought to be in the lubrication of the joints, thus consists of the breakdown products of bone and connective tissue from the joint as well as signalling molecules to regulate the activity of cells (macrophages, osteoclasts and osteoblasts) important in joint homeostasis. In pathological situations such as rheumatoid or osteoarthritis, the composition of the fluid is therefore expected to undergo significant changes. Samples of 500 μL volume can be drawn from the knee joint, and have been found to comprise as many as 135 unique proteins, although 18 of these were keratins that may have resulted from contamination due to skin puncture.[26] Fifteen of these proteins were found to increase in samples from patients with osteoarthritis, while three (aggrecan, cystatin A and dermcidin) were reported to be decreased in this population.[26] Although these data need to be independently validated, synovial joint fluid may be a useful fluid for the detection of musculoskeletal disease.

8.2.1.2.5 Tears. The aqueous humour is a liquid that cleanses and nourishes the cornea and lens of the eye, and is produced from the plasma by a combination of simple diffusion and active transport. The composition of tears may therefore be reflective of the composition of plasma.

8.2.1.2.6 Saliva. The average human produces approximately 1.5 L of saliva a day. The major protein constituents are α-amylase and lingual lipase, as well as glycoproteins called mucins that bind and lubricate both food and bacteria, lysozyme (which attacks the cell walls of bacteria) and the immunoglobulin IgA; saliva contains many of the growth factors that may be expected to promote wound healing. Ionic composition is likely to closely reflect that in plasma. In addition, diseases of the buccal cavity, from dental caries to oral cancer, may be expected to be reflected by changes in the protein composition of saliva.

8.2.2 Biosensors for Clinical Applications

Biosensors comprise three elements: a detector (biological probe), a sensing element that captures the biological, chemical or physical output of the detector and a transducer that transmits and amplifies the signal for interpretation by the human operator.

8.2.2.1 Detectors

Traditionally, detectors have comprised enzymes such as glucose oxidase that metabolise the target molecule and produce a chemical output (such as a change in pH) that is readily sensed by an ion-selective material. Interaction of the ion with the material generates an output (*e.g.* electrical current) that is readily measured, amplified and translated for the operator. To move towards a proteomic biosensor will require the ability to multiplex detectors. In the case of enzymes, the best hope here appears to be the cytochrome P450 family.[27] Alternatively, it is possible to use traditional biological probes such as antibodies, although these need to be selected with great care,[16] or non-traditional biological probes such as recombinant antibodies or engineered proteins such as peptide aptamers (see Chapter 7 by Tkac and Davis) or even non-proteinaceous probes including lectins[28] or chemical entities.[29,30]

8.2.2.2 Sensors

These can be broadly divided into optical and electrical. Optical sensors may rely on the fluorescent labelling of the target proteins, which is expensive, laborious and non-quantitative (it is usually not possible to control the number of dye molecules per protein), either for direct detection or for the more robust, but technically challenging, detection of FRET. These techniques are not readily amenable to high throughput, and are not compatible with the preservation of biological function in the labelled protein. Label-free optical approaches including surface plasmon resonance (SPR; involving a two-dimensional metallic surface) or localised SPR (LSPR, involving three-dimensional maoparticles or nanostructures) are the subject of intensive study, and the latter in particular offer great promise for biosensing.[31] Of similar interest are approaches based on Raman spectroscopy, which has the additional advantage that the Raman spectrum may itself be used for the identification of a target molecule.

8.3 Further Reading

Much of the clinical data mentioned in this text is accepted as given. However, textbooks such as the excellent *Review of Medical Physiology* by W. F. Ganong (Lange/McGraw-Hill), *Kumar and Clark Clinical Medicine* (P. Kumar and M. Clark, eds., Elsevier/Saunders) or the *Oxford Handbook of Clinical Diagnosis* (H. Llewellyn, H. Aun Ang, K. E. Lewis, A. Al-Abdullah, Oxford University Press) serve as useful references and were referred to extensively during the writing of this chapter.

References

1. S. G. Oliver, Q. J. van der Aart, M. L. Agostoni-Carbone, M. Aigle, L. Alberghina, D. Alexandraki, G. Antoine, R. Anwar, J. P. Ballesta and P. Benit, *et al.*, *Nature*, 1992, **357**, 38.
2. A. Goffeau, B. G. Barrell, H. Bussey, R. W. Davis, B. Dujon, H. Feldmann, F. Galibert, J. D. Hoheisel, C. Jacq, M. Johnston, E. J. Louis, H. W. Mewes, Y. Murakami, P. Philippsen, H. Tettelin and S. G. Oliver, *Science*, 1996, **274**, 546, 563.
3. D. A. Lashkari, J. L. DeRisi, J. H. McCusker, A. F. Namath, C. Gentile, S. Y. Hwang, P. O. Brown and R. W. Davis, *Proc. Natl. Acad. Sci. USA*, 1997, **94**, 13057.
4. M. C. Jewett, G. Hofmann and J. Nielsen, *Curr. Opin. Biotechnol.*, 2006, **17**, 191.

5. S. G. Villas-Boas, J. F. Moxley, M. Akesson, G. Stephanopoulos and J. Nielsen, *Biochem. J.*, 2005, **388**, 669.
6. H. Zhu, M. Bilgin, R. Bangham, D. Hall, A. Casamayor, P. Bertone, N. Lan, R. Jansen, S. Bidlingmaier, T. Houfek, T. Mitchell, P. Miller, R. A. Dean, M. Gerstein and M. Snyder, *Science*, 2001, **293**, 2101.
7. J. Burian, L. Guller, M. Macor and W. W. Kay, *Plasmid*, 1997, **37**, 2.
8. J. Burian, S. Stuchlik and W. W. Kay, *J. Mol. Biol.*, 1999, **294**, 49.
9. T. J. Johnson, Y. M. Wannemeuhler, J. A. Scaccianoce, S. J. Johnson and L. K. Nolan, *Antimicrob. Agents Chemother.*, 2006, **50**, 3929.
10. A. Douwe de Boer and P. J. Weisbeek, *Biochim. Biophys. Acta*, 1991, **1071**, 221.
11. T. Jenuwein and C. D. Allis, *Science*, 2001, **293**, 1074.
12. A. J. Ruthenburg, H. Li, D. J. Patel and C. D. Allis, *Nat. Rev. Mol. Cell Biol.*, 2007, **8**, 983.
13. A. K. Leung, L. Trinkle-Mulcahy, Y. W. Lam, J. S. Andersen, M. Mann and A. I. Lamond, *Nucleic Acids Res.*, 2006, **34**, D218.
14. J. Deckert, K. Hartmuth, D. Boehringer, N. Behzadnia, C. L. Will, B. Kastner, H. Stark, H. Urlaub and R. Luhrmann, *Mol. Cell Biol.*, 2006, **26**, 5528.
15. S. Decramer, A. Gonzalez de Peredo, B. Breuil, H. Mischak, B. Monsarrat, J. L. Bascands and J. P. Schanstra, *Mol. Cell Proteomics*, 2008, **7**, 1850.
16. B. B. Haab, M. J. Dunham and P. O. Brown, *Genome Biol.*, 2001, **2**, RESEARCH0004.
17. J. Adachi, C. Kumar, Y. Zhang, J. V. Olsen and M. Mann, *Genome Biol.*, 2006, **7**, R80.
18. N. L. Anderson and N. G. Anderson, *Mol. Cell Proteomics*, 2002, **1**, 845.
19. J. N. Adkins, S. M. Varnum, K. J. Auberry, R. J. Moore, N. H. Angell, R. D. Smith, D. L. Springer and J. G. Pounds, *Mol. Cell Proteomics*, 2002, **1**, 947.
20. J. Ahlqvist, *Scand. J. Clin. Lab. Invest.*, 2003, **63**, 315.
21. G. S. Omenn, D. J. States, M. Adamski, T. W. Blackwell, R. Menon, H. Hermjakob, R. Apweiler, B. B. Haab, R. J. Simpson, J. S. Eddes, E. A. Kapp, R. L. Moritz, D. W. Chan, A. J. Rai, A. Admon, R. Aebersold, J. Eng, W. S. Hancock, S. A. Hefta, H. Meyer, Y. K. Paik, J. S. Yoo, P. Ping, J. Pounds, J. Adkins, X. Qian, R. Wang, V. Wasinger, C. Y. Wu, X. Zhao, R. Zeng, A. Archakov, A. Tsugita, I. Beer, A. Pandey, M. Pisano, P. Andrews, H. Tammen, D. W. Speicher and S. M. Hanash, *Proteomics*, 2005, **5**, 3226.
22. N. L. Anderson, M. Polanski, R. Pieper, T. Gatlin, R. S. Tirumalai, T. P. Conrads, T. D. Veenstra, J. N. Adkins, J. G. Pounds, R. Fagan and A. Lobley, *Mol. Cell Proteomics*, 2004, **3**, 311.
23. S. Schenk, G. J. Schoenhals, G. de Souza and M. Mann, *BMC Med. Genomics*, 2008, **1**, 41.
24. P. Horvatovich, N. Govorukhina and R. Bischoff, *Analyst*, 2006, **131**, 1193.

25. P. E. Cryer, *N. Engl. J. Med.*, 1980, **303**, 436.
26. R. Gobezie, A. Kho, B. Krastins, D. A. Sarracino, T. S. Thornhill, M. Chase, P. J. Millett and D. M. Lee, *Arthritis Res. Ther.*, 2007, **9**, R36.
27. N. Bistolas, U. Wollenberger, C. Jung and F. W. Scheller, *Biosens. Bioelectron.*, 2005, **20**, 2408.
28. J. Hirabayashi, *J. Biochem.*, 2008, **144**, 139.
29. P. Bergese, M. Cretich, C. Oldani, G. Oliviero, G. Di Carlo, L. E. Depero and M. Chiari, *Curr. Med. Chem.*, 2008, **15**, 1706.
30. A. J. Vegas, J. H. Fuller and A. N. Koehler, *Chem. Soc. Rev.*, 2008, **37**, 1385.
31. J. N. Anker, W. P. Hall, O. Lyandres, N. C. Shah, J. Zhao and R. P. Van Duyne, *Nat. Mater.*, 2008, **7**, 442.

Subject Index

102A1 P450 *see Bacillus megaterium* P450

α-fetoprotein 98
α-helix structures 134, 135, 136, 156–7
β-sheet structures 134, 135, 136, 156–7
κ_{el} (electronic transmission factor) 131, 132
λ_D (Debye length) 207, 208
π-donor acceptor complexes 60, 62, 63

1-acetato-4-benzyl-triazocyclononane 134, 135
adhesive forces, AFM 32, 33
adiabatic regimes 132, 142
adverse drug reactions 161–5
affibodies 200
affinity complexes 59, 72–82
AFM *see* atomic force microscopy
AFP (alpha-fetoprotein) 98
agarose gels 195, 196
albumins 104, 107, 109, 235
Alcaligenes faecalis 19–20, 143
alcohol dehydrogenase (AlcDH) 77, 78, 79
aldehyde oxidase 2
alkane dithiols 67–8
alkane hydroxylation 166
alkylthiols 102
allelic variants 164–5
alpha-fetoprotein (AFP) 98
alpha-helix structures 134, 135, 136, 156–7
amicyanin 120, 122
amide linkages 100, 102, 110
amine coupling chemistry 214

amine oxidase 143, 145
amino acid motifs 157, 158
amino-(FAD) (N^6-(2-aminoethyl-flavinadenine dinucleotide)) 60, 61, 67, 69, 70, 71
3-aminophenyl boronic acid 75
amplification strategies, DNA 99
anions, blood 234
antibodies, blood 104, 107, 109, 235, 236
anti-human serum albumin 104, 107, 109
apo-enzymes
    apo-ferritin, electron transport studies 37, 38, 46
    apo-glucose oxidase
        reconstitution on charge-transporting wires 59–63
        reconstitution on cofactor-functionalised nanoparticles/tubes 63–70
        reconstitution in redox polymers 70–2
aptamers
    capture agents 198, 199–200, 201, 208
    scanning probe analyses 40, 42
Arabidopsis thaliana 228
arsenic poisoning 19
arsenite oxidase 2, 19–20
Arthrobacter globiformis 145
atomic force microscopy (AFM) 29–34
    conductive probe 35
    confined metalloproteins 46–7
    constant height/current modes 28, 29
    contact mode 29–31, 32
    direct biomolecule imaging on electrodes 41–3

displacement 31
enzyme-CNT electrodes 69, 70, 71
error signal 31
force aspects 32–4
force spectrum 32
general bioimaging considerations 36–7
imaging case studies 39–41, 42
nanotube assemblies 104–5
technique 203–4
attomolar detection limits 212, 215
attractive regime, AFM 32
azurin 40, 41, 123–5, 143, 144

Bacillus megaterium cytochrome
    P450$_{BM-3}$ (BMP) 176, 178, 181–4
    covalent linkage to cystamine-
        maleimide 180, 181
    heme domain 157, 166
    non-covalent linkage on negatively
        charged electrode surfaces 179
    rational design 166–7, 168
bacteria
    see also Bacillus megaterium
        cytochrome P450$_{BM-3}$
    Alcaligenes faecalis 19–20, 143
    Arthrobacter globiformis 145
    Corynebacterium glutamicum 37
    cytochrome P450 enzymes
        characteristics 155, 156
        coupling/uncoupling 161
        origins 153, 154
    Desulfovibrio vulgaris 181, 182, 183, 184
    Escherichia coli 15, 16, 18, 227–8
    Halobacterium halobium 36–7
    Oligotropha carboxidovorans 9
    Paracoccus pantotrophus 18
    plant P450 fusion protein expression 171
    Pseudomonas
        P. aeruginosa 123–5, 143, 144
        P. putida 172–3
    Rhizobium Spp. 19
    Rhodobacter Spp. 8–9, 15–16, 17, 18
    Rhodococcus 156
    Rhodopseudomonas palustris 173
    Starkeya novella 10, 11, 14
    Sulfolobus tokodaii 178

bacteriorhodopsin 38
ballistic conductance 63
bamboo-like nanotubes 99, 101
basal plane graphite 132, 133
beef products 6
1,4-benzene dithiol cross-linker 64, 66
beta-sheet structures 134, 135, 136, 156–7
beverage industry 10, 15
bias voltage, STM 28
bi-enzyme systems 6, 7
bilirubin oxidase (BOD) 83, 87, 89
bimolecular (mediator–enzyme)
    electron transfer 13
bioaffinity binding 196, 197
bio-barcode assay 204, 205, 206
biofuel cells 57, 58, 82–9
bioimaging
    see also individual techniques
    general considerations 36–7
biological applications of biosensors
    225–40
biological complexity 226–30
biorecognition layers 207, 208, 209–10
biotin 107, 198
bis-bipyridinium cyclophane 60, 62
bis-iminobenzene π donor 60, 63
bis-maleimidomethylether (BMME)
    linkers 124–5
blood
    normal 234
    serum
        albumin 104, 107, 109, 235
        chemical messengers 237
        clotting factors 233–4, 235, 236
        composition 233–8
BMME (*bis*-maleimidomethylether)
    linkers 124–5
BMP *see Bacillus megaterium*
    cytochrome P450$_{BM-3}$
BOD (bilirubin oxidase) 83, 87, 89
body fluids 104, 107, 109, 232–9
boronic acid bridging ligands 60, 61, 75,
    76, 77
bottom-up fabrication methods 214, 215
bovine milk 5–6
bufuralol 170
Butler–Volmer equation 128

Subject Index

cantilevers 29–31
cantilever spring constant (k) 32
capacitance 67, 69, 204, 209–13
capture agents 194, 198–200, 201, 208
carbodiimide chemistry 122
carbon materials surface preparation 132–4
carbon monoxide dehydrogenase 1, 9
carbon nanotubes (CNTs)
   antibody-modified, field effect transistor protein detection 208
   applications 95–112
   arsenite oxidase modified electrode 19
   assembly 95
   defects 70
   definition/characteristics 94–6
   density control 103
   diameter tuning 103
   electrocatalytic properties 96
   electrode preparation 133
   metallic, cofactor-functionalised 63, 67, 70, 78, 87, 89
   modified AFM probes 32, 36
   non-oriented 96–9, 101
   reconstituted protein units assembly 69, 71
   single-wall carbon nanotube forests 94–112
   transistor-based protein detection 204, 213–15
   vertically-aligned forests 99–112
carboxyl groups 100
carboxylic acid-modified carbon nanotubes 67, 69, 70, 71
catalysis
   cycle, cytochrome P450 enzymes 158–61
   electrode modification for P450 immobilisation 176–8
   nanolithography, chemical force microscopy 33–4
   P450 engineering 165–71
   self-sufficient enzymes, P450 reductase fusion proteins 169–70, 171
Catharanthus roseus 171
cationic surfactants 173–8

Cc *see* cytochrome *c*
CcP (cytochrome *c* peroxidase) 120
cells 228, 229–30, 234
cerebro-spinal fluid 238
CFM (chemical force microscopy) 33–4
charge flux mechanisms 42–3
charge transfer mechanisms 47
charge transfer resistance (R_{ct}) 212
chemical force microscopy (CFM) 33–4
chemical messengers, blood plasma 235, 236, 237–8
chemical structures *see* structures
chemical vapour deposition (CVD) 102
chicken liver sulfite oxidase 10–11, 14
chicken liver sulfite oxidoreductase 10
chimeric (fusion) enzymes 168–71
chitosan 173, 178
chlorate reductase 2, 20
chloroplasts 228
chlortoluron 171
cholera toxin sensors 98
chronoamperometry 128–32
classification, P450 enzymes 155–6
clinical applications 231–40
clotting factors 233–4, 235, 236
CNTs *see* carbon nanotubes
co-crystallisation, proteins 120–2
cofactors
   co-crystallisation studies 120
   electrode–enzyme cofactor links using electron transfer mediators 7–9
   enzyme affinity complexes 59, 72–82
   flavine adenine dinucleotide 5, 9
   functionalised metallic nanoparticles/nanotubes 63–90
   heme in sulfite oxidases 10, 11
complexity of biological systems 226–30
Compound 0, definition 160
Compound I, definition 160
compression of biomolecules 38
conductance 7–8, 127, 139–43
conductive-probe atomic force microscopy (CP-AFM) 35
confined molecules *see* immobilisation
conjugated linkers 139–43, 178, 180
conserved motifs 157, 158

continuum model (Dutton) 121
copper 1, 9, 43, 45, 48
Corynebacterium glutamicum 37
coupling/uncoupling 46–9, 159, 160–1
covalent bonds
   P450 immobilisation on gold electrodes 178–80
   protein complexes 122–6
   protein–electrode connections 144–6, 178–80, 195, 196, 197
   protein–protein 120
covalent/non-covalent attachment 195, 196, 197
CP-AFM (conductive-probe atomic force microscopy) 35
CPR (cytochrome P450 reductase) 155, 156, 158, 169–70
cross-linking
   1,4-benzene dithiol 64, 66
   bis-maleimidomethylether 124–5
   boronic acid 60, 61, 75, 76, 77
   conjugated linkers 139–43, 178, 180
   covalent protein complexes 122–6
   cysteines 122, 123, 139, 145
   4,4'-dimercaptobiphenyl 64, 66
   glutaric dialdehyde 77
   surface-confined cofactor–enzyme affinity complexes 72–82
crystal structure
   see also structures
   azurin homodimer 125
   nitrate reductases 18
   protein complexes 120–1
   sulfite oxidoreductases 10, 11
current channel 28
current maps 26–7
current-potential (IE) spectroscopy 44, 46, 47–8, 49
current switching 47–9
current-voltage (IV) spectroscopy 46–8, 49
CVD (chemical vapour deposition) 102
cyclic voltammetry
   electrode modification for P450 immobilisation 176–8
   P450 molcular Lego constructions 184–5
   technique 128, 129

CYP102A1 P450 *see Bacillus megaterium* P450
CYP *see* cytochrome P450 enzymes
cystamine-maleimide 173, 175, 180, 181, 185
cysteamine self-assembled monolayers 100, 101
cysteines
   cross-linked protein complexes 122, 123, 124–5
   cysteine–cysteine bridges, protein immobilisation on gold 139
   gold electrode surface modification 136, 137
cytochrome b_5 161
cytochrome b_5 reductase 156
cytochrome *c*
   cytochrome-c_{551} 120, 122
   cytochrome *c* peroxidase co-crystallisation 120
   gold electrodes
      diffusional cofactor tethered covalently to 77, 80, 81, 82, 84, 85
      immobilisation 139
      potential modulation 43, 45
   redox protein activation 74
   sulfite oxidase mediator 9, 12–13
   trumpet plot 130
cytochrome *c* peroxidase (CcP) 120
cytochrome oxidase (COx) 80, 81, 82, 84, 85
cytochrome P450 enzymes 153–85
   102A1 *see* P450$_{BM-3}$
   catalytic cycle 158–61
   classification 155–6
   CYP prefix 156
   directed evolution 166–7
   drug metabolism 161–5
   electrode interfacing 171–85
   molecular Lego approach 180–2, 184–5
   molecular oxygen 158–9, 160
   mono-oxygenase activity 155, 159
   P450$_{BM-3}$ 157, 176, 178, 181–4
      covalent linkage to cystamine-maleimide 180, 181
      non-covalent linkage on negatively charged electrode surfaces 179
      rational design 166–7, 168

protein engineering 165–71
structure/function 155–8
cytochrome P450 reductase (CPR)
155, 156, 158, 169–70
cytochrome P450$_{cam}$ enzyme 41
cytokines 235

DAPA (DNA array to protein array) 197
DDAB (didodecylammonium chloride)
173, 174, 176, 178
Debye length (λ_D) 207, 208
denaturation/unfolding of proteins 37–8,
231
density control, carbon nanotubes 103
Desulfovibrio vulgaris flavodoxin
181, 182, 183, 184
detection limits 211–15
diagnosis, clinical 231–9
didodecylammonium chloride (DDAB)
173, 174, 176, 178
diethylaniline end groups 143, 145
diffusional cofactors 72–82
4,4'-dimercaptobiphenyl cross-linker
64, 66
dimethyl formamide (DMF) 104, 105
dimethyl sulfide (DMS) 2, 15
directed evolution 166–7
direct processes
  biomolecule imaging on electrodes
  41–3
  electron exchange *see* third
  generation biosensors
  growth of vertically-aligned CNT
  forests 100, 101, 102
  imaging electrochemistry/enzyme
  activity 41–6
  immobilisation of proteins 137–9
  sample labelling 194, 195, 201
displacement, atomic force
  microscopy 31
dithio-bismaleimidoethane (DTME) 180
dithiol cross-linkers 64, 66
DMF (dimethyl formamide) 104, 105
DMSO reductase family of enzymes
  arsenite oxidase 19–20
  chlorate reductase 20

DMSO reductase 15–17
  mechanism of action 2–3
  oxo ligands 3
  perchlorate reductase 20
  structure 2
DNA
  amplification
    barcode-based amplification
    strategies 204, 205, 206
    carbon nanotube biosensors 99
  carbon nanotube–DNA
    electrochemical sensor 110
  electrode surface preparation 134
  human cell nucleus 229–30
  nanostructures, AFM imaging
    considerations 36
  oligonucleotide-modified YCC 98
  protein microarray *in situ* synthesis
    193, 196, 197, 198, 199–200
  recombinant DNA techniques 168–71
drug metabolism 161–5
drug–drug interactions 163–4
DTME (dithio-bismaleimidoethane) 180
dynein 35

E^0 (midpoint potential) 128–31, 172–3
ECSTM (electrochemical scanning
  tunnelling microscopy) 36, 44–9
EDC (1-ethyl-3-(3-dimethylamino-
  propyl)carbodiimide HCl)
  chemistry 122
edge plane graphite electrodes 132, 133
  direct protein immobilisation 138
  DMSO reductase adsorption 17
  electrode modification for P450
    immobilisation 176
  Starkeya novella sulfite
    dehydrogenase 14
  unmodified 172–3
edge plane sites, carbon nanotubes 96
EI/C (electrochemical
  impedance/capacitance) 204
EIS (electrochemical impedance
  spectroscopy) 209
electrochemical impedance/capacitance
  (EI/C) methods 204

electrochemical impedance spectroscopy (EIS) 209
electrochemical scanning tunnelling microscopy (ECSTM) 36, 44–9
electrodes
  see also gold electrodes
  biomolecule electronic coupling, spectroscopic assessment 46–9
  capacitive configurations 211
  confined metalloproteins, atomic force microscopy 39–40, 41
  cytochrome P450 enzyme immobilisation 171–85
  direct biomolecule imaging on 41–3
  enzyme cofactor links 7–9
  redox enzyme electrical interface nanostructures 56–90
  surfaces
    bioelectronic scanning probe analyses 39
    preparation 132–6, 137
electron donors/acceptors 56, 60, 62, 63
electronic transmission factor (κ_{el}) 131, 132
electron self exchange (ese) 123–6
electron transfer (ET)
  carbon nanotubes 95
  co-crystallisation studies 121
  ferritin 37, 38
  immobilised proteins/electrodes 126–32
  mediators see mediators
  protein immobilisation/wiring 119–46
electropolymerised polytyramine 10–12
electrostatic gating 207, 208
ELISA (Enzyme-Linked Immunosorbent Assay) 201, 202, 205
E_m (midpoint potential) 128–31, 172–3
EMs (extensive metabolisers) 165
enantioselective reduction 16
engineering, proteins 165–71, 180–5
environmental monitoring 225
enzyme activity, direct imaging 41–6
enzyme electrodes
  reconstituted 56–90
  monolayers 59–63
Enzyme-Linked Immunosorbent Assay (ELISA) 201, 202, 205

enzymes see individual enzymes
enzymic amplification approaches 205–6
EPG see edge plane graphite electrodes
erythromycin, safety 163
Escherichia coli 15, 16, 18, 227–8
ese (electron self exchange) 123–6
ET see electron transfer
ethanol 89
7-ethoxyresorufin 167, 170
ethylbenzene dehydrogenase 2
exponential decay coefficient 138
extensive metabolisers (EMs) 165

facilitators see promoter molecules
FAD see flavin adenine dinucleotide
families of enzymes 156, 161, 162
Faradaic impedance label-free protein sensing 212–13
feedback electronics
  atomic force microscopy 29, 30, 31
  scanning tunnelling microscopy 26, 27, 29
femtomolar detection 212
FePc (iron(II) phthalocyanine) 102
Fermi function 130
Fermi–Dirac statistics 130
ferredoxin 155
ferritin 37, 38, 46
ferrocene 140, 141, 142
ferrocenium derivatives 8
F-G loop region, cytochrome P450 enzymes 158
fibrinogens 233–4, 235, 236
field effect transistors (FETs) 206–15
first generation biosensors 3, 4
fish spoilage 6, 7
flavin, hot wiring 143
flavin adenine dinucleotide (FAD)
  amino-FAD 60, 61, 67, 69, 70, 71
  biofuel cells 83, 84
  Class I/II P450s 155–6
  glucose oxidase on SWNT mat electrodes 98
  xanthine oxidoreductases 5, 9
flavin mononucleotide (FMN) 155–6
flavodoxin 181, 182, 183, 184, 185
fluorescent labels 201

FM (frequency-modulated) tapping
    mode 29–31
FMN (flavin mononucleotide) 155–6
food industry 6, 10, 15
force spectroscopy 37–8
    see also atomic force microscopy
forests, single-wall carbon nanotubes
    94–112
formate dehydrogenase 2
four-terminal EC-STM 43, 44
frequency-modulated (FM) tapping
    mode 29–31
friction force 33
functionalisation 95, 100, 195
functions
    cytochrome P450 enzymes 155–8
    human cytochrome P450 enzymes 162
    P450 proteins 154
fusion (chimeric) proteins 168–71

gate electrode, FETs 207–8, 213, 214
gating, in situ EC-STM 47–9
GC/polytyramine electrodes 11, 12
GDH see glucose dehydrogenase
gene recombination 166–7
genetic differences 164–5
genomes
    complexity, model organisms 227–8
    mitochondrial 229
    sequences 226
    size, biological complexity 227
ghrelin 236
glassy carbon electrodes
    DMSO reductase/methyl viologen
        adsorption 16
    mercury-coated, sulfite oxidase
        activity monitoring 11, 12
    modification for P450
        immobilisation 176, 184–5
    NADH cofactor–enzyme affinity
        complexes on carbon
        nanotubes 75, 77
    non-oriented carbon nanotubes 97
    osmium-mediated horseradish
        peroxidase reduction 7, 8
    P450$_{cam}$ electrochemistry 173
    perchlorate reductase 20

glucose
    biofuel cells 82–4, 89
    sensors 60, 61, 109–10
glucose dehydrogenase (GDH)
    biofuel cells 83, 84, 85, 87, 89
    derivatised for hot wiring 143
    NADP$^+$ cofactor affinity complexes
        cross-linked with glutaric
        dialdehyde 77, 78, 79
    polyaniline/PQQ-reconstituted
        glucose dehydrogenase
        electrode 73, 74
    PQQ-dependent, assembly on
        functionalised gold
        nanoparticles 65–6
glucose oxidase (GOx)
    biofuel cells 83, 84, 85
    nanotube forests 112
    non-oriented carbon nanotubes 98
    polyaniline-reconstituted glucose
        oxidase electrode 72, 73
    reconstitution as apo-GOx on
        charge-transporting wires 59–63
    reconstitution as apo-GOx on
        cofactor-functionalised
        nanoparticles/tubes 63–70
glutaric dialdehyde 77
glutathione S-transferase (GST) 197, 198
gold
    conductive-probe atomic force
        microscopy 35
    confined metalloproteins 47
    nanoparticles 63–90, 97, 173, 178, 202
    protein microarrays 196, 197
    pyrolytic, nanotube biosensor
        assembly 104
    scanning tunnelling microscopy
        probes 28
gold electrodes
    cysteine links 136, 137, 139
    cytochrome c diffusional cofactor
        tethered to 77, 80, 81, 82, 84, 85
    electron transfer chronoamperometry
        130, 131
    horse heart cytochrome-c
        immobilisation 139, 145

P450 covalent linkages 178–80, 181
pyrroloquinoline quinone assembly on 60–1
spacers 180
surfaces 39, 196, 197
   copper-azurin chemisorption 43, 45
   modification 12–13, 14, 134, 135, 136, 137
GOx *see* glucose oxidase
graphite
   see also edge plane graphite electrodes
   electrode preparation 132, 133
   pyrolytic 19, 39
GST (glutathione S-transferase) 197, 198
guanine oxidation, CNT-modified electrode 99

Halobacterium halobium 36–7
H_{DA} (matrix coupling element) 130, 142
Heck reaction 33
height, constant (STM) 28, 29
heme domain
   bacterial $P450_{BM-3}$ 157, 166
   cofactors
      cytochrome *c*/cofactor enzyme affinity complexes 80, 81
      intermolecular electron transfer 14
      sulfite oxidoreductase 10, 11
   iron, cytochrome P450 enzymes 159, 160
   P450 directed evolution 166
   P450 proteins 154, 157, 158, 166
   thiolate ligand, cytochrome P450 enzymes 157, 158
   zinc ions replacement of heme groups 122
herbicide tolerance 171
heterogeneous rate measurement 126–8
highly oriented pyrolytic graphite (HOPG) 39
high potential redox mediators 8, 13
high-throughput reverse-dot blot assay 195, 201
Hille classification 1–2
histidine-117 ligands 126

histones 230
holoferritin 37, 38, 46
HOPG (highly oriented pyrolytic graphite) 39
hormones, blood plasma 235, 236, 237–8
horse heart cytochrome-*c* 12–13, 139, 145
horseradish peroxidase (HRP)
   osmium-pyridine-based hydrogel polymers 7–8
   protein–electrode connections 144
   SWNT biosensors 107–8, 110, 111
   SWNT forest electrochemical characterisation 105–6
hot wiring, proteins 143–4
HSA *see* human serum albumin
human cells, biological complexity 228–30
human cytochrome P450 enzymes, 2D6 polymorphism 165
human cytochrome P450 enzymes 154
   allelic variants 164–5
   drug metabolism 161–5
   families 162
   hepatic isoforms 163, 164
human serum albumin (HSA) 104, 107, 109, 235
human sulfite oxidase 13
hydrogel polymers 7–8
hydrogen bonds 178
hydrogen peroxide 5–8, 10–11, 105–6
hydrophobic patches 121–2, 123–6
hydrophobins 134
hydroxo ligands 3, 5
hydroxymethylferrocenium 8
hypoxanthine
   fish spoilage 6, 7
   osmium-mediated horseradish peroxide reduction 8
   purine nucleoside phosphorylase 6–7
   ratio determination, purine analysis 7

IC (intermittent contact) mode, AFM 29–30, 32, 36
IE (current-potential spectroscopy) 44, 46, 47–9
imidazole 143, 144

Subject Index 251

immobilisation 46–7
  cofactor–enzyme affinity complexes 72–82
  cytochrome P450 enymes 171–85
    summary 174–5
  molecules for scanning electro-chemical microscopy 44–9
  proteins 136–46, 196–8
  redox enzymes 119–46
immunosensors
  detection principles 108
  prostate antigen 98, 107, 109, 110, 111, 112
  SWNT forests 107–9, 111
impedance 204, 209–13
indium tin oxide (ITO) 132, 134
inhibitor units, tethered to molecular wires 80
inosine 7
in situ protein microarray synthesis 193, 196, 197, 198, 199–200
integrated enzyme electrodes 77–89
interdigitated capacitative electrodes 211
interfacial protein detection 193–4
interleukins, blood plasma 235
intermittent contact (IC) mode, AFM 29–30, 32, 36
intramolecular electron transfer 13
ionic strength 170, 207, 208
ion-pumping membrane proteins 36–7
iridium-platinum STM probes 28
iron-assisted electrostatic SWNT forest assembly 103
iron(II) phthalocyanine (FePc) 102
iron–sulfur proteins
  cytochrome P450 catalytic cycle 159, 160
  ferredoxin 155
  nitrate reductase 18
  xanthine oxidoreductases 5, 9
isoforms, human hepatic P450 163, 164
isosbestic point 160
ITO (indium tin oxide) 132, 134
IV (current-voltage) spectroscopy 46–8, 49

k_0 (standard electron transfer rate) 131, 140, 141
k see cantilever spring constant
ketoconazole, safety 163
kinetic barrier, electron transfer 57, 75

label-free field effect protein sensing 193–215
  capacitance/impedance 209–13
  capture biomolecules 198–200
  detection tools 200–4
  different approaches 206–15
  interfacial protein detection 193–4
  nanoscale devices 213–15
  protein microarrays 194–206
  surface chemistry 196–8
  ultrasensitive protein detection 204–6
label-free protein detection platforms 202–4
labelling
  fluorescent labels 201
  nucleic acid molecular beacons 199
  protein detection 201–2
  sample labelling 194, 195, 201
lactate dehydrogenase (LDH)
  biofuel cells 81, 85, 86, 88
  NADH cofactor–enzyme affinity complexes 75, 77
laser optical systems, AFM 29, 30
lateral forces, AFM 33–4
Laviron's theory 128, 130
layer-by-layer fabrication 99
  see also self-assembled monolayers
LDH (lactate dehydrogenase) 75, 77, 81, 85, 86, 88
leptin 236
linkers 139–43, 178, 180
lipids, blood plasma 235
lipophilic xenobiotics 161–5
lithographically etched patterns 215
liver isoforms 163, 164
long-chain alkane dithiol monolayers 67–8
low potential functioning 8

M44K mutant 123

M64E mutant 123, 125
MAC (magnetic AC) AFM mode 30
MADH (methylamine
    dehydrogenase) 120, 122
magnetic AC AFM mode 30
magnetic fields 85, 86, 88
magnetic nanoparticles 202
malate dehydrogenase (MDH) 75, 76
mammalian cytochrome P450 enzymes
    crystal structures 156–7
    directed evolution 167
    origins 153, 154
Marcus inverted region 131
Marcus theory 13, 56, 129
mass spectrometry (MS) 203, 230
mat electrodes 96, 97–8
matrix coupling element (H_{DA}) 130, 142
MDH (malate dehydrogenase) 75, 76
mediators (of electron transfer)
    cofactors in redox protein assemblies
        57, 58, 59
    electrode–enzyme cofactor links 7–9
    high potential redox 8, 13
    second-generation biosensors 3, 4, 5
    sulfite oxidase by horse heart
        cytochrome c 12–13
    xanthine oxidase cofactor–electrode
        links 7–9
membrane-associated proteins 36–7,
    38, 229
membrane-bound nitrate reductase
    (Nar) 18
meniscus forces, SPL 34
mercury-coated glassy carbon electrodes
    11, 12
metabolism of drugs 161–5
metallic nanostructures 63–90
metalloproteins
    see also individual metalloproteins
    confined, atomic force microscopy
        46–7
    electrode-confined, atomic force
        microscopy 39–40, 41
    metalloprotein-electrode imaging,
        potential control 43–4, 45
metal-molecule-metal (mMm)
    junctions 35

methylamine dehydrogenase (MADH)
    120, 122
methyl viologen 8, 15–16, 18–19
mica surfaces 39, 104
microarrays 193, 194–206, 231
microsomal cytochrome P450
    enzymes 161, 169–70
midpoint potential (E^0/E_m) 128–31,
    172–3
miniaturised xanthine oxidase
    electrodes 6
mitochondrial cytochrome P450
    enzymes 155
mitochondrial genomes 229
mMm (metal-molecule-metal)
    junctions 35
model organisms, genome complexity
    227–8
molecular dipoles 158
molecular landers 134
molecular Lego approach
    cytochrome P450 enzymes 167,
        170–1, 180–2, 184–5
    Desulfovibrio vulgaris flavodoxin
        181, 182, 183
molecular oxygen 158–9, 160
molecular recognition, AFM 38
molecular wires 60, 62, 80
molecule length, linkers 140–2
molybdenum, protonation 16
molybdenum–copper–sulfur moiety 9
molybdoenzymes see mononuclear
    molybdoenzymes
molybdopterin (MPT) ligands 1–2, 5
monolayers
    see also self-assembled monolayers
    1,9-nonanedithiol long chain 67–8
    reconstituted enzyme electrodes
        58–63, 80, 83, 84, 86
mononuclear molybdoenzymes 1–21
    see also individual molybdoenzyme
        families
    mechanism of action 2–3, 4
    molybdenum oxidation states 2
    Oligotropha carboxidovorans carbon
        monoxide dehydrogenase 9
    sulfite oxidases 9–15

three families 1–2
xanthine oxidases 5–9
mono-oxygenase activity 155, 159
motifs, conserved 157, 158
MPT (molybdopterin) ligands 1–2, 5
mRNAs 227
MS (mass spectrometry) 203, 230
multi-wall carbon nanotubes (MWNTs)
  definition/characteristics 94–5
  mat electrodes 97, 98
  non-oriented 96–9, 101
mutagenesis
  hot wiring 143, 144
  random 166–7
  site-directed
    cytochrome P450 enzymes 167–8
    hot wiring 143, 144
    hydrophobic patches 123–6
  steroid metabolism 168
myocardial cell culture 6
myoglobin 105–6, 144
myosin 35

N^6-(2-aminoethyl-flavinadenine
  dinucleotide) (amino-FAD) 60,
  61, 67, 69, 70, 71
N42C azurin dimers 125
NADPH see nicotinamide adenine
  dinucleotide phosphate
Nafion® 8, 20
  -FeO(OH)/FeOCl layers 100, 101,
    102, 103, 104
  non-oriented carbon nanotubes 97
n-alkanes
  conducting properties 139–43
  gold electrode SAMs 138–9
  gold electrode surface modification 134
  hot wiring 143, 144
nano-dissection, AFM 37
nanografting 33
nanolithography, SPL 33, 34
nanoparticles (NPs)
  gold 63–90, 97, 173, 178, 202
  magnetic 202
  metallic, cofactor-functionalised
    63–7

nanopore method of conducting
  properties measurement 141, 142–3
nanoscale devices, label-free field
  effect protein sensing 213–15
nanostructures, bioelectrocatalytic
  56–90
nanowires
  polymer-carbon nanotube coaxial
    nanowire 110
  semiconducting 204, 213–15
  single-wall carbon nanotube forests
    94–112
NAPPA (nucleic acid programmable
  protein array) 196, 198
Nap (periplasmic nitrate reductase) 2,
  17–19
Nar (membrane-bound nitrate
  reductase) 2, 17–19
National Centre for Biological
  Information (NCBI) 226
negative differential resistance (NDR)
  48
neomycin B 133
nervous system studies 6–7
nicotinamide adenine dinucleotide
  phosphate (NADPH)
  coupling/uncoupling of P450
    substrate turnover 159, 160–1
  cytochrome P450 enzymes 155, 159
  P450 reductase fusion proteins 169–70
  redox protein activation 74–9
  screening method 166
Nile blue 75, 77, 78, 79, 89
nitrate reductase 80–2
  membrane-bound (Nar) 2, 17–19
  periplasmic (Nap) 2, 17–19
nitrite reductase 138, 143
nitrocellulose 196
NMR (nuclear magnetic resonance) 121
nomenclature systems 155, 156
non-adiabatic regimes 129, 132
1,9-nonanedithiol long chain
  monolayers 67–8
non-Faradaic capacitance/impedance
  label-free protein sensing 209–12
non-oriented carbon nanotubes 96–9, 101

NP (purine nucleoside phosphorylase) 6–7
NPs *see* nanoparticles
NR *see* nitrate reductase
nuclear magnetic resonance (NMR) 121
nucleic acid programmable protein array (NAPPA) 196, 198
nucleic acids
　see also DNA
　aptamers 40, 42, 198, 199–200, 201, 208
　assays 207, 208
　RNA 200, 227
nucleus, human cells 229–30

oligonucleotide-modified YCC 98
oligophenylacetylenes 80
oligo-polyphenyl-ethynylenes (OPEs)
　conducting properties 139–40, 141, 142
　protein–electrode connections 143, 144, 145
oligo-polyphenyl-vinylenes (OPVs)
　conducting properties 139–40, 141, 142
　protein–electrode connections 146
Oligotropha carboxidovorans 9
open-reading frames 230
optical sensors, clinical applications 240
OPVs (oligo-polyphenyl-vinylenes) 139–40, 141, 142, 146
organelles 228, 229–30
orientation
　carbon nanotube properties 96
　chemisorption
　　direct biomolecule imaging on electrodes 41–3
　　peptide aptamers 40, 42
　　redox enzymes by surface reconstitution 56–90
osmium-mediated reduction 7–8
osmium-pyridine-based hydrogel polymers (Heller) 7–8
overpotentials 128–9, 131
oxidation states, molybdenum 2, 16
oxo ligands 3
oxygen, molecular 158–9, 160
oxygen activation motif 158
oxygenation, carbon nanotubes 96

P450s *see* cytochrome P450 enzymes
PAA (polyacrylic acid) 70, 72
PAn (polyaniline) 70, 72, 73, 74
pAPEC01-R plasmid 228
Paracoccus pantotrophus 18
parallel metal plate capacitative electrodes 211
partner proteins 136, 138
patches
　hydrophobic 121–2, 123–6
　positively charged on P450 2E1 179
pathway model of electron transfer 121
patterns 102, 106, 215
PDDACl (polycationic poly(diallyldi-methyl-ammonium chloride)) 101
PDDA (poly-(dimethyldiallyl-ammonium chloride))
　Desulfovibrio vulgaris flavodoxin 181, 182, 183
　P450 immobilisation 173, 174, 177, 178, 185
PECVD (plasma-enhanced chemical vapour deposition) 110
PEI (polyethylenimmine) 173, 175, 177
PEO (poly-ethylene oxide) 173, 174, 178
peptides
　aptamers 198, 199
　　oriented chemisorption 40, 42
　　protein microarrays 200, 201, 208
　electrode surface preparation 132, 133, 134, 136, 137, 138
perchlorate reductase 2, 20
periplasmic bacterial dimethylsulf-oxidoreductase 15–16, 17
periplasmic nitrate reductase (Nap) 18
peroxidase enzymes 6, 105–6
peroxide shunt reaction 160–1, 166, 167
personalised medicine 231–9
pH, blood 234
Phase I/II drug metabolism 163
phase lags, AFM 31
phenacetin 167
phenytoin, safety 164, 165
phosphate determination 7
photolithography 215
Photosystem I (PSI) 122

phylogenetic trees 153, 154
pi-donor acceptor complexes 60, 62, 63
pKL1 plasmid 228
planar glass 194
plants 153, 154, 171, 228
plasma composition 233–8
plasma-enhanced chemical vapour deposition (PECVD) 110
plasma membrane 229
plasmids 198, 227–8
plastocyanins 122
platinum 28, 35, 63, 97
PMMA (polymethylmethacrylate) 102
PMs (poor metabolisers) 165
polyacrylamide gels 196
polyacrylic acid (PAA) 70, 72
polyaniline (PAn) 70, 72, 73, 74, 102
polycationic poly(diallyldimethyl-ammonium chloride) (PDDACl) 101
polycations 101, 173–8
poly-(dimethyldiallylammonium chloride) (PDDA) 173, 174, 177, 178
  Desulfovibrio vulgaris flavodoxin 181, 182, 183
poly-ethylene oxide (PEO) 173, 174, 178
polyethylenimmine (PEI) 173, 175, 177
poly(mercapto-*p*-benzoquinone) 7
polymers
  see also individual polymers
  electrode modification for P450 immobilisation 173–8
  electron transfer mediation 7–8
  polymer-carbon nanotube coaxial nanowire 110
  redox-active
    apo-enzymes reconstituted in 70–2
    biofuel cells 83
polymethylmethacrylate (PMMA) 102
polymyxin B1 133
poly-*p*-benzoquinone 7
poly-pyrrole matrix 18–19
poly(sodium 4-styrenesulfonate) (PSS) 173, 175, 178
polytyramine 10–12
polyvinylidene fluoride (PVDF) membranes 196

poor metabolisers (PMs) 165
porphyrin rings 159
positively-charged patches 179
positively-charged surfactants 173–8
post-translational modifications 227
potential control 43–4, 45
potential step amperometry 209–10, 211
power output, biofuel cells 82, 83, 85, 86, 88
PQQ *see* pyrroloquinoline quinone
pregnancy, plasma composition 237, 238
probe construction 28, 32–4, 46
prognosis, clinical 231–9
prokaryotes *see* bacteria
promoter molecules 132, 133
propane hydroxylation 166
prostate specific antigen 98, 107, 109, 110, 111, 112, 236
protein detection
  amplification methodologies 205
  concentration ranges vs. techniques 204, 205
  label—free field effect methods 193–215
  SWNT biosensors 107, 109
protein engineering 165–71, 180–5
protein microarrays 194–206, 231
protein unfolding 37–8, 231
proteomics, biosensor applications 225–6
protonation, molybdenum 16
pseudo-azurin 138
Pseudomonas Spp.
  P. aeruginosa 123–5, 143, 144
  P. putida 172–3
PSS (poly(sodium 4-styrenesulfonate)) 173, 175, 178
pterindithioline *see* molybdopterin
pure research 225–31
purine nucleoside phosphorylase (NP) 6–7
purines 5–9
purple membrane of *Halobacterium halobium* 36–7
PVDF (polyvinylidene fluoride) 196
pyridine 143, 145
pyridine-terminated undecanethiol linker 145

pyrolytic graphite electrodes 19, 39
pyrroloquinoline quinone (PQQ)
  apo-enzymes reconstitution in redox-active polymers 72, 74
  assembly on gold electrode 60–1
  biofuel cells 83–6
  GDH assembly on functionalised gold nanoparticles 65–6
  relay unit for NAD(P)H electrocatalysed oxidation 75

QD (quantum dots) 202
quality control, SWNT assembly 104, 105
quartz plates 102

random mutagenesis, directed evolution techniques 166–7
rat 169, 170
rate-limiting step, biofuel cells 85
rational design, cytochrome P450 enzymes 167–71
R_{ct} (charge transfer resistance) 212
recombinant DNA techniques 168–71
reconstituted systems 56–90, 170
redox active matrices 7–8
redox enzymes
  electrically contacted by crossing of surface-confined cofactor–enzyme affinity complexes 72–82
  ET kinetic barrier at redox sites 57
  immobilisation/wiring 119–46
    co-crystallisation 121–2
    ET at electrodes 126–32
    immobilisation processes 136–46
    protein complexes 120–6
    surface preparation 132–6
  reconstitution on cofactor-functionalised metallic nanoparticles/nanotubes 63–70
  redox state tuning 43, 44–6
  surface reconstitution methods 56–90
redox polymers 70–2, 83
reduced CO complex 155
refective coatings 31
relay-cofactor wire 58, 59
relay units, PQQ 75

repulsive regime, atomic force microscopy 32
resolution, scanning tunnelling microscopy 27
Resonance Raman spectra, nanotube assemblies 104–5
respiration, bacteria 20
respiratory chain proteins 40, 41
reverse-phase sample blot 195
Reverse Type I binding spectra 160
Rhizobium Spp. str. NT-26 19
Rhodobacter Spp. 15–16, 17, 18
  R. capsulatus 8–9
Rhodococcus 156
Rhodopseudomonas palustris 173
ribose-1-phosphate 6–7
rifampicin, safety 163
RNA 200, 227
rotaxane configurations 60, 62
roughness, electrode surfaces 39

Saccharomyces cerevisiae genome sequence 226
safety, drugs 161–5
saliva 239
SAM *see* self-assembled monolayers
sandwich format
  bio-barcode assay 204, 205, 206
  immunoassays, SWNT biosensors 107, 109
  protein microarrays 194, 195
saturated linkers 139–43
scaffold proteins 40, 42, 200
scanning electrochemical microscopy (SECM) 44–6
scanning probe analyses 25–50
scanning probe lithography (SPL) 33, 34
scanning probe microscopy (SPM) 25–39
scanning tunnelling microscopy (STM) 26–9
  direct biomolecule imaging on electrodes 41–3
  electrochemical 44–9
  two-layer junction model 140
screen-printed carbon electrode composites 13

Subject Index

screen-printed xanthine oxidase electrodes 6
SECM (scanning electrochemical microscopy) 44–6
secondary structure 157–8
second generation biosensors 3, 4
selenate reductase 2
self-assembled monolayers (SAM)
  atomic force microscopy 33
  cysteamine 100, 101
  direct protein immobilisation 138
  gold electrode surface modification 134, 135
  wiring proteins 144, 145, 146
self-assembled protein immobilisation 198
self-assembled vertically-aligned CNT forests 100, 101, 102
self-sufficient enzymes 169–70
semiconducting nanowires 213–15
semiconducting quantum dots 202
semi-rotaxane configurations 60, 62
sensitivity, capacitance/impedance label-free protein sensing 211–15
serum albumin 104, 107, 109, 235
serum composition 233–8
silane chemistry 134, 135
silicon chips 194, 196
silicon dioxide 134, 135, 136
silicon wafers 104
silver 6, 8, 63, 100
single-wall carbon nanotubes (SWNTs)
  biosensor applications of forests 107–12
  definition/characteristics 94–6
  forests, biosensors 94–112
  non-oriented 96–9, 101
  quality control 104, 105
single working capacitative electrodes 211
site-directed mutagenesis
  cytochrome P450 enzymes 167–8
  hot wiring 143, 144
  hydrophobic patches 123–6
sol-gel techniques 6
solvent stability 166

spacers 139–43, 178, 180
species differences 156
specific/non-specific attachment 195, 196, 197
spectrophotometric properties
  cytochromes, substrate binding spectra 159–60
  electrode-biomolecule electronic coupling 46–9
  P450 proteins 153–4, 155
SPL (scanning probe lithography) 33, 34
SPM (scanning probe microscopy) 25–39
SPR (surface plasmon resonance) 40, 42, 203
standard electron transfer rate (k_0) 131, 140, 141
Starkeya novella 10, 11, 14
Stefin Triple Mutant (STM) 200, 201
steroid hormones 163, 235, 237–8
steroid metabolism 162, 168
STM (scanning tunnelling microscopy) 26–9, 41–3
STM (Stefin Triple Mutant) 200, 201
stoppering 60, 62
structures
  α-helix 134, 135, 136, 156–7
  β-sheet 134, 135, 136, 156–7
  azurin homodimer 125
  cytochrome P450 enzymes 155–8
  DMS dehydrogenase 2
  DMSO reductase family of enzymes 2
  ethylbenzene dehydrogenase 2
  formate dehydrogenase 2
  mammalian cytochrome P450 enzymes 156–7
  mononuclear molybdoenzyme families 2
  nitrate reductases 18
  Oligotropha carboxidovorans carbon monoxide dehydrogenase 9
  perchlorate reductase 2
  periplasmic nitrate reductase 18
  protein complexes 120–1
  selenate reductase 2
  sulfite oxidases 2, 10
  sulfite oxidoreductases 10, 11

sub-families of enzymes 156, 161, 162
substrate recognition sequences (SRS) 167
substrates
  cytochrome P450 enzymes 159–60, 168
  label-free field effect protein sensing 194–6
  P450 proteins 154
  protein immobilisation on microarrays 194–6
sulfite dehydrogenase 10, 11, 14
sulfite oxidases 9–15
  cytochrome c electron transfer mediation 12–13
  eukaryotic 9
  horse heart cytochrome c electron transfer mediation 12–13
  mechanism of action 2–3, 4
  oxo ligands 3
  structure 2, 10
sulfite oxidoreductases
  characteristics 10–15
  crystal structure 10, 11
  distribution 9
  third generation biosensors 14
Sulfolobus tokodaii 178
surface adsorption, proteins 196
surface chemistry, protein immobilisation 196–8
surface-confined *see* immobilisation
surface modification, gold electrodes 134, 135, 137
surface plasmon resonance (SPR)
  long-chain alkane dithiol monolayers 67, 69
  oriented peptide aptamers 40, 42
  technique 203
surface preparation, electrodes 132–6, 137
surface raster scans, STM 26–7
surface reconstitution methods 56–90, 170
surfactants 173–8
SWNTs *see* single-wall carbon nanotubes
synovial joint fluid 238–9
systems biology applications 225–6

tapping mode, AFM 30, 31, 104–5
tears 239
temporal resolution, SPM 34–5
TEM (transmission electron microscopy) 69, 70, 71
terfenadine 163
thermal equilibrium, AFM 31
thin films, redox polymers, apo-enzymes reconstituted in 70–2
thiolated diethylaniline oligophenyl-acetylene monolayers 80
thiol-modified gold electrodes 12–13
third generation biosensors
  arsenite oxidase 19
  definition 4, 5
  sulfite oxidoreductases 14
Thlaspi arvensae 171
threaded redox-active rings 60, 62
three-electrode electrochemical cell 127
thyroid hormones 236
time-lapse SPM imaging 34–5
tip convolution, AFM 32
TMAO (trimethylamine N-oxide) reductase 2
top-down methods 215
top gate method 213, 214
topography channel, STM 28
total prostate specific antigen (T-PSA) 98, 107, 109, 110, 111, 112
toxicity, drugs 161–5
transient complexes, inter-protein 119, 121–2
transmission electron microscopy (TEM) 69, 70, 71
trimethylamine N-oxide (TMAO) reductase 2
trumpet plots 128, 129, 130
tuning
  nanoscale devices for label-free field effect protein sensing 213
  nanotube diameter control 103
  redox state 43, 44–6
tunnelling set-point 28
two-layer junction model 140
Type I/II binding spectra 160

Subject Index

ultra-rapid metabolisers (UMs) 165
ultrasensitive protein detection 204–6
UMs (ultra-rapid metabolisers) 165
unfolding of proteins 37–8, 231
unmodified electrodes, P450 cytochrome immobilisation on 172–3
uric acid 5, 6, 7
urine composition 233

vertically-aligned carbon nanotube forests 99–112
vertically-aligned CNT forests, fabrication methods 100, 101, 102
video-imaging 34, 35
viologens 8, 15–16, 18–19
voltammetry *see* cyclic voltammetry

warfarin 164
water 124, 126, 209, 233
weak electronic coupling 129
wiring
   process, redox protein assemblies 58, 59
   proteins 143–6
   protein–electrode contact 139–43
   redox enzymes 119–46

xanthine dehydrogenase 8–9

xanthine oxidases
   bi-enzyme systems 6–7
   electrode–enzyme cofactor links using electron transfer mediators 7–9
   ferrocenium artificial electron acceptors 8
   mechanism of action 2–3, 4
   sensor applications 5–9
   structures 2
   substrates/products 6
xanthine oxidoreductases
   electrochemical studies 8–9
   hydroxo ligands 3, 5
   iron–sulfur clusters 5
   sensor applications 5–9
xenobiotics *see* drug metabolism
X-ray structures
   see also structures
   azurin homodimer 125
   cytochrome P450 enzymes 156–7
   nitrate reductases 18
   protein complexes 120–1
   sulfite oxidoreductases 10, 11
p-xylene dithiol crosslinker 64, 66

yeast cytochrome *c* (YCC) 40, 45, 48, 49, 98
yeast genome sequence 226

zinc ions 122